U.S. Marine Corps Physical Readiness Training for Combat

Plus

Military Training for Urban Operations

Department of Defense

Pentagon
Publishing

UNITED STATES MARINE CORPS
Marine Corps Combat Development Command
Quantico, Virginia 22134-5001

29 January 1988

FLEET MARINE FORCE REFERENCE PUBLICATION............0-1B

MARINE PHYSICAL READINESS TRAINING FOR COMBAT

1. PURPOSE

Fleet Marine Force Reference Publication (FMFRP) 0-1Marine Physical Readiness Training for Combat is intended for use by all Marines. It provides the information and references necessary to establish and conduct physical conditioning programs to prepare Marines for the physical demands of combat.

2. APPLICABILITY

Although not directive, this handbook is intended for use as a reference by all Marines in developing unit physical readiness training for combat. This FMFRP does not establish policy for the conduct of physical fitness training. Refer to Marine Corps Order 6300.3_ for applicable Marine Corps policy on physical fitness. It may also be used as a basis for instruction conducted by unit leaders or as a reference for instruction or training.

3. RECOMMENDATIONS

To improve this manual, the MAGTF Warfighting Center seeks comments and recommendations from the user. Officers and noncommissioned officers with constructive comments are encouraged to respond. Lack of response will be considered as concurrence. The User Suggestion Form which follows can be detached and forwarded to - -

Commanding General
Marine Corps Combat Development Command (WF 06)
Quantico, Virginia 22134-5001

4. DISTRIBUTION

Initial distribution of this FMFRP 0-1B is shown below. Addressees on this list should ensure that appropriate subordinate organizations are provided copies.

JOHN I. HOPKINS
Brigadier General, U.S. Marine Corps
Assistant Chief of Staff, Warfighting
Marine Corps Combat Development Command
Quantico, Virginia

DISTRIBUTION: TDJ

From:

To: Commanding General, Marine Corps Combat Development Command (WF 06), Quantico, Virginia 22134-5001

Subj: FMFRP 0-1B, MARINE PHYSICAL READINESS TRAINING FOR COMBAT, RECOMMENDATIONS CONCERNING

1. In accordance with the Promulgation Page to FMFRP 0-1B, which invites individuals to submit suggestions concerning this FMFRP directly to the above addressee, the following unclassified recommendation is forwarded:

Page________Article/Para No.________Line No.________Figure/Table No.________

Nature of Change:_____Add_____Delete_____Change_____Correct

2. Proposed New Verbatim Text: (Verbatim, double spaced; continue on additional pages as necessary.)

3. Justification/Source: (Need not be double spaced.)

NOTE: Only one recommendation per page.

MARINE PHYSICAL READINESS TRAINING FOR COMBAT

Chapter 1
PHYSICAL READINESS LEADERSHIP
Section I. Role of Physical Fitness in Combat

1101. PURPOSE OF PHYSICAL FITNESS TRAINING

Physical fitness training in the Marine Corps has one purpose: to prepare Marines to physically withstand the rigors of combat. All other goals of physical fitness training are subordinate to and must support attainment of this goal. The idea that only infantry or reconnaissance units and their attachments normally face physically demanding combat is wrong. This error must not influence the priority commanders of combat support, combat service support, aviation, and headquarters units give to physical fitness for combat. Physical fitness for combat has a high priority for all Marines. A sound, effective unit program, requiring limited time and material, offers a greater payoff in combat than many more expensive and time-consuming training programs. This handbook describes the concept and provides the ingredients of a program for commanders to use in carrying out one of their most serious responsibilities: ensuring that their Marines are physically ready for combat.

1102. SCOPE OF THE MANUAL

This handbook provides guidance for all leaders, trainers, and planners of physical training programs. It describes unit physical fitness for combat training in the following chapters:

a. Physical Readiness Leadership. Chapter 1 provides guidance to leaders in the conduct of physical combat readiness training.

b. Physical Readiness Training Programs. Chapter 2 discusses how to structure programs to meet physical readiness goals in a variety of combat training situations.

c. Physical Conditioning Activities. Chapter 3 discusses the primary physical conditioning activities which commanders may use to attain readiness for combat goals. Sections in this chapter address foot marches under load, strength-building activities, and activities which build confidence and the aggressive spirit needed in combat.

d. Combat Water Survival. As an amphibious force, Marines cannot ignore the importance of combat water survival training. Chapter 4 describes a program for commanders to conduct this training.

e. Competitive Conditioning Activities. Chapter 5 describes competitive activities which stress the relationship to combat actions, featuring team-building types of competition.

f. Evaluation of Performance During Training. Chapter 6 describes tests which measure individual and unit physical fitness for combat.

g. The Human Body. Chapter 7 covers the structure and functioning of the body.

1103. COMMANDER'S ROLE

Major General Commandant John A. Lejeune in the 1921 edition of Marine Corps Manual expressed our philosophy of leadership. Among other things, General Lejeune stated that "it will be necessary for officers ... to endeavor to enlist the interest of their men in building up and maintaining their bodies in the finest physical condition." It is significant that the Commandant prefaced these writings in 1921 with the statement that his thoughts were greatly influenced by the experience of World War I. We have never varied as a Corps from this belief that one of a commander's most serious responsibilities is to have Marines physically prepared for combat. Senior commanders and their staffs carry out this role primarily by providing command emphasis, including planning, support, and supervision. Commanders at the battalion/squadron and company/battery level execute a program of physical fitness for combat by close supervision and, most importantly, leadership by example.

1104. HISTORY OF MILITARY PHYSICAL READINESS

Every war has revealed our military physical deficiencies during the initial periods of mobilization. This realization followed the Civil War and has recurred regularly with each national emergency.

a. Training programs in each war were geared to the physical need of the era. Success was dependent upon the amount of time available during training to physically prepare Marines for battle conditions. Frequently, casualties in initial engagements were attributed to the inability of our Marines to physically withstand the rigors of combat over rugged terrain and under unfavorable climatic conditions. With adequate preparation, Marines have always handled the test of battle.

b. During World War II, the first physical conditioning doctrine that could be scientifically justified by testing was introduced. As the war progressed, this program was effective in the physical conditioning of millions of Marines for combat.

c. Postwar periods have traditionally been a time of consolidation. Unfortunately, some leaders considered the conditioning phase of training to be a wartime tool. With such a philosophy prevalent between wars, physical readiness was relegated to secondary importance resulting in a serious lowering of combat effectiveness. The initial commitment of Marines in Korea dramatically displayed this failure to recognize the extreme physical demands of warfare. Call-up of large numbers of reserve Marines and pressing them into the war within weeks after call-up mainly caused the loss in effectiveness.

d. Costly lessons learned from our military experiences over a period of years and the course of several wars led to an increasing interest in the physical conditioning of the individual Marine. No longer can we afford to emphasize physical fitness during wartime and de-emphasize it during peacetime. It is evident that, in spite of increased mechanization and modern weapons, physical readiness retains a vital place in the life of each individual Marine and in every unit within the Marine Corps regardless of mission.

e. Commanders are well aware of the need for rugged and well conditioned Marines. The daily demands of housekeeping, maintenance, support, training, operations, and other time-consuming tasks make it necessary for commanders to set aside time for frequent, regular, and vigorous exercise periods.

1105. PHYSICAL DEMANDS OF COMBAT

There are three primary elements of effective physical fitness for combat: lower body strength and stamina; upper body strength and stamina; and a competitive, combative spirit. Cardiovascular functioning is not treated as a separate element since it is a necessary foundation for the other three. Exercises and activities designed to improve Marines' condition in the three primary elements will foster good cardiovascular and respiratory condition.

a. Lower Body Strength and Stamina. Some common demands that combat places on lower body strength and endurance are marching long distances under load and functioning effectively at the destination; moving quickly and evasively under fire; and carrying wounded Marines to safety.

b. Upper Body Strength and Stamina. Some common demands that combat places on upper body strength and stamina are rapidly emplacing crew-served weapons; handling large-caliber ammunition for extended periods; climbing walls, cliffs, and other high obstacles; and performing field maintenance on air craft or heavy machinery.

c. Competitive, Combative Spirit. While part of this critical aspect of combat readiness is mental and emotional, robust physical condition and a training program which promotes physical aggressiveness greatly enhance a competitive and combative spirit. Activities which force Marines to overcome natural physical fear, which directly leads to fatigue, are particularly valuable.

1106. PHYSICAL EFFECTS OF COMBAT STRESS

It has been thoroughly documented that the added uncertainty and stress of combat have a major physical effect on Marines as well as the acknowledged psychological effect. In other words, in combat, fear equals fatigue. In training, we cannot easily reproduce this combat stress which reduces the effectiveness of individuals and units. However, we can produce fatigue and psychological doubt in training situations by developing a challenging physical training program which stresses the participants. This compels Marines to exercise their ability to continue to function under difficult and trying conditions. Such activities as long foot marches under load and difficult confidence courses train Marines to overcome their own fear and fatigue. Rugged competitive activities such as martial arts and pugil stick fighting prepare Marines psychologically to overcome an opponent as well as their own fear and fatigue. This sort of physically demanding training builds self-confidence and unit morale. The adage that "the unit that sweats more in peacetime will bleed less in battle" certainly applies to its physical training program. Athletic coaches have long followed this approach by such practices as having teams shoot foul shots after wind sprints in order to strengthen their mental conditioning and toughness. Although pushing Marines to their limit is necessary, commanders often set lower standards initially than ultimately desired. The program should then call for steadily increasing the difficulty. A good guide for a starting point is a standard that will be physically and psychologically demanding for no less than three-fourths of the unit. Expressed another way, no more than one-fourth of the unit should be working at less than their limit. The program should progress quickly, augmented, if needed, by remedial training for those unable to meet the unit standards, until the training is challenging for all but the most exceptional Marines in the command. Commanders must never forget that Marines expect to be challenged and thrive on being pushed to their limit.

1107. LIMITS OF PHYSICAL READINESS

A well-conditioned Marine unit has a significant advantage in combat, but this does not mean that physical conditioning can substitute for technical knowledge or good planning. For instance, units must, as discussed in chapter 3, pursue a continuing and challenging program of marches under loads. This does not disprove the axiom that fighting men, if required to carry over one-third of body weight, will generally become ineffective in battle. Leaders and planners must decide what is essential to be carried and must use an effective logistic distribution system, such as unit trains, to make remaining items available when they are needed. In a similar vein, the valid requirement to foster a competitive and combative spirit in Marines does not excuse leaders from the responsibility to practice tactics which will not needlessly risk lives. Physical readiness for combat is a vital goal, but it is only one part of a unit's overall readiness for the lest of combat.

Section II. Fundamentals of Physical Fitness

1201. FITNESS FOR MARINES

Total combat readiness includes technical proficiency and mental, emotional, and physical fitness. If any one of these attributes is lacking, combat effectiveness suffers proportionately. Without technical fitness, Marines lack the knowledge and skill to fight; without mental and emotional fitness, they lack the incentive and will to fight; and without physical fitness, they lack the physical ability and confidence to fight. Physical fitness in a Marine means a whole and healthy body, the capacity for skillful and sustained performance, the ability to recover from exertion rapidly, the desire to complete a designated task, and the confidence to face any eventuality.

1202. COMPONENTS OF PHYSICAL FITNESS

A sound body, free of disease and defect, does not in itself constitute physical fitness. Before an untrained Marine can be considered physically fit for combat, the following traits must be developed:

a. Strength. Every Marine needs enough strength to perform the heaviest task encountered in routine and emergency activities. The basic areas where strength is required are in the arm and shoulder girdle, abdomen, back, and legs. Muscles increase in size, strength, and firmness with regular and strenuous exercises. Without work, they grow flabby and weak.

b. Endurance. Each Marine needs sustaining power to maintain the maximum ability without undue fatigue. There are two types of endurance:

(1) Muscular Endurance. The Marine needs muscular endurance to fight the enemy under the most tiring combat conditions. Muscular endurance is the ability to perform continuous work over long periods of time. Endurance depends on the bloodstream's ability to deliver large amounts of oxygen and nutrition to the muscle masses and then carry away the waste products quickly.

(2) Cardiovascular-Respiratory Endurance. The development of cardiovascular-respiratory endurance ("wind") is necessary to maintain muscular endurance. Cardiovascular-respiratory endurance depends on the efficiency of the blood vessels, lungs, and heart. The maximum effort a Marine can exert over a period of time is limited by the capacity to absorb oxygen and expel carbon dioxide. The average Marine's cardiovascular-respiratory capacity can be greatly increased by exercise.

c. Agility. A Marine must be able to change direction quickly and as faultlessly as possible. The ability to react instantly and to maintain orientation during rapid changes of body position is important to survival. This important ability may be developed by conditioning exercises that require varied and rapid changes of body position on the ground and in the air, such as obstacle courses.

d. Coordination. Coordination is the ability to move all parts of the body in a smooth, efficient, and concerted effort (commonly called timing). A well-coordinated individual does not make useless movements. An individual who moves with precision and accuracy saves energy. Coordination is best developed by practicing diversified muscular activities and skills affecting all body parts.

1203. TYPES OF EXERCISE

Basically, there are two forms of exercise: isotonic and isometric. Both forms are contained in the activities found in this manual.

a. Isotonic. Isotonic exercises are those in which the expenditure of energy is regulated and released during concentrated efforts. The regulated expenditure of energy is controlled by both the mode of exercise and the individual's effort. This type of effort is common to most exercise and sports. To develop endurance, coordination, and flexibility, isotonic exercise must be used. Strength can also be developed or increased through this type of exercise.

b. Isometric. Isometric exercises are those in which maximum effort is applied and held until the engaged muscle energy is depleted during a single contraction effort. The individual exerts full force against an immovable object for a relatively short period of time (6 to 10 seconds) and then repeats the contraction several times with a short period of rest between each contraction. This type of exercise develops only strength; therefore, it has limited application.

1204. PRINCIPLES OF PHYSICAL CONDITIONING

Men and women vary in their physical makeup. Physiological (body) function and reaction also vary in proportion to heavy demands placed upon the body. To attain the maximum benefit without sacrificing Marines' welfare, fitness must grow with a careful program of conditioning. To allow for adjustments in body functioning as the conditioning program progresses and to ensure attainment of objectives, the following principles of physical conditioning must be applied:

a. Overload. As strength and endurance increase, the physical load must be increased until the desired level of fitness is reached.

b. Progression. In beginning stages, the load must be moderate. Gradual progression from this low state of fitness to a higher state is possible through application of a progressive program.

c. Balance. An effective program utilizes various types of activities and provides for the concurrent development of strength, endurance, and coordination as well as basic physical skills.

d. Variety. Some programs fail because the routine becomes boring. The most successful programs always include conditioning activities, competitive events, and military physical skill development.

e. Regularity. There is no easy or occasional way to develop physical fitness. Regularity of exercise is a must, with daily exercise preferred.

1205. THREE STAGES OF PHYSICAL CONDITIONING

Unconditioned or poorly conditioned Marines pass through the following stages in reaching the desired state of physical condition.

a. Toughening Stage. This stage is approximately 2 weeks long and is usually characterized by muscular stiffness and soreness followed by recovery.

b. Slow Improvement Stage. This stage is approximately 6 to 10 weeks long and is characterized by slow and steady improvement until the desired level, or a high level, of fitness is attained.

c. Sustaining Stage. This stage goes on indefinitely in order to maintain the level of conditioning achieved by passage through the previous stages.

1206. EFFECTS OF CLIMATIC CONDITIONS

Temperature, both atmospheric and body, affects the physical performance of Marines. Proper maintenance of body temperature through warm up exercise, proper dress in cold weather, and removal or adjustment of clothing in hot weather is necessary for effective performance and health. Climatic factors to be considered are the following:

a. Exercise In High Temperatures and High Humidity. Marines can endure strenuous physical activity in extremely hot temperatures if they are given an opportunity to become acclimated and if they take enough salt and water. It is essential to continue physical training programs in hot climates. Marines can better withstand high temperatures when they are well-conditioned. High humidity combined with high temperatures presents a serious danger. These conditions prevent the natural cooling of the body by the evaporation of perspiration. Training schedules should conform to the provisions of the base commander in respect to wet bulb conditions. Those who conduct training under these conditions should monitor weight loss and be careful to make fluid replacement adjustments.

b. Exercise at High Altitudes. Certain problems are encountered in conditioning Marines stationed in high altitudes because the heart undergoes greater exertion during exercise. It is particularly important that only light exercise be given initially at such altitudes. A Marine's body gradually adjusts to high altitudes within a few weeks. After this adjustment, progressively greater amounts of exercise are possible.

c. Exercise in Arctic Regions. Military duty in the arctic is so arduous that a high level of physical conditioning is essential. Because of the difficulties of carrying on physical conditioning exercises in extreme cold, Marines should be conditioned to the highest level possible before they arrive. A sustaining program will then maintain that level. When exercising in cold weather, Marines should be required to remove excess clothing to prevent them from becoming damp with perspiration.

1207. WARMING UP AND COOLING OFF

It is a fundamental physiological principle that Marines should warm up gradually before taking strenuous exercise. Such action speeds up the circulation to prepare the body to take an overload and helps to prevent injury to muscles and joints after exercising. Marines should be kept mildly active, walking, stretching, or performing some other mild muscular activity until breathing and temperature have returned to normal. Marines should never be allowed to cool off too rapidly; in cool or cold weather, they should put on additional clothing during the cooling-off period.

1208. PHYSICAL ACTIVITY AS AGE INCREASES

Combat makes severe physical demands on Marines. All Marines, regardless of age, must be physically ready to meet these demands. There is no physiological reason to cease exercise or exertion with age. Increased age usually brings increased responsibility which, in many instances, leads to a routine that can become almost devoid of physical activity. The key to fitness with increased age is to continue exercising at a reasonable level and to include exercise of a vigorous type in the daily routine. Older persons who have not regularly maintained a reasonable state of physical fitness require more time than younger persons to become fit. Such individuals usually require a longer period of time to recover from physical effort than younger Marines. If general health is good, evidence shows that older Marines can develop and maintain a degree of fitness which permits vigorous activity and proper performance of duties. It is both the individual's and the commander's responsibility that all Marines continue a daily sustaining exercise program.

Section III. Goals of Physical Readiness Training

1301. NECESSITY FOR PHYSICAL READINESS TRAINING

A very important objective of training is attainment and maintenance of operational readiness. Marines must be physically ready for operations at any time and under all conditions of climate and environment. A combination of training to develop proficiency in physical skills and conditioning to improve strength and endurance results in physical readiness for combat. The degree of physical fitness required of Marines can be acquired through physical exercises in a normal well-coordinated and closely monitored training routine. The performance of purely military duties, such as drills and marching, is not enough to build all the desired areas of fitness. Few recruits are physically fit for the arduous duties ahead of them. The softening influences of our mechanized civilization and the difficulties of conditioning Marines make physical fitness more important than ever before. If Marines are to be fully developed to and maintained at the desired standard of physical fitness, a well-conceived plan of physical readiness training must be a basic part of every training program. Marines cannot be adequately prepared in any other way for the hard work and arduous demands required on the battlefield.

1302. OBJECTIVE OF PHYSICAL READINESS TRAINING

The overall objective of the physical readiness training program is to develop individuals and units who are physically able and ready to perform their duty assignments or missions during training and in combat. To attain the objective of physical readiness, exercise activities must develop the following areas:

a. Strength and Endurance. Developing adequate strength to perform required duties and adequate endurance to sustain activity over a long period of time.

b. Muscle Tone. Developing muscle tone adequate to maintain proper posture and reasonable weight control.

c. Skills. Developing military physical skills which are essential to personal safety and effective combat performance. As skill is developed, agility and coordination are attained. The essential skills are-

(1) Marching Under Load. Marching with individual and unit weapons and equipment.

(2) Running. Distance and sprint running on roads and cross-country.

(3) Jumping. Broad jumping and vertical jumping downward from a height.

(4) Dodging. Changing body direction rapidly while running.

(5) Climbing and Traversing. Vertical climbing of ropes, poles, walls, and cargo nets. Traversing horizontal objects such as ropes, pipes, and ladders.

(6) Crawling. High crawling and low crawling for speed and stealth.

(7) Throwing. Propelling objects, such as grenades, for a distance with accuracy.

(8) Vaulting. Surmounting low objects, such as fences and barriers, by use of hand assists.

(9) Carrying. Carrying objects and employing man-carries.

(10) Balancing. Maintaining proper body balance on narrow walkways and at heights above normal.

(11) Falling. Contacting the ground from standing, running, and jumping postures.

(12) Surviving in Water (Or Other Specialized Situations) Using water survival techniques.

d. Character Traits. Jnstilling character traits which help accomplish military missions to include-

(1) Confidence. Confidence develops through achieving progressively more difficult tasks as physical ability grows.

(2) Aggressiveness. Participation in combative activities and contests develops desire and willingness to overcome an opponent.

(3) Reaction Under Pressure. Competitive contests and game situations train Marines to think and to act quickly while under pressure.

(4) Teamwork. Teamwork develops through competitive events in which a number of Marines must coordinate their efforts to accomplish a physical task.

1303. BENEFITS OF EXERCISE

The benefits of exercise are not always understood. Some of the more important results of exercise are listed below:

a. Improved Muscle Tone. Muscular tone improves and, at the same time, muscular strength and endurance are built up.

b. Cardiovascular-respiratory Endurance. Cardiovascular-respiratory endurance, or wind, improves through a process of opening up dormant lung capacity to absorb greater amounts of oxygen.

c. Circulation. Circulation of the blood speeds up and extends to a greater portion of the body as exercise forces the blood to service all parts of the body. The efficiency and effectiveness of the heart, lungs, and blood vessels improve.

d. Flexibility. A wider range of muscular movement is possible and rapidity in physical skills grows.

e. Elimination of Body Waste. Bending and twisting the body and the general speedup of body processes caused by exercise regulate and help eliminate body wastes.

f. Tension. Working off excess nervous energy and relief from daily worries and cares relieve tension. Participation in exercises leaves little time for worry.

g. Sleep. Sleep improves because muscles are healthfully tired after about of exercise. A by-product of sound sleep is relief of tension.

h. Obesity Control.. Control of obesity (fat) is made possible by using up excessive amounts of fat-producing food elements.

i. Injury Susceptibility. Susceptibility to injury is reduced through exercise. Muscles, tendons, and joints are strengthened. Injuries such as hernia, back strain, and joint sprains are less likely to occur if muscles are maintained in proper tone.

Section IV. Leadership Roles

1401. PSYCHOLOGICAL LEADERSHIP

The full development of a Marine's resources is not all physical. To be effective in developing physical readiness, leaders must realize that mind and attitude are also important to success. The more important psychological considerations are to—

a. Promote Understanding of the Value of Physical Readiness. A desire to be physically ready should be created in all Marines. Motivation is increased and Marines take greater interest in their individual physical fitness if they understand the value and benefits of vigorous exercise. When Marines realize their efforts are an investment in their own personal welfare, it should not be difficult to obtain their cooperation. Marines should understand the objectives, the benefits, and the value of each type of exercise activity in their program. They should also understand the relation of physical readiness to survival in combat.

b. Maintain a Positive Approach. Physical readiness training for combat is strenuous and demanding. It is a responsibility of leadership to create an atmosphere where all desire to participate fully. This attitude should be fostered. A negative approach must not be identified with physical readiness training, even with those having difficulty. Only in unusual cases should fear of punishment be the motivating factor behind good performance. For those few who cannot keep up or attempt to malinger, an effective remedial program is essential.

c. Seek Cooperation and Develop Morale. In a program placing maximum physical stress upon individuals, it is necessary to gain their cooperation. Favorable reaction is enhanced by proper planning and organization, challenging requirements, use of competition, and application of a progressive program resulting in physical fitness. As physical fitness grows, morale also grows.

1402. COMMAND AND SUPERVISORY FUNCTIONS

a. Command Functions. Commanders should take the following actions to support physical readiness training:

(1) Lead by personal example.

(2) Instill command interest and indicate to subordinate personnel the importance of this training to the welfare of the organization.

(3) Allot sufficient time for the achievement of objectives and monitor the use of such allotted time. The substitution of other training or routine duties for scheduled physical readiness training is unsound and unwise.

(4) Assign and properly utilize qualified personnel to supervise and conduct physical readiness training. If leaders are not competent, take action to ensure they become competent quickly. Reassign those who do not meet this standard to other duties.

(5) Make necessary facilities and funds available to support a program to develop physical readiness within all personnel.

(6) Measure the physical fitness of individuals and units in order to evaluate progress and to determine if the program is successful.

b. Supervisory Functions. Leaders responsible for planning, conducting, and supervising physical readiness training should take the following actions:

(1) Prepare physical readiness training schedules which apply the principles of physical conditioning and which aim for a particular type of program plan.

(2) Provide for wide participation of as many Marines as possible. All Marines, regardless of position or age, will benefit from regular exercise. In some instances, special efforts are necessary to overcome obstacles to regular and frequent training. Special effort is also necessary to ensure remedial conditioning. Such conditioning should occur for those who are physically substandard and after extended absence due to leave, sickness, injury, and travel.

(3) Prevent waste or unwise use of time allotted for physical readiness training. Time-wasters include unprepared instructors; assignment of one instructor to a group larger than a platoon; progression which does not keep pace with the physical development of the Marines; extreme formality; inadequate equipment or facilities which require waiting turns to exercise; and lengthy rest periods between exercises which interfere with the application of overload.

(4) Ensure that the program contains vigorous physical activity. Such activity places progressively greater demands upon the body during each exercise session and also over the duration of the training program. To be of benefit, exercise must tire the muscles and cause the heart to increase its rate of beat.

(5) Set an overall objective for each physical fitness program.Observe the training as necessary to ensure that the established objectives are being achieved.

(6) Observe physical readiness training to insure the use of a positive approach. To implement a positive attitude, small-unit leaders and instructors should personally set the example; have an understanding, fair, and sympathetic attitude; recognize individual differences; and motivate Marines toward their best effort.

(7) Guide and inform small-unit leaders and instructors concerning approved techniques, directives, and literature. As necessary, arrange for local training of instructors to include clinics, conferences, schools, and demonstrations.

(8) Determine the effectiveness of physical readiness training by personal participation in and observation of training, analysis of field inspection reports, and analysis of individual physical fitness test scores. Scores may be combined to reflect the fitness of the unit.

1403. SMALL-UNIT LEADERS AND INSTRUCTORS

a. Responsibility. The instruction and conduct of physical readiness activities are the function of company/battery commanders, platoonleaders and persons assigned as instructors. Experience has proved the effectiveness of physical fitness development when conducted in company-and platoon-size units under direct control of the leader with overall supervision by the parent-unit commander. For example, all the platoons of a company may exercise at the same time under the general supervision of the company commander, with each platoon conducting the assignment separately and under its own leadership.

b. Leader's Assignment. A small-unit leader or an instructor in a school or training activity is assigned to a combat unit or to a support unit. In this assignment, the leader is responsible for all training to include physical readiness training. In a different situation, the noncommissioned officer (NCO) or officer may beassigned as a full-time physical readiness instructor. This contrasts to a unit leader assignment where only part of the time is devoted to such training. In either case, the leader will hold an important and vital position for the physical fitness of Marines.

c. Leader's Training. Leaders may come to the assignment either fully or partially trained. It may be their first responsibility for the development of physical fitness. If they have had previous training through experience, make certain that their information is supplemented with study of this manual. If they have had professional training in physical education during civilian life, but no military experience, they should also use this manual to learn the methods used by the Marine Corps. A new leader should take advantage of various ways to learn including attendance at leader training courses, self-study, practice, and discussion with more experienced leaders.

d.Leader's Objective. As a physical readiness training instructor, the leader has two general objectives. The first is to motivate Marines to want to be physically fit. The second is to conduct a program that will develop a high degree of physical fitnss. Motivated Marines will react enthusiastically to such a program. It aids greatly in achieving local program objectives.

e. Leader's Personal Fitnes. A unit leader who must instruct and demonstrate physical activities must be in physical condition to do the job without undue physical stress.The leader should be able to do those things that must be demonstrated. The leader's strength, endurance, posture, and skill should set the example. This does not mean that the leader must excel, as other Marines do not expect championship performance. However, they do expect, and deserve, a creditable showing of fitness for the job.

f. Leader's Knowledge. The leader must have three types of knowledge to properly administer physical readiness training. Theyare- -

(1) Knowledge of Marines. The leader must understand Marines, know how to lead and motivate them, understand how they learn, and apply this knowledge wisely in the day-to-day training situation.

(2) Understanding of Body Functioning. A more intelligent exercise program results from understanding and applying the principles which govern physical conditioning of the body. The leader with such knowledge can better prescribe, adjust, and regulate exercise types, amounts, and progression to attain fitness.

(3) Understanding Exercise Activities. The leader needs to understand the contribution each type of physical activity makes to physical fitness, and how to use each activity to develop fitness. Skill to demonstrate and lead the various activities is a necessary part of technique and is invaluable to the instructor or small-unit leader.

Chapter 2
PHYSICAL READINESS TRAINING PROGRAMS
Section I. Development of a Program

2101.GUIDANCE FOR PLANNERS

This chapter instructs planners on physical readiness training procedures. It contains program planning guidance including factors to consider when developing programs; steps in assembling a program; definitions of activity packages and systems of exercise; selection of activity packages; and selection of systems used in implementing those packages.

2102. MARINE CORPS MISSION

The mission of the Marine Corps is to seize and defend advanced naval bases and to perform other missions as the President of the United States may direct. Traditionally, to perform this mission, Marines have been projected into the area of operations by Navy vessels and have attacked hostile beaches over the shore in landing craft, assault amphibian vehicles, and helicopters. Recently, performance of this task has been somewhat complicated by the introduction of the concepts of maritime prepositioning and the air-landed Marine expeditionary-brigade. Under these concepts, Marines, landed at airfields in foreign countries, will join their equipment at commercial ports. The distance between these airfields and ports is often 10 or more miles. In time of international crisis, it can be expected that transportation assets will not be adequate. Marines from all elements of a Marine Air-Ground Task Force may have to march from the airfield to the site of their equipment. All Marines must be physically conditioned to rigors of conducting foot marches with individual weapons and equipment.

2103. EVALUATION OF FITNESS

The Marine Corps mission is such that all Marines must constantly and consistently achieve a high level of fitness that prepares them for the demands of combat. While Marines have many different MOS's, all Marines must be prepared for the demands of marching under load and performing basic infantry tasks, such as rear security and patrolling. While the physical fitness test remains our universal measure of individual fitness, it is the commander's responsibility to observe and evaluate the unit's ability to perform effectively in combat. Often the commander's most effective evaluation tool is weekly physical training sessions. The commander should participate in these sessions and see that they are fully integrated into the unit physical conditioning program.

2104. INTEGRATION OF TRAINING

Training time must be used efficiently and wisely. Every opportunity to integrate physical training into other training activities should be seized. For example, foot marches under load can be integrated into the normal daily activities as a way to move from one training or work site to another. This integration reflects realism in training and should be used to maintain overall proficiency. To accomplish this requires imagination and enthusiasm on the part of the commander. The area of operations in which the training is conducted determines what can be integrated and how. If possible, training should also consider the terrain and climate of the area in which the unit will subsequently conduct training or operations. Training should include familiarization with special equipment and the application of specialized techniques to tactical principles. Activities such as gun drills and command post emplacement drills also have physical conditioning value, particularly if supervised and done competitively or measured against time standards.

2105. ACTIVITIES AT THE UNIT LEVEL

Physical readiness training is a command responsibility and is generally conducted in Marine units at the company and platoon levels. Battalion/squadron commanders must constantly supervise, making sure that all Marines are physically prepared for combat.

a. Physical readiness training should usually be conducted weekly by each battalion/squadron-level unit, using events designed for all personnel (e.g., marches under load, battalion/squadron runs, military field days, etc.).These events enable the commanders to demonstrate personal leadership, and observe and evaluate the physical condition and combat physical readiness of their Marines. This exercise period should be designed to elevate the unit's morale and emphasize unit identity through the wearing of uniforms and the carrying of battalion and company guidons. The battalion/squadron commander should make every effort to be a visible participant in the unit's physical readiness program on a weekly basis. These activities could be conditioning marches under load or a weekly calisthenics/unit run.The suggested times for these activities are early on Monday morning or late on Friday afternoon after the training week has been completed. The uniform may be physical training gear orutilities and boots.

b. Commanders should be cautious of long runs in boots and utilities. This is not to suggest that units should not run in utilities, but to remind commanders that boots are designed for marching, not running. If commanders choose boots and utilities, they should select off-road routes which will provide a cushioning effect.

c. Units should be made well aware of the commander's intent before the unit falls out for exercise. If the exercise period is conducted early in the morning, it need not necessitate an extensive shower/clean-up period afterward, so long as all Marines from the commander on down remain dirty. Commanders should inform their Marines not to fall out for these sessions in clean, pressed uniforms.

d. The physical readiness training of headquarters units is often difficult to manage. Often physical training is left for individuals or sections to develop and manage on their own. This approach can be effective, but if not supervised, it can result in a "paper" program. Headquarters unit commanders should endeavor to conduct weekly integrated training which will enable them to evaluate the physical readiness of their personnel. Often this training has to be scheduled before or after normal working hours to maximize participation of the entire unit.

2106. STEPS IN PLANNING

To implement workable and effective programs (as directed by CMC ALMAR 261/87) the planner must-

a. Determine the Type of Program Needed. Marine units are inherently different in organization and mission. The physical readiness program must be tailored to the mission and to the current physical condition of most unit personnel. Programs to meet this need are of the following types:

(1) **Developmental Programs.** Marines in a beginning or poor state of physical readiness need a program which will develop strength, endurance, physical skills, and character traits which are beneficial to successful accomplishment of military missions. Such programs should be applied progressively to rise gradually to a peak of fitness and skill.

(2) Maintenance Programs. Once Marines reach the sustaining stage of conditioning, their goal is then to maintain this level by participation in a maintenance program.

(3) Remedial Programs. The term "remedial" is usually applied to those individuals or groups who possess substandard physical fitness. For example, a remedial physical conditioning program could be applied to persons who are overweight, who fail to reach physical fitness test standards, or who have missed extended periods of conditioning due to illness, injury, extended hospitalization, or other absence. Saturday training sessions are designed as remedial sessions for all individuals who either miss the daily training or have failed to meet adequate standards. Note that personnel in light duty or no-duty status should be expectedto attend all training sessions, observing, supervising, or assisting as needed.

b. Determine the Time Required. The amount of time for training operations varies considerably. However, every unit can find time to conduct physical readiness training. Frequent (i.e.,daily) physical training of short duration-30 to 60 minutes-is preferred over occasional longer periods. Other demands for training time are so urgent that every minute of time allotted for physical readinesstraining should be used. Determine the time required per week anddivide it into daily blocks.

c. Organize for Various Group Size. It is essential to stress exercise rather than formality. Marines must complete the program where they are- -on the training field, in the motor pool, on the range, next to the classroom, in the office area, in the shop, aboard ship, or elsewhere. Although desirable, it is not always possible to assemble company-size units for physical training. Platoon-size groups are appropriate for the proper conduct of physical conditioning activities. Certain situations may require exercise programs for section- or squad-size units. This manual outlines programs for all situations and types of organizations.

d. Allow for Weather and Exercise Area. In programing and scheduling, the climate and terrain often govern the selection of activities.

(1) Weather changes cause differences in temperature, rainfall, wind chill, and snow. These changes should be anticipated as they dictate the type of program. Alternate plans should be part of the schedule. Nevertheless, normal weather changes and rain should not drive training inside.
(2) Local terrain and available exercise areas may also influence the selection of activities and the type of program which it is possible to support. Some activities can be completed in nearly any area.
e. Plan for Seasonal Change. As most physical readiness training is conducted outdoors, it is necessary to recognize seasonal change. A program should be divided into fall, winter, spring, and summer parts. Seasonal change also causes change in light. For example, an early morning program started in the summer will have ideal lightconditions, yet in the fall or winter, darkness will occur at that same hour and interfere with the conduct of the program. Develop programs in seasonal blocks, and make provisions for anticipated changes.

f. Consider Needed Facilities. An excellent program can be conducted with practically no facilities since there are exercises which require no equipment. However, a better program can be developed when supported by certain facilities and items of equipment. Proper command support, plus ingenuity, will solve this problem. (Items of equipment, when necessary to support the recommended exercises, are included in the chapters on exercise activities.)

g. Specify Appropriate Uniform. The uniform worn for exercising depends upon the season of the year, the state of the weather, and local regulations. Whenever possible, Marines should be dressed alike. Undershirts are preferred as the upper garment when the weather permits. A uniform that restricts the free movement of the body should not be worn when exercising.

h. Consider Availability of Instructors. Leaders who can lead and direct the scheduled activity must be available. Organizational units should train junior officers and noncommissioned officers down to squad or section leaders to instruct and lead the various activities.

i. Select Activity and System. With the type of program needed and the objectives in mind, select an activity package or an integrated training package for each day's scheduled physical training. At this same time, the system to be used in employment of the selected activity or activities must be determined. This selection affects equipment, areas, instructors, transportation, and other support requirements.

j. Secure Command Participation and Support. Prepare and brief the commander to assure full understanding of the objectives and administration of the program. The full participation and support of the commander will greatly improve the level of success.

k. Supervise Execution. Determine needs, publish the program, and supervise its execution as a necessary part of developing the unit's training schedule.

2107. SELECTION OF ACTIVITIES AND SYSTEMS

a. Exercise Activities. The planner must determine exercise activities which will be appropriate to include in the program. An exercise activity is a single means of exercise usually identified by the name applied; for example, running, log exercises, and obstacle course. Many exercise activi ties can become part of activity packages.

b. Activity Packages. Many of the physical activities described in this manual are arranged in prescribed sequences and are known as activity packages. An activity package is a number of exercises of the same type, assembled as a group or a set, and arranged in a specific sequence. Exercise packages are organized in such manner that not more than 15 minutes will be required to complete the execution of any package. Each type of activity is explained in laterchapters. The number of available drills, tables, or circuits; the manner of organization; and the contribution each makes to the total program are covered. Full understanding of this information will greatly assist in developing effective programs. Various designations are used to identify exercise packages; for example, conditioning exercises when arranged in a set order are known as drills, and other packaged activities are designated as tables or circuits. The following are activity packages:

(1) Activity packages for groups:

o Conditioning Drill Two
o Conditioning Drill Three
o Rifle Drill
o Log Drill
o Grass Drill
o Running Tables
o Guerrilla Tables
o Circuit-Interval Table
o Combatives Tables
o Relay Tables

(2) Strength Circuits:

o Fixed Circuit
o Movable Circuit
o Simplified Circuit (Circuit Interval Table)

(3) Activity packages for individuals:

o The 6-12 Plan
o Weight Training
o Isometric Exercise

c. Advantages of Using Packages. These of exercise packages simplifies scheduling and conducting of exercise and results in the following benefits:

(1) Schedule development is simplified as the planner assembles packages which will satisfy the training objective. There is no need to deal with selection of individual activities or to be concerned about the amount of time to be expended on each.

(2) Any 15-minute period, and in some cases less time, can be scheduled or used to perform an activity.

(3) If longer periods of time are available or if the objective demands, several packages can be assembled to provide a more complete period of activity.

(4) Marines are assured a balanced set of exercises or activities as each package is carefully arranged to reach all muscle groups.

(5) The instructor can concentrate on the conduct of a vigorous workout as the type and duration of the activity have already been determined.

d. Nonpackaged Activities. Several types of activity are not packaged. Activities in this category are conditioning marches, unit runs, obstacle courses, combat water survival swimming, teamcontests, and team sports. These activities can be scheduled in combination with packaged activities, or they may be scheduled separately. Often they require a longer period of time. Most nonpackaged activities require a 50-minute period to satisfactorily complete their objective. For conditioning marches, a few hours are required. The benefits of these activities should not be overlooked as some desirable objectives cannot be attained without them.

e. Systems of Exercise. Several methods or systems of organizing exercise and activity packages can be used. Each system is based upon a specific organization as follows:

(1) Single Activity System. The unit leader immediately assumes command of the unit at the beginning of the exercise period. The leader moves the unit to a predesignated exercise site at double time, forms the unit in a circle around the leader, groundsclothing and equipment as appropriate, and quickly moves into the exercise routine. There is usually no time to teach; therefore,the Marines must know the activity to be used. At the conclusion of a 5- to 15-minute period, the unit leader returns the unit to the instructional area at double time and releases the unit for the next scheduled activity.

(2) Progressive Activity System. All Marines (company or platoon) complete activities in the same order during the period. For example, Drill One is followed by dual combatives, and finally a 1-mile run. This system is usually progressive from a warm up activity, such as Drill One, to an activity which contributes in a major way to one of the objectives, such as aggressiveness development through combatives or cardiovascular development, such as running.

(3) Rotating Activity System. Set up the same number of activities or stations as there are platoons in the company. Each platoon rotates through each station in turn. With four platoons in a 50-minute period, about 10 minutes can be devoted to each station. With three platoons, approximately 15 minutes may be spent at each station. Activities must be of a type that can be covered in the time allotted. For example, with three platoons, Station 1 could be Conditioning Drill One; Station 2, running; and Station 3, a team contest.

(4) Circuit System. Set up a number of stations to provide various types of exercise equipment and/or items of apparatus. The idea is to keep all Marines busy and exercising vigorously for a short period of time at each station. The fixed strength circuits and the movable strength circuits are examples of this system. Station changes must be rapid, and the exercise must be started quickly after each change. Since the objective is to exercise at top speed, the motivation comes from frequent change of activity by moving to another station. Rotation by station groups continues until all Marines have covered all stations.

(5) Interval System. This system stresses the development of strength and endurance. It involves heavy work for a given distance within a specified time, alternated with lighter work and recovery, but never stopping during the workout. This procedure is repeated, and the intensity is increased gradually as physical condition improves, but always with adequate recovery. The important concept is stress, recover, stress, recover, and so on. This system is often applied through running, but other activities of a continuous nature may also be used. An example is the Circuit-Interval Table.

f. Steps in Selecting Activities and Systems. The planner should follow these steps.

(1) Evaluate the needs of one unit with emphasis upon unit mission, objective, and time available.

(2) For each day, select an activity package or a combination of activities which will contribute to the objective.

(3) Then determine the system to be used in implementing the selected activities. There are several possible choices and much opportunity for flexibility in program development. For example, the weekly program may contain various exercise packages, systems, and time periods. Additional guidance is contained in following chapters.

2108. SAMPLE WEEKLY PROGRAMS

The variety of weekly programs given shows available alternatives. Programs and schedules other than those illustrated may be assembled to provide schedules to fit any situation. Many other scheduling combinations are possible through use of the packaged activities, varied time periods, and different types of assigned duty.

a. Sample A. Figure 2-1 illustrates a 60-minute time allotment in which the single activity system is used with activity packages that vary from day to day. In this program, 5 hours of training are scheduled for the week. This program needs some type of integrated unit training to supplement it each week. A bimonthly conditioning march of 10 miles in 3 hours is suggested to round out this program. The Saturday session is designed for remedial physical training.

NOTE: A physical training activity must be conducted at a minimum of once every 72 hours. After 72 hours, muscles begin to atrophy.

Figure 2-1 is available in the illustration section at the end of the book

b. Sample B. A more comprehensive weekly program (fig. 2-2) illustrates the scheduling of various-length periods, a variety of activities, and the use of three systems during the week. In this schedule, 3.5 hours are included for the week. The following should be noted concerning this schedule:

Figure 2-2 is available in the illustration section at the end of the book

(1) On Monday, a 60-minute period is available for physical readiness training. Three 20-minute packages are scheduled. Each platoon, under its own leadership, will progress through each scheduled activity in turn.

(2) On Tuesday, Wednesday, and Thursday, only 30-minute periods are available. Here each platoon leader supervises each platoon in a single activity system.

(3) On Friday, the company commander directs training to exhibit personal leadership example and supervision. In this example, only one strength circuit and one cross-country course are available, and each will accommodate only one platoon. For this reason, Conditioning Drill One station is included. A platoon rotates to each station within the 60-minute period.

(4) The Saturday session once again is designed for remedial physical training.
(5) Bimonthly, a 10-mile/3-hour march under load will be necessary to round out this program.
NOTE: Cardiovascular/respiratory exercise to improve endurance must be of long enough duration (20 minutes or more) to induce and maintain oxygen debt.

c. Sample C. A third sample schedule (fig. 2-3) illustrates the use of the single activity system for 5.25 hours of training. This schedule includes sustaining-type activities for a unit that has passed through both the toughening and slow improvement stages of conditioning. The use of competitive activities is featured to holdinterest and provide self-motivation. A bimonthly 10-mile/3-hour march under load will be necessary to round out this program. The Saturday session once again is designed forremedial physical training.

Figure 2-3 is available in the illustration section at the end of the book

NOTE: Thc body will respond to hard work or physical training without soreness and stiffness provided the work or physical training is conducted regularly.
NOTE: Excrciscs of short duration increase strength and bulk but do little to effect endurance.
NOTE: Thc times shown in the charts arc participation times.

Section II. Remedial Physical Training

2201. NEED FOR ATTENTION

Remedial physical training is a process by which physically substandard individuals are conditioned to meet standard levels of performance. To achieve a full degree of operational readiness, it is necessary to bring all Marines up to the prescribed standard. Experience has demonstrated that some Marines have difficulty due to a poor state of fitness, obesity, or lack of motivation. Attention to these individual deficiencies will help improve unit combat readiness.

2202. IDENTIFICATION OF INDIVIDUALS

The company or battery commander identifies persons who cannot achieve the prescribed level of proficiency through the physical fitness test or as a result of observation during training. The commander notes particular weakness of body parts as indicated by failure of certain test events. These Marines are then placed in a special remedial program either at platoon, company/battery, or battalion/squadron level. Extra time is devoted to overcoming the weakness. These extra periods of conditioning may be during or after duty hours, as determined by the commanders. Other persons need reconditioning as result of hospitalization. Restoring physical fitness to damaged areas of the body is accomplished through progressively graded physical activities under professional supervision, not by the physical fitness trainers.

a. Types of Deficiencies. Physical deficiencies which can be corrected by exercise fall into several categories.

(1) Lack of Strength in One or More Body Parts. The major muscle areas concerned are the arms and shoulder girdle, back, abdomen, and legs.

(2) Lack of Overall Endurance. Usually, there is deficient muscular and cardiovascular endurance.

(3) Deficiency in Coordination and Agility. In these cases, physical skill is not developed to a satisfactory degree in activities such as crawling, running, jumping, climbing, traversing, vaulting, pushing, pulling, lifting, and carrying.

(4) Overweight or Underweight. Either condition may interfere with physical fitness and mission accomplishment. Lack of exercise is not always the cause. The cause may be malfunction of normal physiological functions or it may be poor health habits such as over- or under-eating, lack of adequate rest, or overconsumption of alcohol.

(5) Lack of Motivation. Not all persons are motivated to attain or maintain **a** desirable state of fitness. Some Marines do not understand the importance of physical fitness, some find proper exercise too difficult, and others find it inconvenient.

b. Causes of Deficiencies. There are several causes for Marines being deficient in physical fitness:

(1) Absence of exercise.

(2) Exercise which fails to develop all muscle groups and components of fitness.

(3) Exercise which is not vigorous enough or which lacks progression.
(4) Injury or illness which depletes fitness.
(5) Inadequate amounts of sleep or rest.

c. Methods of Detection. The commander has several means by which to detect physical deficiencies:
(1)
from those who pass. Make an analysis to determine the cause of failure as revealed by test scores.
(2) Observe Marines as they perform physical tasks-
o Marines who have difficulty during training or in physical types of work,
o Marines who have become obese and therefore experience difficulty.
(3) Be alert to those Marines who are often on sick call, returnees from hospitalization, or newly assigned.

2203. ADMINISTRATION OF REMEDIAL ACTION

a. Group. Attitude Marines who are singled out as being physically deficient are self-conscious and are not always convinced they need extra help. Within a deficient group, motivation may be low. These Marines must be convinced that a remedial program which is tailored to their needs will help them and will eliminate their deficiencies. Encouragement is often needed and desirable.

b. Leadership. The leader of this group must study individuals and know their deficiencies. The unit leader should counsel Marines individually, observe them closely as they progress through the remedial program, maintain records, and adjust the program as required.

c. Measurement.. Whenever improvement in performance is noted, Marines should be measured by the physical fitness test either individually or by group. When testing reveals an individual to be satisfactory, that individual should be released from the remedial group. There may be exceptions to this policy in the case of Marines who are overweight or in the execution of an exercise program prescribed for some special purpose.

d. Organization of a Remedial Group
(1) A remedial group is usually a small group of Marines within a company or battalion. In some situations, it may be a larger group numbering 50 to 75 Marines. In the case of a larger group, divide the Marines initially into subgroups according to ability. Prescribe exercise loads commensurate with their ability. General conditioning is sufficient in the early part of the program to qualify Marines who are on the borderline. Qualification will reduce the size of the group, permitting more individual and specialized attention to those who remain.
(2) As the program progresses, regroup individuals who have like deficiencies in order that they can concentrate on their weakness. For example—

(a) One group may be weak in the arms and shoulders as revealed by failure in the chin up event. This group, with an assistant instruction charge, can work on pull-ups, rope climb, pushups, rifle or logexercises, horizontal ladder, and similar types of exercise.

(b) Another group may be weak in general endurance as revealed by the3-mile run. This group could profit from participation in conditioning drills, running, grass drill, and strength circuit.

(3) It may be necessary to form some groups to overcome weaknesses in skills, such as an inability to throw, to quickly change direction while running, to crawl rapidly, or to carry a load. Lack of coordination or lack of practice may cause these deficiencies. In this instance, instructors must provide an opportunity to practice and correct poor form and other errors as they are noted.

Chapter 3
PHYSICAL CONDITIONING ACTIVITIES
Section I. Marching Under Load

3101. GENERAL

Few physical fitness activities are as directly related to readiness for combat as foot marches under load. In addition to the obvious physical conditioning and unit cohesiveness benefits, marches under load prepare most Marines for numerous foreseeable tasks in combat. The idea that only infantry and reconnaissance units actually require training marches to prepare them for combat is clearly a mistake. In addition to the very real possibility of Marines from all types of units being used to fill combat-depleted infantry units, it is likely in maritime prepositioning force operations that aviation, combat service support, and command element personnel will have to march with weapons and equipment, from the arrival airfield to the equipment marshalling site. In American coastal cities, the harbor is nearly always 10 or more miles from the airport. It is difficult to imagine any different condition in a lesser developed country .A further combat-related benefit is that foot marching under load is the most effective way to develop leg strength in the context of a unit training program. In his classic study of troop performance in World War II, Men Against Fire, S.L.A. Marshall noted that the greatest single weakness of replacements of all occupational fields was leg strength. Truck drivers must push trucks which are stuck in mud, and headquarters communicators must climb hills to emplace antennas. The demands on the lower body in combat apply to all Marines, and foot marching is one activity which can help build these muscles while being conveniently integrated with the rest of the training program.

3102.TRAINING GOALS

The standard for success of a foot march is very simple to measure: did the unit arrive at the destination at the prescribed time with Marines in condition and required equipment present to accomplish the mission?A progressive program can increase a unit's readiness. It can instill pride by increasing distances and rates of march and by selecting routes over increasingly challenging terrain as the Marines become better conditioned. However, the ability to execute the mission at the conclusion of the march must remain the standard of success.

3103.MOVEMENT PLANNING

The basic considerations in planning a foot march are the mission, tactical situation, terrain and weather, and the units to participate. The success of the march will depend largely upon the thoroughness with which it is planned. A successful march is characterized by adherence to prescribed routes and time schedules, the efficient employment of the means available, and the ability of the unit to accomplish its assigned mission upon arrival at the destination.

a. Movement Orders. Movement planning culminates in the preparation and issuance of an operation order prepared in the standard, five-paragraph format. Necessary annexes are attached to furnish detailed information required for the movement. Written movement orders are rarely prepared at company level. A discussion of the annexes is contained in this chapter.

b.March Planning. March planning, as discussed here, is the planning conducted at battalion level. March planning may be organized into the following steps:

(1) **Preparation and Issuance of the Warning Order.** In order to afford subordinate units the maximum possible time to prepare for a pending move, a warning order containing all available information about the march is issued. The amount of planning time available will determine the time of issuance and the content of the warning order.

(2) Estimate of the Situation. In his estimate, the commander considers the mission, terrain, weather, time, and space factors, available routes, available transportation for the movement of equipment and/or shuttling of Marines, enemy capabilities, disposition of own forces, physical condition and training of Marines, and courses of action available to the command.

(3) Organization and Dispatch of a Reconnaissance. Party very march plan is based on as thorough and complete a ground reconnaissance as time and the situation will permit. Map and aerial reconnaissance are valuable in formulating a plan, but are not a substitute for ground reconnaissance. Route reconnaissance is accomplished by a reconnaissance party which usually consists of a reconnaissance element, an engineer element from the attached or supporting engineer unit, and a traffic control element. Unit standing operating procedures generally establish the basic composition of the reconnaissance party. It is modified as necessary to meet the requirements of a particular march. A recommended method of reporting information obtained by the route reconnaissance party is shown in figure 3-1.The minimum information required from the reconnaissance party is—

(a) Available routes and conditions.(Routes may be specified by higher headquarters.)

(b) Recommended rate of march.

(c) Selection of start point and release point, or confirmation of the suitability of start point and release point previously selected by map reconnaissance.

(d) Confirmation of location of the assembly or bivouac area.

(e) Location of critical points on the route.

(f) Distance between critical points on the route and total distance from start point to release point.

Figure 3-1 is available in the illustration section at the end of the book

(g) Location of obstacles and estimation of necessary Marines and equip- ment needed to repair and maintain routes.

(h) Number of guides required and their location on the route.

(4) Development of Detailed Movement Plan. Some of the elements included in a detailed movement plan are-

(a) Organizing the Column. To facilitate control and scheduling, units will be organized into serials and march units and given an order of march. In determining the order of march, the march planner must consider the enemy situation and the desirable order of arrival of the units at the destination. Where dispersion is required, a unit may be organized into two or more columns, each assigned a different route.

(b) Using Reconnaissance Information. Results of the route reconnaissance will be used to select the route(s); determine the start point, critical points along the route(s), and the release point; and select the rate of march.

(c) Determining March Computations. March unit pass time is based on the strength, formation, and rate of march. The pass time of the marching columns, plus necessary time distance computations, will be used to determine the completion time of the march.

(d) Drafting of Road Movement Table. Using the completed march computations, a draft road movement tabic is compiled.

(e) Checking the Plan. Using the draft road movement table and a road movement graph, the movement plan is checked to ensure that it conforms to the directive of the higher headquarters and the battalion commander's instructions.

(5) Preparation and Issuance of the Road Movement Order. After the plan has been checked and approved by the commander, an operation order is prepared and issued. The operation order may be in written form or issued orally, and is accompanied by a road movement table, overlay and/or strip map, and appropriate administrative details.

(6) Road Movement Table. A road movement table, prepared as an annex to the operation order, provides serial commanders with arrival and clearance times at critical points along the route of march. It also provides the column commander with information as to the proposed location of elements of the column at various times.

(7) Overlay/Strip Map. An overlay serves the normal purpose and should show, as a minimum, the present location of units, route of march, critical points, and the new location of units at the destination. A strip map is a schematic diagram of the route of march and shows landmarks and critical points with the distances between them. A strip map may be issued as an annex to the road movement order, in addition to or in lieu of an overlay.

c. March Computations. Prior to issuance of the operation order, the S-3 must verify time and space computations as they provide him with the necessary data for the preparation of a road movement table.

(1) Time-Distance. Time distance (TD) is determined by dividing the distance to be traveled (D) by the rate of inarch (R):

TD = D/RWhere TD = hours D = distance in kilometers R = kilometers per hour

(2) Length of ColumnThe length of column (LC) is used to determine the pass time of a column. The sum total of the following two parts determines the length of column. (See pars, (a) and (b)).The two parts are the space occupied by the Marines alone (including the distance between Marines) and the sum of the distances between the units of the column (column gap).

(a) The length of column of Marines alone is determined by multiplying the number of Marines by the appropriate factor selected from the table below. The length of column does not include distances between units. LC Marines = No. of Marines x factor.

(b) The total distance in meters between units is determined as follows:

Step One. Determine the number of gaps between serials (total serials minus one).

Step Two. Multiply the number of serial gaps from Step One by the length (in meters) between respective units.

Step Three. Determine the number of gaps between march units (total march units minus one, minus the number of serial distances).

Step Four. Multiply the number of march unit gaps from Step Three by the length (in meters) between respective units.

Step Five. Add the totals from Step Two and Step Four to get the total meters for the column gap.

EXAMPLE:A battalion foot column is organized into 12 platoon-sized march units and 3 company-sized serials. REQUIRED: total column gap distances when there are 100 meters between serials and 50 meters between march units.

(3) Pass-Time.Pass-time (PT) is the time a unit takes to pass a specified point. For foot columns, the pass-time is determined by applying the following formula:PT (minutes) = LC x FACTOR (for appropriate rate of march).

PASS-TIME FACTORS - FOOT TROOPS

.0150 for 4.0 km/h

.0187 for 3.2 km/h

.0250 for 2.4 km/h

.0375 for 1.6 km/h

EXAMPLE: Determine the pass-time of a unit whose length of column is 1,500meters and is marching at a rate of 4 km/h.PT (min) = 1,500 x .0150 (the factor for 4.0 km/h) = 22.5 mm.

(4) Completion Time. Completion time is the time of day that a march will be completed. Completion time is determined by using the following formula: Completion Time = SP time + TD + PT + Scheduled Halts (other than normal hourly halts).

EXAMPLE: A column's starting point (SP) time is 0700.The time-distance is6 hours and 40 minutes. Pass-time of the column is 30 minutes. A 35-minutelunch halt has been scheduled. What is the completion time of the march?Employing the 24-hour clock system, the formula can be applied as follows for simplified addition of the times:

HrMin SP time................0700

TD.......................0640

PT.......................0030

Lunch halt..............0035

Completion time.........13 105or 1445

The march will be completed at1445.

(5) Experience Tables. Based on previous movements made by a unit, data is accumulated to facilitate march planning. Such data includes approximate pass-times lor various elements of the battalion. The S-3 can utilize these data rather than computing them each time a march is scheduled. Such experience tables serve to reduce the time required to complete the computation phase of march planning. Matter appropriate to the unit standing operating procedure should be integrated therein.

d. Road Movement Graph

(1) A road movement graph is a time-distance diagram used in planning, preparing, or checking road movement tables, and for controlling marches. The graph provides a visual representation of a march plan so that conflicts and discrepancies may be prevented in the planning stage, before congestion occurs on the route. It is not usually issued as a part of the order. Road movement graphs may be applied to small units, to movements of a single column, or to a large organization scheduling separate elements, moving by various means, with different rates of march, over one or more routes.

(2) To construct a road movement graph, use the following steps: (See fig. 3-2.)

Figure 3-2 is available in the illustration section at the end of the book

Step One. Determine the number of hours available for the march. Designate the lower left corner of a sheet of graph paper as the SP time or the earliest even hour before the march is to begin. Select a scale and plot the hours available in sequence from the left on the horizontal axis.

Step Two. Determine the distance to be moved in kilometers. Indicate the SP at the lower left corner of the graph sheet and, using an appropriate scale, plot the number of kilometers on the vertical scale from the SP to the release point. Indicate the location of critical points on the vertical scale.

Step Three. At the proper distance from the SP, draw a horizontal line indicating the location of the RP. Indicate by a vertical line the hour when the movement must be completed. Plot lines representing route restriction, if any, at the proper distances and times on the graph.

Step Four. Determine the pass-times of foot and motor elements of the column.

Step Five. Starting at the SP at the specified hour, plot the movement of the head of the leading element. If the rate of march is 3 kilometers per hour, the lead element will have moved 1.5 kilometers in 30 minutes, 3 kilometers in 1 hour, etc. Plot the trace of the lead man to the RP. The last Marine will cross the SP in pass-times after the lead man. Measure this time on the graph and plot the trace of the last Marine of the column. The lines describing the head and tail of a march serial are parallel. Indicate the time subsequent serials reach the SP and plot the trace of the head and tail. Check to see that the plan complies with all restrictions and orders. If there are any violations or conflicts, the plan may be made to conform by changing the order of march, the starting time of the serials, the planned rate of march, or the organization of the column.

e. Conversion to the Metric System. The adoption of the metric system for expressing linear distances requires a simple means of conversion.

(1) To accomplish this, use the following tables:

o Multiply inches x 2.54 to obtain centimeters. Multiply centimeters x .39 to obtain inches,

o Multiply yards x .91 to obtain meters. Multiply meters x 1.1 to obtain yards,

o Multiply miles x 1.6 (or 8/5) to obtain kilometers. Multiply kilometers x .62 (or 5/8) to obtain miles.

(2) Linear distances can be accurately measured on maps which have been converted to the metric system by using a straightedge graduated in centimeters (cm).The centimeter scale can be used in lieu of the graphic scale when the scale of the map is shown, as follows:

Map Scale:	One cm Equals:
1:500,000	5,000 meters
1:250,000	2,500 meters
1:100,000	1,000 meters
1:50,000	500 meters
1:25,000	250 meters

3104. INDIVIDUAL LOAD

Since the backbone of Marine ground combat forces is the infantry, tactical mobility is largely dependent on the marching ability of the individual Marine. The load that each Marine must carry is the greatest impediment to mobility. In general, the commander must make every effort to reduce the individual load of his Marines to the absolute minimum. While the load has the greatest effect on Marines marching on foot, the value of carrying the minimum essential supplies and equipment applies for all Marine forces.

a. Fighting Load. The fighting load consists of items of clothing, equipment, weapons, and ammunition that are carried by, and essential to, the effectiveness of the combat Marine and the accomplishment of the immediate mission. The fighting load should not weigh more than 40 pounds. The commander must ensure that his Marines' loads- -and his own- -be stripped to the bare minimum. The addition of small, light-weight items in excess of the limit should not be tolerated, since cumulatively, these items will take a toll in energy. Every extra pound a Marine must carry decreases his combat effectiveness.

b. Existence Load. The existence load consists of items other than those in the fighting load that are required to sustain or protect the combat Marine, that may be necessary for increased personal and environmental protection, and that are not normally carried by the individual. The existence load is normally carried by the Marine's unit. Marching, but not engaged in combat, a Marine may be expected to carry a maximum of 50 pounds of supplies and equipment.

c. Considerations

(1) The primary consideration is not how much a Marine can carry, but how much he can carry without impaired combat effectiveness-moral or physical. The combat strength of a unit is not counted simply in numbers of Marines, but in the numbers of willing and physically able Marines.

(2) It is generally better to risk temporary inconvenience from lack of health and comfort items than to exhaust Marines due to overloading.

(3) It is a fundamental truth that men become physically exhausted more quickly when under the stress of combat. Marines must be required to carry less into battle than they are conditioned to carry in training. Marines should be conditioned for carrying weight, but should be equipped in combat for fleetness of foot.

(4) A common mistake is to base the fighting load on the gear and supplies necessary to meet every contingency. The commander cannot reasonably expect the Marines to carry enough gear for every possible eventuality. The items to be carried must be based on reasonable expectations.

(5) It is the commander's responsibility to procure the transport to carry any additional gear. As a rule of thumb, a rifle company, or unit of similar size, requires one 5-ton truck and trailer in support to carry additional gear. In cold weather, or other conditions in which the necessary personal gear increases, this requirement will increase.

(6) The commander must ensure that the supply system provides, on a dependable and timely basis, the balance of essential supplies and equipment not carried by the unit. Marines must feel confident that they will be supported with the necessary supplies and equipment.

(7) In training, commanders must implant in their Marines the pride in operating under austere conditions. To effectively demonstrate the need for this spartan mentality, the commander must personally set the standard. Marines should be trained in field craft, foraging techniques, and the use of caches and field expedients. Maximum use should be made of captured stores.

3105. MARCH TECHNIQUES AND PROCEDURES

a. Organization. A command executing a march is organized into march units, march serials, and march columns, as necessary for control.

(1) March Units. A march unit is a unit of command which moves and halts at the command of a single commander. The march unit normally corresponds to one of the smaller Marine units such as a platoon or company.

(2) March Serials. A march serial consists of one or more march units organized under the senior officer and given a numerical or alphabetical designation to facilitate control. A serial is usually a battalion, but may be a company if the battalion is marching alone.

(3) March Columns. A march column is composed of elements of a command moving over the same route. It may be composed of one or more serials. To facilitate control, a column commander is designated. A column is normally a regiment or larger unit, but may be a battalion if marching alone.

b. Order of March. The order of march is determined by the mission, terrain, and the commander's desired order of commitment of units into action.

c. Control and Coordination Measures. The commander establishes initial control of the march by designating control measures in his march order. The most frequently used measures include:

- Start point and release point.
- Time at which head or tail of column passes the start point.
- Rate of march.
- Order of march.
- Route of march.
- Location of command post.
- March communications.

d. The Pacesetter. The pace-setter is an experienced individual carrying the same load as the majority of the Marines and marching from 4 to 10 meters at the head of the column. (See fig. 3-3.)The pace-setter's primary duty is to maintain the rate of march ordered by the column commander.

(l) Setting the Pace. The pacesetter does this by establishing the pace (length of step) and cadence (steps per minute) to obtain the prescribed rate of march. He should be of medium height so normal strides will be taken. Overstriding or understriding tends to tire leg muscles quickly and affects the combat efficiency of marching Marines. The officer marching at the head of the column supervises the pacesetter to ensure that he takes normal strides and maintains a uniform cadence.

Figure 3-3 is available in the illustration section at the end of the book

(2) Reducing Accordion Effect. Accordion effect occurs when the column alternately lengthens and shortens, causing the rear elements of the column to double time to maintain distance. It is caused by changes in the rate of march by the lead elements, usually after passing a slowing obstacle. To reduce this effect as much as possible, lead elements should slow their rate of march for a sufficient time after crossing obstacles to allow the rear elements to maintain the prescribed distance without running. The rear elements can signal the head when the last man has cleared the obstacle and the rate can be resumed. Since some accordion effect is nearly inevitable, the order of march should be periodically rotated to prevent units in the rear from becoming physically exhausted at the conclusion of the march.

e. Length, Rate and Pace

(1) **Length of March.** The length of march varies depending on terrain and weather, enemy situation, and physical and mental condition of Marines. The normal length of march for a 24-hour period is from 20 to 32 kilometers (12 to 20 miles) marching from 5 to 8 hours at a rate of 4 kilometers (2.5 miles) per hour (km/hmi/h).A march in excess of 32 kilometers in a 24-hour period is considered a forced march. Well-trained units that have been progressively trained will be able to achieve a unit goal of 60 miles in a 3-day period.

(2) Rate of March. The same factors of terrain, weather, load to be carried, and condition of Marines affect the rate of march. The commander considers these and selects a rate which will place his unit at the destination in the shortest possible time in condition to accomplish the mission. Unit standing operating procedure should state the standard rate of march for that unit over normal terrain for both road and cross-country movement and in both day and night. The commander then modifies this rate if the situation requires. Normally prescribed standards are as follows:

Figure is available in the illustration section at the end of the book

(3) Pace and Cadence. The normal pace is 30 inches. A pace of 30 inches and a cadence of 106 steps per minute result in a speed of 4.8 kilometers per hour or 3 miles per hour and a rate of 4 kilometers per hour or 2.5 miles per hour if a 10-minute rest halt per hour is taken. Since the pace of each man may vary, the cadence may have to be adjusted to maintain the prescribed rate. Slope of ground and surface footing affect pace. A normal cadence is relatively simple to maintain on level or gently rolling terrain on firm footing. Mud, sand, loose gravel, and hills all greatly affect pace.

f. Halts. Halts during a day's march are taken at regular intervals to rest personnel and adjust loads .Halts are specified by standing operating procedures or march order. Such factors as midday heat or enemy action may force the adoption of long midday halts or night marches. At long halts, each unit disperses to a previously selected location near the route of march.

(1) Normal Time and Frequency. Under normal conditions, a 15-minute halt is made after the first 45 minutes of marching. Following the first halt, a 10-minute halt is made after each 50 minutes of marching. Variations of this schedule are made when a scheduled halt time occurs when passing through a built-up area or when cover and concealment are required by the tactical situation and none is available. Observation posts may be established, if required for security of the unit during halts.

(2) Actions at Halts. All units in the column should be halted at the same time. At the halt signal, Marines should move to the side of the road, staying within the immediate vicinity of their unit. Marines should remove or loosen gear and sit or lie with feet elevated. Commanders inspect Marines and equipment, and corpsmen administer medical treatment, as required.

g. Communications. The four primary means of communications used in controlling foot marches are radio, visual, sound, and messenger.

(1) Radio. Radios are normally used for communications between platoons and higher headquarters in the march column. Radio transmissions should be held to the absolute minimum necessary for control and should be at minimum power required.

(2) Visual.. Visual communications may include flashlights, luminous markers, panel sets, pyrotechnics, and hand and arm signals. When contact with the enemy is not expected, the loss of light discipline from some of these signals may be a lesser security risk than the risk of disclosure from radio transmissions.

(3) Sound. Sound communications include voice and such signaling devices as horns and whistles. Some of their uses include assembling Marines at the conclusions of halts and providing warnings of air or chemical attacks.

(4) Messengers. Messengers are particularly effective during periods of radio silence and reduced visibility. Messages should be simple and brief to preclude mistakes by the messenger.

h. Security. Both active and passive measures are used against attack by enemy aircraft and indirect fire weapons during movement. Active measures include the use of air sentries and organic and attached weapons in accordance with the unit air defense standing operating procedure. Passive measures include the use of concealed routes and assembly areas, night marches, and extended interval between elements of the column. With imagination and planning, all of these measures can be practiced in training without detracting from the physical conditioning benefit of the march. In fact, such drills as rapidly taking cover during simulated attacks and then quickly resuming the march can prove physically challenging.

i. Reconnaissance. Reconnaissance determines in advance critical points along the march route such as bridges, fords, and obstacles in order that congestion or delay may be held to a minimum and local security provided. Paragraph 3103.C describes organization of the reconnaissance party.

3106. TACTICAL MARCHES

These marches are movements of Marines and equipment not in direct ground contact with the enemy but expecting early ground contact either en route or upon arrival at the objective. Some characteristics of this type of movement include security elements to the front, flanks, and, if necessary, to the rear of the march column. Additional considerations include the selection of concealed areas on defendable terrain for start and release points and rest stops. In training, infantry units of regimental size or smaller will normally derive the greatest benefit from tactical marches. The security requirements usually dictate a slower rate of march and, therefore, reduce the physical conditioning value for other types of units which would seldom move in combat with the expectation of early ground contact with the enemy. Two key points should be remembered regarding tactical marches in a training program.

a. Security. Although security requirements often slow the rate of march, it is a valid and important training objective for infantry units to work toward reducing march times while still practicing sound tactical security. March standing operating procedures, good land navigation skills, and simple but complete march orders contribute as much to this goal as physical conditioning. The idea that proceeding slowly enhances security is seldom true. Keeping the enemy off balance by rapid movement frequently is the best guarantee of security.

b. Training. In peacetime training, it is critical that support personnel who normally are located with infantry units in combat, participate completely with infantry units in forced march training. Not only must they not slow the infantry rate of march because of inferior conditioning, but they must be thoroughly familiar with the infantry unit's standing operating procedure for security on the march. Personnel from within the ground combat element such as artillery forward observer teams and combat engineer units, can be easily identified for this training and should be habitually associated with the same infantry units. Personnel from other elements, such as low-altitude air defense teams and radio battalion detachments, are more difficult to identify and earmark for support of the same infantry unit. Their normal mode of employment is in vehicles and their units frequently provide general support vice direct support. This does not, however, lessen their requirement to be prepared to provide support to dismounted infantry over terrain or in a tactical situation not conducive to vehicles. Such training needs to take place, and planning and cooperation among the elements of the MAGTF can overcome the obstacles.

3107. ADMINISTRATIVE MARCHES

These marches are movements of Marines and equipment made when no enemy interference is expected except from aircraft or indirect fire. During these marches, units are administratively grouped for ease of control and speed of movement. The term administrative foot marches should not be taken to imply that tactical considerations are disregarded. There are simply less stringent security measures used than when ground contact with the enemy is a distinct possibility. The normal formation is the route column with one file moving on either side of the road and with negligible security. Air sentries are used and dispersion between individuals and units is practiced. Although roads usually provide the fastest route, there may be reasons to conduct these marches cross-country (particularly as part of a training program). Administrative marches are the type of foot marches most practical for training programs for non-infantry units. These marches may also play a key role in the physical conditioning program of infantry units, because of the greater speeds possible due to lessened security requirements.

3108. FORCED MARCHES

A forced march is a march which covers a greater distance than 32 kilometer (20 miles) in a 24-hour period. Normally the extra required distance should be achieved by increasing the number of hours marched in a day beyond the norm (8 hours) rather than increasing the rate of march beyond the norm (4 kilometers per hour/2.5 miles per hour). However, there will be occasions when the situation or mission demands an increase in the rate of march.

a. Maximum Recommended Distances. The maximum recommended distances for forced marches are-56 kilometers (35 miles) in 24 hours; 96 kilometers (60 miles) in 48 hours; or 128 kilometers (80 miles) in 72 hours.

b. Sample Time Breakdown. A sample time breakdown for a forced march of 52 kilometers, beginning at first light is as follows:

Figure is available in the illustration section at the end of the book

3109. NIGHT MARCHES

Night marches are characterized by closed formations, more difficult control and reconnaissance, and a slower rate of march than day marches.

a. Control. Control is increased by reducing the distance between individuals and units, and by using connecting files to maintain contact between platoons and companies. Connecting files are normally constituted from personnel from the rear most march unit.

b. Safety on Roads. Because of reduced visibility, night marches on roads used by vehicles require attention to safety procedures to prevent accidents. Commanders should not think these safety measures apply only to peacetime training. In combat, nearly all vehicles operate with blackout lights at night, making visibility even more difficult than in garrison driving. If consistent with the tactical situation, the following measures can reduce the possibility of marchers being struck by vehicles:

(1) Use off-road trails and routes as much as possible.

(2) Guards to the front and rear of march units should be marked with strips of luminous tape and carry red-filtered flashlights. These measures are generally consistent with light discipline, while still affording reasonable warning to approaching drivers.

Section II. Conditioning Drills (One, Two, and Three)

3201. GENERAL

a. Description and Objective. Conditioning drills are calisthenic exercises. Each drill contains seven exercises organized and numbered in a set pattern. Each drill takes 15 minutes to complete. The objective of conditioning drills is to exercise all major muscle areas in order to develop strength, endurance, coordination, and flexibility.

b. Area and Equipment. Any level area is satisfactory for conduct of the drills. Drills One and Two contain ground exercises. If ground conditions are unsatisfactory, Drill Three should be used as it contains no ground positions. Usually, no equipment is required;however, if the group exceeds a platoon in size, an instructor's stand is necessary.

c. Formation. The extended rectangular formation is used in this drill. (See app. A.)

d. Starting Level and Progression. The starting level is six repetitions of each exercise. An increase of one repetition for each three periods of exercise in which the drill is performed is an acceptable rate of progression. This rate is continued until 12 repetitions can be completed. To maintain, continue the drill at 12 repetitions. To progress, move to 6 repetitions at a more difficult drill which exercises the same muscle group. Progression can also be gained by moving from Drill One to Drill Two, as Drill Two is more demanding.

e. Starting Positions. Starting positions vary with the exercise and are explained in each exercise.

f. Leadership. A principal instructor demonstrates and leads the drills. The instructor must be familiar with leadership techniques peculiar to conditioning drills to include the exercises, commands, counting cadence, cumulative count, formation, method of teaching the exercises, and utilization of assistant leaders.

g. Use With Other Programs. Conditioning Drills One, Two, and Three reach all major muscles of the body. They are easy to learn and to perform, and they are simple to administer and supervise. These features, coupled with the short time required for completion, the fact that no equipment is necessary, and adaptability to most areas of execution, make these drills possible in any programs.

3202. CONDITIONING DRILL ONE

This conditioning drill is similar to the calisthenics drill commonly known throughout the Marine Corps as the "Daily Seven."The "Daily Seven" may be used as a substitute for Conditioning Drill One.

a. Exercise l:High Jumper

(1) Starting Position. Feet separated shoulder width, knees flexed, body bent forward at the waist, arms aligned with the trunk and hips, elbows locked, palms facing, fingers extended and joined, head and eyes to the front.(See fig. 3-4, A.)(Elbows remain locked throughout the exercise.)

Figure 3-4 is available in the illustration section at the end of the book.

(2) Cadence. Moderate.

(3) Movement. A four-count exercise: at the count of—

(a) ONE -- Take a slight jump into the air, swinging the arms forward and up to shoulder level.

(b) TWO -- Take a slight jump into the air and swing the arms downward and back, returning to the starting position.

(c) THREE -- Take a vigorous leap into the air, swinging the arms forward and up to an overhead position, momentarily looking skyward, on returning to the ground the knees are flexed, head and eyes return to the front.

(d) FOUR--Repeat the action of count two.

b. Exercise 2: Bend and Reach

(1) Starting Position. Feet spread more than shoulder width, arms overhead, elbows locked, palms facing, fingers extended and joined, head and eyes to the front.(See fig. 3-4, B.)

(2) Cadence. Moderate.

(3) Movement. A four-count exercise: at the count of—

(a) ONE-Bend at the knees and waist, swing the arms straight downward and reach between the legs. Touch the ground as far to the rear as possible and look to the rear. (Elbows remain locked throughout the exercise).

(b) TWO-- Recover sharply to the starting position.

(c) THREE--Repeat the action of count ONE.

(d) FOUR-- Repeat the action of count TWO.

c. Exercise 3: Pushup

(1) Starting Position. Front leaning rest position. To assume this position there is a silent one-two count:on the silent count of one, assume the squatting position, heels together, elbows locked inside the knees, hands flat on the ground directly beneath the shoulders; on the silent count of two, thrust the legs to the rear, toes and heels together, body straight from head to heels.(Seefig. 3-4, C.)

(2) Cadence. Moderate.

(3) Movement. A four-count exercise: at the count of—

(a) ONE -- Flex the elbows lowering the body until the thick portion of chest touches the ground.

(b) TWO -- Raise the body until elbows are straight and locked.

(c) THREE-- Repeat the action of count ONE.

(d) FOUR -- Repeat the action of count TWO. (On returning to position of attention, the silent one-two count is used in reverse).

d. Exercise 4: Trunk Twister

(1) Starting Position. Feet are spread more than shoulder width apart, fingers placed behind neck, thumbs pointing downward, elbows back. (See fig. 3-4, D.)(Elbows remain well back throughout the exercise).

(2) Cadence. Slow.

(3) Movement. A four-count exercise: at the count of-

(a) ONE-- Keeping the knees locked and the back straight, bend forward at the waist sharply, with a slight recovery.

(b) TWO-- Twist the trunk to the left vigorously at the waist, keeping the elbow back. The left elbow is higher than the right.

(c) THREE-- Twist vigorously to the right, so the left elbow comes under the right.

(d) FOUR-- Straighten sharply to the starting position. NOTE: Do not attempt to touch the elbows to the knees on counts two and three.

e. Exercise 5: Squat Bender

(1) Starting Position. Feet are spread less than shoulder width apart, hands on hips, thumbs in small of back, elbows back. (See fig. 3-4, E.)

(2) Cadence. Moderate.

(3) Movement. A four-count exercise: at the count of—

(a) ONE-- Assume the squatting position, maintain balance on the balls of the feet, with trunk erect thrust arms forward to shoulder level, elbows locked, palms down.

(b) TWO-- Recover to starting position. Elbows are well back.

(c) THREE-- Keeping the knees locked, bend forward at the waist, touching the ground in front of the toes.

(d) FOUR-- Vigorously recover to the starting position.

f. Exercise 6: Body Twist

(1) Starting Position. On the back, arms extended sideward on the ground, palms down. The legs are raised to a near vertical position, feet together, knees locked.(See fig. 3-4, F.)

(2) Cadence. Slow-fast.

(3) Movement. A four-count exercise: at the count of-

(a) ONE-- Lower legs slowly to the left until they touch the ground near the left hand, keeping the knees straight and shoulders on the ground.

(b) TWO-- Recover the starting position by quickly raising the legs, keep knees straight and feet together.

(c) THREE-- Repeat movement of count ONE, except the movement is to the right side.

(d) FOUR-- Recover sharply to the starting position.

g. Exercise 7: Stationary Run

(1) Starting Position. Position of attention. (See fig, 3-4, G.)

(2) Cadence. Fast.

(3) Movement.

(a) At the command of execution, start running in place at double-time cadence, lifting the left foot first time cadence. Follow the instructor as he counts two repetitions of cadence;

e.g., 1, 2, 3, 4--1, 2, 3, 4. The instructor then gives informal commands such as FOLLOW ME, running on the toes and balls of the feet, keeping the back straight, speeding up the cadence to a sprint, raising the knees high, leaning forward at the waist, and pumping the arms vigorously.

(b) To halt the exercise, the instructor will count two repetitions of cadence as the left foot strikes the ground: 1, 2, 3, 4 - 1, 2, 3, HALT.

NOTE: When counting cadence, the instructor counts only as the left foot strikes the ground. The duration of the exercise is approximately 1 1/2 minutes.

3203. CONDITIONING DRILL TWO

a. Exercise 1: Jumping Jack

(1) Starting Position. Feet separated more than shoulder width, arms overhead. (See fig. 3-5, A.)

(2) Cadence. Moderate.

(3) Movement. A four-count exercise: at the count of~

(a) ONE-- Jump to position with the feet together and assume the squatting position, swinging the arms sideward and downward, placing the hands palms down on the ground, elbows locked inside the knees.

(b) TWO-- Recover to the starting position by jumping to the side straddle and swinging the arms sideward overhead.

(c) THREE-- Repeat the action of count ONE.

(d) FOUR-- Recover to the starting position.

b. Exercise 2: Turn and Bend

(1) Starting Position. Side straddle, arms overhead. (See fig. 3-5, B.)

(2) Cadence. Moderate.

(3) Movement. A four-count exercise: at the count of-

(a) ONE-- Turn the trunk to the left and bend forward over the left thigh, attempting to touch the fingertips to the ground outside the left foot. Keep the left knee straight. On successive repetitions attempt to touch farther and farther to the side.

(b) TWO-- Recover to the starting position.

(c) THREE-- Turn the trunk to the right and bend forward over the right thigh, trying to touch the hands to the ground outside the right foot. Keep the right knee straight.

Figure 3-5 is available in the illustration section at the end of the book.

(d) FOUR-- Recover to the starting position.

c. Exercise 3: Eight-Count Pushup

(1) Starting Position. Position of attention. (See fig. 3-5, C.)

(2) Cadence. Moderate.

(3) Movement. An eight-count exercise: at the count of—

(a) ONE-- Assume the squatting position, palms on the ground directly beneath the shoulders, elbows locked inside the knees.

(b) TWO-- Thrust the legs to the rear assuming the front leaning rest position.

(c) **THREE**-- Flex the elbows until the thick portion of the chest touches the ground.

(d) **FOUR**-- Raise the body on a straight plane until the elbows are locked.

(e) **FIVE**-- Repeat the action of count THREE.

(f) **SIX**-- Repeat the action of count FOUR.

(g) **SEVEN**-- Recover to the squatting position as in count ONE (elbows locked inside the knees).

(h) **EIGHT**—Return sharply to the position of attention.

d. Exercise 4: Turn and Bounce

(1) **Starting Position.** Feet spread more than shoulder width apart, arms sideward at shoulder level, palms up. (See fig. 3-5, D.)

(2) **Cadence.** Slow.

(3) **Movement.** An eight-count exercise: at the count of—

(a) **ONE**-- Turn sharply to the left as far as possible, then recover slightly.

(b) **TWO**-- Again turn to the left as far as possible and recover as in ONE.

(c) **THREE**-- Repeat the action of count TWO.

(d) **FOUR**-- Recover sharply to the starting position.

(e) **FIVE**—Turn sharply to the right as far as possible, then recover slightly.

(f) **SIX**—Again turn to the right as far as possible and recover as in FIVE.

(g) **SEVEN**-Repeat the action of count SIX.

(h) **EIGHT**-Return to the starting position.

NOTE: The head and hips remain to the front throughout the exercise and the knees and elbows are locked at all times.

e. Exercise 5: Squat Stretch

(1) **Starting Position**. Attention. (See fig. 3-5, E.)

(2) **Cadence**. Moderate.

(3) **Movement**. A four-count exercise: at the count of—

(a) **ONE**-- Squat, placing the hands on the ground about 12 inches in front of the feet.

(b) **TWO**— Keeping the fingertips on the ground, straighten the knees completely and raise the hips.

(c) **THREE**-- Recover to position ONE.

(d) **FOUR**-- Recover to the starting position.

f. Exercise 6: Leg Circular

(1) **Starting Position.** On the back, arms stretched sideward, palms down, feet raised foot from ground, knees straight. (See fig 3-5,F.)

(2) **Cadence.** Slow.

(3) **Movement** A four-count exercise: at the count of—

(a) **ONE--** Swing the legs as far as possible to the left, keeping the knees straight and the legs together.

(b) **TWO--** Swing the extended legs over head with the thighs as close as possible to the trunk.

(c) **THREE--** Swing the legs as far as possible to the right.

(d) **FOUR--** Recover to the starting position.

g. Exercise 7: Stationary Run

(1) Starting Position. Position of attention. (See fig. 3-5, G.)

(2) Cadence. Fast.

(3) Movement

(a) At the command of execution, start running in place at double-time cadence, lifting the left foot first. Follow the instructor as he counts two repetitions of cadence; for example: 1, 2, 3, 4--1, 2, 3, 4. The instructor then gives informal commands such as FOLLOW ME. Running on the toes and balls of the feet, keeping the back straight, speeding up the cadence to a sprint, raising the knees high, leaning forward at the waist, and pumping the arm vigorously.

(b) To halt the exercise, the instructor will count two repetitions of cadence as the foot strikes the ground: 1, 2, 3, 4-1,2, 3, HALT.

NOTE: When counting cadence, the instructor counts only as the left foot strikes the ground. The duration of the exercise is approximately 1 1/2 minutes.

3204. CONDITIONING DRILL THREE

a. Exercise 1: Side Straddle Hop

(1) Starting Position. Position of attention. (See fig. 3-6, A.)

(2) Cadence. Moderate.

(3) Movement. A four-count exercise: at count of-

(a) ONE-- Take a slight jump into the air, moving the legs sideward (more than shoulder width apart); at the same time, swing the arms overhead (to an overhead position) clapping the palms together.

Figure 3-6 is available in the illustration section at the end of the book.

(b) TWO-- Take a slight jump into the air, swing the arms sideward and downward returning to the starting position.

(c) THREE-- Repeat the action of count ONE.

(d) FOUR-- Repeat the action of count TWO.

b. Exercise 2: Back Bender

(1) Starting Position. Standing, feet 12 inches apart, fingers placed behind the head. (See fig. 3-6, B.)

(2) Cadence. Slow.

(3) Movement. A four-count exercise: at the count of-

(a) ONE-- Bend the upper trunk backward, raising the chest high, pulling the elbows back, and looking upward. Keep the knees straight.

(b) TWO-- Recover to the starting position.

(c) THREE-- Repeat the action of count ONE.

(d) FOUR-- Recover to the starting position.

c. Exercise 3: Squat Thrust

(1) Starting Position. Position of attention. (See fig. 3-6, C.)

(2) Cadence Moderate.

(3) Movement. A four-count exercise: at the count of—

(a) ONE-- Assume the squatting position; heels together, placing the hands flat on the ground, shoulder width apart, elbows locked and inside the knees.

(b) TWO-- Thrust the legs to the rear, assuming the front leaning rest position, body in line from head to toe, heels and toes together.

(c) THREE-- Return to the squatting position as in ONE.

(d) FOUR-- Return to position of attention.

d. Exercise 4: Side Bender

(1) Starting Position. Feet are spread more than shoulder width apart, arms are raised sideward and overhead, thumbs interlocked palms to front, fingers extended and joined, elbows locked. (See fig. 3-6, D.)

(2) Cadence. Slow.

(3) Movement. An eight-count exercise: at the count of—

(a) ONE-- Bend to left as far as possible, then recover slightly.

(b) TWO-- Again bend to the left as far possible, then recover slightly.

(c) THREE-Repeat the action of count TWO.

(d) FOUR-- Recover sharply to the starting position.

(e) FIVE-- Bend to the right as far as possible, then recover slightly.

(f) SIX-- Again bend to the right as far as possible, then recover slightly.

(g) SEVEN-- Repeat the action of count SIX.

(h) EIGHT-- Recover sharply to the starting position.

NOTE:Keep the elbows and knees locked throughout the exercise. The bend should occur to the side and not the front.

e. Exercise 5: Knee Bender

(1) Starting Position. Feet are spread less than shoulder-width apart, hands on hips, thumbs in small of back, elbows back. (See fig 3-6, E.)

(2) Cadence. Moderate.

(3) Movement. A four-count exercise: on the count of—

(a) ONE-Do a knee bend, lean trunk forward at the waist, thrust arms between legs until the extended fingers touch the ground palms to the ground, hands 6 inches apart.

(b) TWO~Recover sharply to the starting position.

(c) THREE-Repeat the action of count ONE.

(d) FOUR-Repeat the action of count TWO.

f. Exercise 6: Bottoms Up

(1) Starting Position. Front leaning rest position. A silent one-two count is used as in the pushups. (See fig. 3-6, F.)

(2) Cadence. Moderate.

(3) Movement A four-count exercise: at the count of-

(a) ONE-- With the weight on the hands, and knees locked, jump forward bringing the feet as close to the hands as possible; look to the rear.

(b) TWO--Keeping the knees locked, thrust the legs backward assuming the front leaning rest position.

(c) THREE-- Repeat the action of count ONE.

(d) FOUR-- Repeat the action of count TWO.

g. Exercise 7: Stationary Run

(1) Starting Position. Position of attention. (See fig. 3-6, G.)

(2) Cadence. Fast.

(3) Movement

(a) At the command of execution, start running in place at double time, lifting the left foot first. Follow the instructor as he counts two repetitions of cadence; for example: 1, 2, 3, 4--1, 2,3, 4. The instructor then gives informal commands such as FOLLOW ME, running on the toes and balls of the feet, keeping the back straight, speeding up the cadence to a sprint, raising the knees high, leaning forward at the waist, and pumping the arms vigorously.

(b) To halt the exercise, the instructor will count two repetitions of cadence as the left foot strikes the ground: 1, 2, 3, 4 - 1, 2, 3, HALT.

NOTE: When counting cadence, the instructor counts only as the left foot strikes the ground. The duration of the exercise is approximately 1 1/2 minutes.

Section III. Rifle and Log Drills

3301. RIFLE DRILL

a. Description and Objective. Rifle exercises are conditioning exercises performed with a rifle. Each drill contains six exercises (fig. 3-7) and they are numbered in a set pattern. The drill takes 15 minutes to complete. The objective of rifle drills is to exercise the arms, shoulders, and back muscles in order to develop strength and endurance, particularly in the upper body. In units without rifles, log drills may be substituted.

b. Area and Equipment. Any level area is satisfactory for conducting this drill. Each Marine completes these exercises from a standing position and no ground contact is required. Each Marine will need a rifle and, if the group exceeds a platoon in size, then the instructor will need an instructor's stand.

c. Formation. The extended rectangular formation is used in this drill. (See app. A.)

d. Starting Positions. Starting positions vary with the exercises and are explained in each exercise. As in all set conditioning drills, the command used to start the exercise is STARTING POSITION, MOVE. The following directions apply to rifle drill.

(1) In those exercises which start from the rifle downward position, on the command MOVE, execute port arms as prescribed in NAVMC 2691, Drill and Ceremonics Manual, and then assume the starting position. The command to return to the men to the position of attention at the conclusion of the exercise is POSITION OF ATTENTION, MOVE.

(2) In exercises which terminate in the rifle downward position, on the command MOVE, execute the position followed by order arms as prescribed in NAVMC 2691.

(3) In the exercises which terminate in a position other than the rifle downward position, Marines first assume the rifle downward position before executing port arms and order arms.

(4) These movements are executed without command. This procedure promotes uniformity, but precision is not expected. To be effective, rifle exercises must be strenuous enough to tire the arms, but not to the point where the arms cannot move with precision.

e. Leadership. A principal instructor demonstrates and leads the drill. He must be familiar with leadership techniques for conditioning exercises and the peculiar techniques for rifle drill.

3302. EXERCISES PERFORMED WITH RIFLES

The exercises of rifle drill are outlined in the following paragraphs,

a. Exercise 1: Foreup, Behind Back

(1) Starting Position. Rifle downward, feet together.(See fig. 3-7, A.)

Figure 3-7 is available in the illustration section at the end of the book.

(3) Movement. A four-count exercise: at the count of—

(a) ONE-- Swing the arms forward and upward to the overhead position. Inhale.

(b) **TWO**-- Lower the rifle to the back of the shoulders. Exhale.

(c) **THREE**-- Recover to position ONE and inhale.

(d) **FOUR**-- Recover to the starting position and exhale.

b. Exercise 2: Lunge Side, Turn and Bend

(1) Starting Position. Rifle downward, feet together.(See fig. 3-7, B.)

(2) Cadence. Moderate.

(3) Movement. An eight-count exercise:at the count of—

(a) **ONE**-- Lunge sidewards to the left, swing the rifle forward and upward to the overhead position.

(b) **TWO**-- Turn the trunk to the left and bend forward over the left hip. At the same time, swing the rifle to a **low** horizontal in front of the left ankle.

(c) **THREE**-- Recover to position ONE.

(d) **FOUR**-- Recover to the starting position.

(e) **FIVE, SIX, SEVEN, and EIGHT**-- Repeat on the right side.

c. Exercise 3: Foreup, Back Bend

(1) Starting Position. Rifle downward, feet together.(See fig. 3-7, C.)

(2) Cadence. Moderate.

(3) Movement. A four-count exercise:at the count of—

(a) **ONE--** Swing the arms forward and upward to the overhead position.

(b) **TWO**-- Bend backward, emphasizing the bend in the upper back. The face is up.Keep the knees straight.

(c) **THREE**-- Recover to position ONE.

(d) **FOUR**-- Recover to the starting position.

d. Exercise 4:Up and Forward

(1) Starting PositionRifle downward, feet together.(See fig. 3-7, D.)

(2) Cadence. Fast.

(3) Movement. A four-count exercise:at the count of—

(a) **ONE**-Swing the arms forward and upward to the overhead position.

(b) **TWO**- Swing the arms forward to shoulder level.

(c) **THREE**-Recover to position ONE.

(d) **FOUR**-Recover to the starting position.

e. Exercise 5: Foreup, Full Squat

(1) Starting Position. Rifle downward, feet in narrow stance. (See **fig.** 3-7, E.)

(2) Cadence. Moderate.

(3) Movement. A four-count exercise: at the count of—

(a) **ONE**-- Swing the arms forward and upward to the overhead position.

(b) **TWO**-- Swing the arms down to shoulder level and assume the squatting position.

(c) **THREE**-- Recover to position ONE.

(d) **FOUR**-- Recover to the starting position.

f. Exercise 6: Arms Forward, Side Bend

(1) Starting Position. Sidestraddle, regular stance, rifle forward.(See fig. 3-7, F.)
(2) Cadence. Moderate.
(3) Movement. A four-count exercise:at the count of-

(a) ONE-- Bend the trunk to the left. Keep the knees straight.
(b) TWO-- Recover to the starting position.
(c) THREE-- Bend the trunk to the right. Keep the knees straight.
(d) FOUR-- Recover to the starting position.

NOTE:Keep the rifle on the same level as the shoulders throughout the exercise.

3303. LOG DRILL

a. Description and Objective. Log exercises are conditioning exercises performed with a log. Each drill contains six exercises and they are numbered in a set pattern. The drill takes 15 minutes to complete. The objective of log drills is to develop strength and muscular endurance and, in this instance, under maximum loads. Log exercises also develop teamwork. Log may be used in lieu of conditioning drills after the Marines have become somewhat conditioned.

b. Area and Equipment. Any level area is satisfactory for conducting this drill. Each Marine completes these exercises from a standing position and no ground contact is required. If the group exceeds a platoon in size, then the instructor will need instructor's stand. Each six-person group or eight-person group will need a log. The logs should be from 6 to 8 inches in diameter. They may vary in length from 14 feet (for 6 people) to 18 feet (for 8 people).The logs should be skinned, smoother, and dried. The 14-foot logs should weigh approximately 300 pounds and the 18-foot logs, approximately 400 pounds. Rings should be painted on the logs to indicate each person's position. When not in use, the logs should be stored on a rack to keep them off the ground.

c. Formation. All the Marines assigned to the same log team should be about the same height at the shoulders. The recommended method of dividing the platoon is to have the Marines form a single file or column with short people to the front and tall people to the rear. Have the Marine assume their positions in the column according to shoulder height, not head height. When they are in position, they are given the command COUNT OFF BY SIXES (OR EIGHTS), COUNT OFF, to divide them into six- or eight-person log teams. Each team in turn, can then proceed to the log rack, shoulder a log, and carry it to the designated exercise area. The log teams form in columns in front of the instructor. With the Marine holding the log in the chest position, have them face the instructor and ground the log at least 10 yards between columns and 10 yards between log teams within the columns.

d. Starting Positions. The Marines fall in, facing the log, with their toes about 4 inches from it. The basic starting positions and commands are as follows:(See fig. 3-8.)
(1) **RIGHT-HAND STARTING POSITION, MOVE.** It the command MOVE, move the left foot 12 inches to the left, and lower the body into a flat foot squat. Keep the back straight, head up, and arms between the legs. Encircle the far side of the log with the left hand. Place the right hand underneath the log.(See fig. 3-8, A.)

(2) LEFT-HANDED STARTING POSITION, MOVE. These commands are executed in the same manner as in paragraph a. except that the left hand is underneath the log and the right hand encircles its far side. (See fig. 3-8, B.)

Figure 3-8 is available in the illustration section at the end of the book.

(3) RIGHT SHOULDER POSITION, MOVE. At the command MOVE, pull the log upward in one continuous motion to the right shoulder. At the same time, move the left foot to the rear and stand up, facing left. Balance the log on the right shoulder with both hands.(See fig. 3-8, C.)This movement cannot be performed from the left-hand starting position because of the position of the hands.

(4) LEFT SHOULDER POSITION, MOVE. These commands should be given from the left hand starting position. At the command MOVE, pull the log upward in one continuous motion, to the left shoulder. At the same time, move the right foot to the rear and stand facing right. Balance the log on the left shoulder with both hands.(See fig. 3-8, D.)This movement cannot be performed from the right-hand starting position.

(5) WAIST POSITION, MOVE. From the right hand position pull the log waist high. Keep the arms straight and fingers laced underneath the log. The body is inclined slightly to the rear and the chest is lifted and arched. (See fig. 3-8, E.)

(6) CHEST POSITION, MOVE. This command should be given after the waist position has been assumed .On the command MOVE, shift the log to a position high on the chest, bring the left arm under the log and hold the log in the bend of the arms.(See fig. 3-8, F.)Keep the upper arms parallel to the ground.

(7) To move the log from the right shoulder to the left shoulder, the command is: LEFT SHOULDER POSITION, MOVE. On the command MOVE, push the log overhead and lower it to the opposite shoulder.

(8) To return the log to the ground from any of the above positions, the command is: STARTING POSITION, MOVE. At the command MOVE, slowly lower the log to the ground. The hands and fingers must be kept from under the log.

e. Leadership. A principal instructor demonstrates and leads the drill. He must be familiar with the leadership techniques for conditioning exercises and the peculiar techniques for log drill.

3304. EXERCISE PERFORMED WITH LOGS

The exercises of log drill are outlined in the following paragraphs. Figure 3-9 graphically explains log drill.

a. Exercise l:Two-Arm Pushup

(1) Starting Position. Right or left shoulder position. Regular stance. (See fig. 3-9, A.)

(2) Cadence. Moderate.

(3) Movement. A four-count exercise:at the count of—

- **(a) ONE--** Push the log overhead until the elbows lock.
- **(b) TWO--** Lower the log to the opposite shoulder.
- **(c) THREE--** Repeat the action of count ONE.
- **(d) FOUR--** Recover to the starting position.

b. Exercise 2: Forward Bender

(1) Starting Position. Chest position. Regular stance.(See fig. 3-9, B.)

(2) Cadence. Moderate.

(3) Movement. A four-count exercise:at the count of-

(a) ONE-- Bend forward at the waist, keeping the back and legs straight.

(b) TWO-- Recover to the starting position.

(c) THREE-- Repeat the action of count ONE.

(d) FOUR-- Recover to the starting position.

c. Exercise 3: Straddle Jump

(1) Starting Position. Right or left shoulder position, feet together, fingers interlaced on top of the log.(See fig. 3-9, **C.)**

(2) Cadence. Moderate.

(3) Movement. A four-count exercise:at the count of—

(a) ONE-- Jump to a side-straddle. Pull down on the log with both hands to keep it from bouncing on the shoulder.

(b) TWO-- Recover to the starting position.

(c) THREE-- Repeat the action of count ONE.

(d) FOUR-- Recover to the starting position.

Figure 3-9 is available in the illustration section at the end of the book.

d. Exercise 4: Side Bender

(1) Starting Position. Right shoulder position, feet regular stance.(See fig. 3-9, D.)

(2) Cadence. Moderate.

(3) Movement. A four-count exercise:at the count of—

(a) ONE-- Bend sideward to the left as far as possible, bending the left knee.

(b) TWO-- Recover to the starting position.

(c) THREE-- Repeat the action of count ONE.

(d) FOUR-- Recover to the starting position.

(4) After completing the required number of repetitions, change shoulders and execute an equal number of repetitions to the other side.

e. Exercise 5: Knee Bend

(1) Starting Position. Right or left shoulder position. Narrow stance. Fingers interlocked on top of the log.(See fig. 3-9, E.)

(2) Cadence. Slow.

(3) Movement. A four-count exercise:at the count of—

(a) ONE-- Flex the knees to a quarter-squat position.

(b) TWO-- Flex the knees to a half-squat position.

(c) THREE-- Lower the body to a three-quarter squat position.(Lean slightly forward.)

(d) FOUR-- Recover to the starting position.

NOTE:Pull forward and downward on the log throughout the exercise.

f. Exercise 6: Overhead Toss

(1) Starting Position. Right or left shoulder position, regular stance. The knees are bent to a quarter-squat.(See fig. 3-9, F.)

(2) Cadence. Moderate.

(3) Movement. A four-count exercisc:at the count of-

(a) ONE-Straighten the knees and toss the log into the air approximately 12 inches overhead. Catch the log with both hands and lower it toward the opposite shoulder. As the log is caught, lower the body into a quarter-squat.

(b) TWO-- Again toss the log into the air and when caught return it to the original shoulder.

(c) THREE-- Repeat the action of count ONE.

Section IV. Grass Drills

3401. GENERAL

a. Description and Objective. Grass drills are extremely strenuous exercises and are performed at top speed for only short periods of time. No cadence is counted but the Marines continue to execute the multiple repetitions of the command until the next command is given. The grass drill consists of two drills:Drill One and Drill Two. Each drill contains six exercises. The objective of the drills is to decrease reaction time, to develop cardiovascular endurance, and to provide a vigorous workout for all major muscles.

b. Area and Equipment. Any level area suitable for ground contact and of a size to accommodate the group is adequate. No equipment is needed.

c. Formation. All movements are executed in place. The extended, rectangular formation is recommended for a platoon- or company-size unit. The circle formation is suitable for groups of squad or section size. At the beginning of an exercise program, 2 to 3 minutes of grass drills will insure a good workout.

d. Progression. Progression is gained by gradually increasing the length of time devoted to the drills. As the physical condition of the Marines improves, the periods should be gradually lengthened to 5 minutes. As the second drill is more difficult than the first, some progression can be attained by initially executing Drill One; then as the program and the Marines progress, introduce Drill Two. To extend the duration of the drill, it may be necessary to repeat the drill.

e. Starting Position

(1) The drills are started from the GO position. Other basic positions are FRONT, BACK, and STOP. (See fig. 3-10, A.)

(a) GO. Running in place (top speed): on the toes and balls of feet, knees raised high, arms pumping, body bent forward at waist.

(b) FRONT. Prone position: elbows bent (along body), palms flat on ground directly under the shoulders, legs together and straight.

(c) BACK. Supine position (flat on back): arms extended near side on ground with palms down, legs together and straight, feet toward the stand or instructor.

(d) STOP. Football lineman stance: feet spread and staggered, left arm across left thigh, right arm straight, knuckles on ground, head up, back parallel with ground.

(2) To assume the FRONT or BACK position from the STANDING, GO, or STOP position, vigorously get into the prescribed position as quickly as possible. (See fig. 3-10, B.)

(3) To change from the FRONT to the BACK position, quickly do a pushup, move the feet several short steps to the right or left, lift the arm on the side toward which the feet move, and thrust the legs vigorously to the front. (See fig. 3-10, C.)

Figure 3-10 is available in the illustration section at the end of the book.

(4) To move from the BACK to the FRONT position, sit up quickly, place both hands on the ground to the right or the left of the legs. Move the feet several short steps to the rear on the side opposite the hands. When the feet are opposite the hands, thrust the legs vigorously to the rear and lower the body to the ground.(See fig. 3-10, D.)

f. Leadership. A warm up activity of lesser intensity should proceed grass drill. During the instructional phase and conduct of these drills, the following points should be applied.

(1) The instructor executes only GO and STOP with the Marines. This allows the instructor to supervise the drill.

(2) The commands for grass drills are given in rapid succession without the usual preparatory command.

(3) To prevent confusion, the instructor should give the commands sharply to distinguish them from comments or encouragement.

(4) As soon as the Marines know the drill, they should respond to the instructor's commands and perform all exercises vigorously and as rapidly as possible. All exercises are executed continuously until the next command is given. Insist on top speed performance; anything less is not effective.

(5) The commands peculiar to each exercise are identical to the name of the exercise.

(6) Marines are not to be required to assume the position of attention once the drills are started. To halt the drill for instructions or for rest, the command UP is used. At this command, the Marines assume a relaxed standing position. Do not demand formality. At the conclusion of a fast and vigorous 5-minute grassdrill, it is physically impossible for people to stand at attention.

(7) The sequence of commands for the execution of grass drills should occur as follows:Drill One. GO, FRONT, Bouncing Ball; GO, BACK, Bicycle; GO, Full Squatter; GO, BACK, Situps; GO, FRONT, Mountain Climber; GO, FRONT, Roll Left; GO, STOP, UP.

g. Use With Other Programs. Since grass drills can be executed in a short period of time, they may be executed where only a few minutes are available for exercise or in conjunction with another type of activity. Grass drills are an excellent substitute for running when time is a factor.

3402. GRASS DRILL ONE AND TWO

a.Grass Drill One

(1) Bouncing Ball. From the FRONT position, push up, supporting the body on the hands (shoulder-width apart) and feet. (See fig. 3-11.)Keep the back and legs in line and the knees straight. Bounce up and down by a series of short, upward springs from the hands, hips, and feet simultaneously,

(2) Bicycle. From the BACK position, raise the legs and hips. Keep the elbows on the ground and support the hips with the hands. Move the legs vigorously as if pedaling a bicycle.

(3) Full Squatter. From the STOP position, assume a full knee bend, the feet on line, hands on hips. Bounce up and down in place by short, bouncing jumps.

Figure 3-11 is available in the illustration section at the end of the book.

(4) **Situps.** From the BACK position and with arms stretched overhead, sit up, reach forward, and touch toes. Return to the supine position.

(5) **Mountain Climber.** From the STOP position, place both hands on the ground directly under the shoulders. Thrust the right leg to the rear, knee straight. The left foot should be close to the left hand, the left knee outside the left arm. Shift the weight to the hands, thrust off with the rear (right) foot and bring that foot up close to the right hand, the right knee outside the right arm. At the same time, thrust the left leg vigorously to the rear, knee straight. Continue at a fast cadence, alternating the legs.

(6) **Roll Left.** From the BACK or FRONT position, make one complete roll in the direction commanded. On completing the roll, return to the FRONT or BACK position.

b. Grass Drill Two

(1) **Legs Over.** .From the BACK position and with arms stretched overhead, palms up, raise the legs upward and then swing them backward over the head until the toes touch the ground behind the head. Return legs to the starting position. (See fig. 3-12.)

(2) **V-Up and Touch Toes.** .From the BACK position, raise the legs with the knees straight, sit up until the trunk and legs form a V, and touch the toes with the hands. Return to the BACK position.

(3) **Rocker.** .In the FRONT position, clasp the hands behind the back, arch the body, holding the head back. Start rocking, using the front part of the trunk as a rocker.

(4) **Bounce** and **Clap Hands**. The procedure is the same as for bouncing ball, but while in the air, clap the hands. This requires a more vigorous bounce or spring. (See fig. 3-12.)

(5) **Leg Spreader.** .From the BACK position, raise the legs so that the heels are 10 to 12 inches from the ground, spread them apart as far as possible, then close them together. Open and close legs as rapidly as possible.

(6) **Forward Roll.** .For forward roll from the STOP position, place both hands on the ground, tuck the head, and do one complete forward roll, keeping the legs tucked as you roll, and come back to the STOP position.

Figure 3-12 is available in the illustration section at the end of the book.

Section V.Guerrilla Exercise

3501.GENERAL

a. Description and Objective. Guerrilla exercises are individual exercises of various basic skills that are performed rapidly while moving forward in a circle formation. There are two tables of guerrilla exercises. Each table takes 15 minutes to complete. The objective of this exercise is to increase strength and endurance, aid flexibility, and develop coordination.

b. Area and Equipment. Any level area is suitable for conducting guerrilla exercises. No ground contact is required other than the hands. There is no equipment requirement.

c. Formation
(1) The circle formation (app. A) is used for guerrilla exercises. Each platoon forms its own circle and engages in guerrilla exercises under a platoon instructor. If the platoon exceeds 30 Marines, double or concentric circles may be used.
(2) When the circle is formed, the instructor steps into the center of the circle and moves clockwise in a small circle. He commands:QUICK TIME, MARCH, 1-2-3-4. (Rapid cadence of approximately 130 counts per minute. Cadence and step are maintained between exercises.)
(3) To reform the platoon after completing guerrilla exercises, the instructor halts the Marines and places the base man or platoon guide where he wishes and commands:
  (a) BASE MAN (or platoon guide),POST.
  (b) FALL OUT AND FALL IN ON THE BASE MAN (or platoon guide).

d. Progression. Progression may be attained by moving from table I to table II or by shortening the quick time marching periods between exercises and performing all exercises a second time.

e. Leadership
(1) To execute the exercises, the Marines continue at quick time while the instructor simultaneously explains and demonstrates the exercise to be performed, and then commands the Marines accordingly. In each instance, the preparatory command will be the name of the exercise and, in all instances, the command of execution will ***MARCH**.. To* terminate each exercise, the command *is* ***QUICK TIME, MARCH.*** *The* Marines immediately pick up the step as the instructor counts cadence.
(2) Unless specified differently, each exercise should be continued for 20 to 40 seconds depending upon the vigor of the exercise. The leader can determine the duration of each exercise by observing its effect upon the Marines.
(3) To form for double guerrillas, the commands for pairing the Marines (who are in circle formation) are-
(a) PLATOON, HALT.
(b) FROM (designate an individual), BY TWO'S, COUNT OFF(.Example1-2; 1-2; 1-2; etc.)
(c) EVEN NUMBERS MOVE UP BEHIND ODD NUMBERS(At this time, adjust pairs according to height and weight.)

(d) YOU ARE NOW PAIRED UP FOR DOUBLE GUERRILLAS. To change the Marines position, merely command CHANGE.)

(e) FORWARD, MARCH.

f. Place in the Program. Many Marines have not had the opportunity to perform the simple skills involved in guerrilla exercises. The conduct of these exercises is a simple matter since they can be performed easily and quickly in almost any situation. The tables of exercise are applicable to all personnel. The tables can constitute a station within a 1-hour period or be completed within a separate 15-minute period.

3502.GUERRILLA TABLES

a. Table I

(1) Double Time.(See fig. 3-13, A.)Hold arms in the thrust position. Execute a double-time run, maintaining the circle formation and the prescribed distance between your fellow Marines. Duration-1 minute.

(2) All Fours.(See fig. 3-13, B.) Face downward. Support the body with the hands and feet. Walk forward hands first.

(3) Crab Walk.(See fig. 3-13, C.) Get in the sitting position, face upward, and lift the hips. Support the body with the hands and feet. Walk forward feet first.

(4) Squat Walk.(See fig. 3-13, D.) Assume a full knee bend position. Grasp the ankles (left ankle with the left hand, right ankle with the right hand). Walk forward.

(5)Broad Jump.(See fig. 3-13, E.) Jump forward on both feet in a series of broad jumps. Swing the arms vigorously to assist the jumps.

(6)Toe-Touch Walk.(See fig. 3-13. F.) Walk forward, bending at the waist and touching one hand to the toe of the opposite foot while it is on the ground. Raise the trunk to the vertical position between steps. Keep the knees straight.

(7)Bottoms-Up Walk.(See fig. 3-13, G.) Assume the front leaning rest position and move the feet toward the hands in short steps, keeping the knees locked. When the feet are as close to the hands as possible, walk forward on the hands to the front leaning rest position.

(8) Straddle Run.(See fig. 3-13, H.) Run forward, leaping to the right from the left foot and to the left from the right foot.

(9) Fireman's Carry .(See fig. 3-13,1.)

(IO)Single-Shoulder Carry(See fig. 3-13, J.) Two men execute the carries as indicated by the diagram. No. 1 man executes one type; No. 2 man executes the other.

Figure 3-13 is available in the illustration section at the end of the book.

b. Table II

(1) Double Time.(See fig. 3-14, A.)Hold arms at the thrust position. Execute a double-time run, maintaining the circle formation and the prescribed distance between your fellow Marine.Duration-1 minute.

(2) Toe-Grasp Walk.(See fig. 3-14, B.) Bend forward and grasp toes. With knees slightly bent, walk forward.

(3) Hand-Kick Walk..(See fig. 3-14, C.) Walk forward, kicking the moving foot upward on every step. At the same time, lean forward and touch the elevated toe with the hand of the opposite arm.

(4) Pike Juinping.(See fig. 3-14, D.) Jump forward and upward from both feet, keeping the knees straight, and at the same time, swing the legs forward and touch the toes with the hands at the top of each jump.

(5) Squat Jump.(See fig. 3-14, E.) Leap forward from the squatting position, with the hands on the ground and the arms between the legs. Land on the ground with hands and legs extended. Bring up the legs to the squatting position.

(6) Steam Engine..(See fig. 3-14, F.) Lace the fingers behind the neck and walk forward in the following manner: as the left leg moves forward, raise the knee high, bend the trunk forward, and touch the outside of the right elbow to the outside of the knee. Then lower the left leg and step forward on the left foot and raise the right leg. Repeat with the right leg and left elbow.

(7) Knee-Touch Walk.(Scc fig. 3-14, G.) Walk forward, bending the knees and touching the ground on each step. The knees are bent and straightened on each step.

(8)Hobble Hopping.(See fig. 3-14, H.) Hold foot behind back with opposite hand and hop forward. On the command CHARGE, grasp the opposite foot with opposite hand and hop **forward.**

(9) Cross Carry.(See fig. 3-14,1.)(10)Saddle Back Carry. (See fig. 3-14, J.) Two Marines execute the carries as indicated in the diagram. No. 1 Marine executes onetype;No. 2 Marine executes the other.

Figure 3-14 is available in the illustration section at the end of the book.

Section VI. Running and Orienteering

3601. RUNNING

a. General

(l)Description. The general form and technique for all types of running are fairly constant. (See fig. 3-15.)The head is erect, body slightly forward without bending at the waist, and the arms are at a loose thrust position alternating from front to rear in straight planes. A cross-body arm movement wastes energy .The movement of the legs and feet will be discussed in subsequent paragraphs dealing with the different types of running. Of primary importance is the fact that in all types of running, the toes should be pointed straight ahead. Toeing out is a common error in both running and walking and should be an item of individual correction.

(2) Objective. The objective of running is to develop cardiovascular endurance. Despite the fact that Marines have developed their muscle structure and the strength of their muscle tissue, unless they have developed cardiovascular endurance to a satisfactory degree, they are not entirely physically fit or combat ready. Cardiovascular endurance (wind) depends on the efficiency of the lungs and heart. The efficiency of the lungs and heart depend on the amount of oxygen the lungs can absorb with each breath inhaled and the amount of carbon dioxide the lungs can expel. The process of absorbing oxygen and expelling carbon dioxide (cardiopulmonary process) is performed by the blood that circulates through the lungs. The condition of this process will determine the amount of effort a person can exert over a period of time. Running is one of the best activities to develop this vitally important endurance.

b. Running Skills. In the development of running skills, individuals may require instruction to improve their proficiency. Some important skills to consider are-

Figure 3-15 is available in the illustration section at the end of the book.

(1) Action of the Arms. Arm action is important. Check to see that arms are held loosely and that the action is relaxed. The faster the run, the more rapid the arm action.

(2) Breathing. Allow the individual to breathe through the mouth as the body demands a large supply of oxygen. Oxygen can be inhaled in greater quantities through the mouth.

3602. TYPES OF RUNNING

a. Double Time

(1) Description and Objective. Double timing is marching at the rate of 180 steps per minute, each step being 36 inches in length. It takes practice to double time with precision in formation. The Marines should keep in step, placing their feet flat on the ground. This, however, should not be a stamping motion, but should be done with as slight a jolt as possible. Double timing is like a jog, the difference being that in a jog the feet are lifted well off the ground and the running motion is bouncy. In double timing, the feet skim the ground and there is no bounce to the ran. Double timing is a vehicle for teaching proper running form and for the development of the cardiovascular system.

(2) Area and Equipment. This type of running can be completed over a variety of surfaces. Usually a Marine uses a field or road. There is no equipment requirement.

(3)Progression. There is no set standard for alternating quick time and double time in the early conditioning of Marines. A general rule is to begin with enough quick-time marching to ensure a thorough warming up, then double time about 100 paces. Change again to quick time until the individuals have made a reasonable recovery from the running, then double time another 100 paces. The amount of double time can be increased and the quick time decreased from week to week, until the individuals are double timing about 1800 yards. This type of training should be given at least twice a week, but by no means is it adequate as the sole means of conditioning.

(4) Leadership

(a) The instructor should be to one side of the column or group and toward the rear so there is a full view of all Marines. Inexperienced instructors have a tendency to supervise from a position too far forward.

(b) Select an individual who can maintain the proper cadence to act as the guide during double-time running.

(c) There are several ways for the instructor and group to count cadence while double timing. If not contrary to local policy, learn several methods and use them for variety.

b. Wind Sprints

(1) Description and Objective. This type of running involves a series of 30- or 40-yard dashes, usually conducted in successive waves of squads. Each squad is in line and the squad leader on the right flank. Wind sprints assist indeveloping speed and cardiovascular endurance. Any flat and level area may be used which will permit the squad to form a line and run the required distance.

(2) Progression. One or two 30-yard sprints will be adequate at the beginning. As time passes, sprints can be lengthened and up to six or seven sprints may be used.

(3) Leadership

(a) At the command READY (given by squad leader), each runner assumes the sprinter's starting position. At the command GO, the squad sprints approximately 30 yards, takes 10 yards to stop, and lines up immediately with the squad leader who repeats, READY, GO, and again the squad sprints. At the conclusion of the third sprint, the squad waits until all the squads of the platoon have made three sprints. Then they all line up and the squad leaders conduct three more wind sprints in the opposite direction.

(b) Valuable time is gained by having each squad ready to go when the preceding squad has moved off its second sprint mark.

c. Cross-Country Running

(1) Description and Objective. Cross-country running is a distance run conducted on a course laid out along roads, across fields, over hills, through woods, and on any irregular ground. The cross-country ran may be utilized as a conditioner or as a competitive event; the objective is to cover the distance in the shortest possible time. The course should be 2 to 2.5 miles in length and be laid out to avoid heavy vehicular traffic. The course should be marked by directional arrows until the runners know the course. These runs build leg muscles, increase lung capacity, and develop endurance. Any local area of varied terrain is suitable.

(2) Progression. In the mass training of a large group, leaders should be stationed at the head and the rear of the column and should make every effort to keep the individuals together .After determining the abilities of the individuals in cross-country running, it is advisable to divide the unit into three groups. The poorest conditioned group starts first, and the best conditioned group, last. The starting time of the groups should be staggered so that all of them finish about the same time. In preliminary training, the running is similar to ordinary road work in that it begins with rather slow jogging, alternating with walking. The speed and distance of the run is gradually increased. As the condition of the individual improves, occasional sprints may be introduced. At first the distance run is from 1/2 to 1 mile. It is gradually increased to 2 or 2.5 miles. Well-conditioned personnel can run 2 to 2.5 miles within a 15-minute period.

(3) Leadership. Marines should not be required to take part in distance running until they have been through a progressively scheduled training program which requires a considerable amount of running. Cross-country runs should be scheduled occasionally to provide variety in the program. Cross-country running has the advantage of allowing mass participation. Interest can be stimulated by putting the runs on a competitive basis. As a single activity, short cross-country runs can be scheduled once a week, gradually increasing the distance as the physical conditioning improves.

(4)Use With Other Programs. Cross-country running can be combined with other activities such as conditioning exercises.

d. Fartlek Training

(l)Description and Function. Fartlek training is various running exercises conducted along an intense combat-like course. The running is conducted to overwork the lungs, allowing only partial recovery which is followed by another intense period of overload. This sequence is repeated for the duration of the workout. Fartlek training is a useful combat training method and general cardiovascular conditioner. The entire unit must be in uniformly good to excellent condition to qualify for this training.

(2) Progression. The following is an example of one fartlek session:

(a) Warm up by stretching 3-5 minutes.

(b) Warm up running easily 5-10 minutes.

(c) Run at a fast, steady pace for 3/4 to 1 3/4 miles (dependent on terrain).

(d) Walk/jog at a moderate pace for 5 minutes (recovery).

(e) Run easily sprinting 15 to 20 meters occasionally.

(f) Run full speed uphill for 175 to 200 meters.

(g) Warm down by running easily for 1/2 to 1 mile.

(h) Warm down by stretching 3 to 5 minutes.

(3)Leadership. To add variety, try incorporating combat movements, terrain association, a series of exercises or carrying table of equipment weapons (hit and roll). This will take some initiative in setting the course, but it allows for variety. In this manner, the fartlek course can be used as a training session or part of one. It works extremely well with highly motivated, competitive Marines.

3603.ORIENTEERING

Orienteering is land navigation over a prescribed course as a timed, competitive event. It is an excellent way to integrate land navigation training with physical conditioning and requires Marines to be skilled in both areas. Commanders adjust the training as needed to change the degree of difficulty or to emphasize either navigation or conditioning. Taken to the extreme, an orienteering exercise is conducted in full combat gear over rugged terrain against aggressors. Orienteering requires the following physical skills: endurance, speed, strength, and all-around conditioning.

Section VII. Circuit Training

3701. GENERAL

a. Description and Objective. A strength circuit consists of a series of stations where individuals in small groups exercise vigorously for a short period of time and then move (on signal) to the next station where a different form of exercise is conducted. This rotation continues until all groups move through all stations. Strength circuits contain no set or specific types of exercise stations within the circuit. The objective of circuit training is to develop strength. There are three general types of circuits.

(1) Fixed Circuit. This is a circuit in which apparatus of an immovable type (fixed into the ground) is used.

(2) Movable Circuit. This circuit consists of individual exercise apparatus which is portable and can be moved to and from the training area.

(3) Simplified Circuit. This circuit requires no equipment or apparatus.

b. Formation. The exercises are done at will, but rapid, steady, and continuous work is required of all. Each Marine's nervous and muscular system reacts differently to timed vigorous exercises. His performance should be measured on how many movements per exercise he can complete as an individual. For example, one Marine may be able to complete 5 movements, while another may be able to complete 20, and yet each is receiving the maximum benefit. All three circuits contained in this chapter are designed for platoon-sized groups. Expansion beyond this capacity requires a large amount of equipment, as each Marine in the fixed and movable types of circuits must have an item of equipment available for exercise at each station. A group larger than a platoon could be exercised through use of the simplified type of circuit; however, the group would be unwieldy and control could be a problem.

c. Place in the Program. All circuits illustrated can be completed in a 15-minute period. This feature allows the exercise of a platoonor smaller group on the circuit for a single 15-minute period, or the scheduling of the circuit as a 15-minute period within a longer period. A circuit can thus be utilized within a rotating activity system of scheduling. Choice of a circuit by the unit depends upon area, facilities, and other local factors; however, there is a circuit for every need.

3702. FIXED STRENGTH CIRCUIT

a. Description and Objective. The strength circuit is an arrangement of various types of exercise apparatus which are fixed in position. (See fig. 3-16.)Seven basic exercises are used and each exercise requires an apparatus. All apparatus of one type are positioned together to constitute a station. Each station will accommodate 10 Marines. The circuit is designed to be accomplished in 15 minutes when conducted on a time-rotation basis, as normally executed for unit training. For individual use, a Marine may complete the circuit by executing a specific number of repetitions for each exercise. In this case, the time required for completion of the circuit would vary slightly depending on the number of repetitions the Marine completed. The objective of this circuit is to provide a series of exercises which will improve and maintain the strength of the body's major muscle groups.

Figure 3-16 is available in the illustration section at the end of the book.

b. Starting Level. Marines must be thoroughly warmed up prior to participating in the circuit system. If Marines have not engaged in vigorous exercise immediately prior to starting the circuit, then an instructor should execute the following warm up exercises. These exercises should be conducted in the normal formation for set drills. Seven repetitions of each exercise will normally provide sufficient warm up. The exercises are-

(1) High jumper, Conditioning Drill 1.

(2) Bend and reach, Conditioning Drill 1.

(3) Squat bender, Conditioning Drill 1.

c. Progression. The instructor can adjust the circuit system through several methods to accommodate participants of varying physical ability. One method of adjusting an exercise is to change the method of executing the leg lift or to select a heavier weight for the barbell curls. When it becomes apparent during a unit program that the overall fitness of the group has Improved, then the exercise may be made more strenuous in two additional ways. First, the instructor can have one or more of the supplementary stations added. Second, the time spent exercising at each station can be increased in 5-second increments to a maximum of 60 seconds. Also, as Marines learn the circuit, the 45-second period for movement and instruction between stations can be eliminated, thus allowing only enough time to change stations.

d. Starting Positions. The Marine starts on any station, exercises steadily for a certain period (45 seconds initially), then moves on command to the next type of apparatus where he again exercises steadily for an equal period. The Marine continues until the required exercise is completed at each station. If it is desired to expand the number of stations in the circuit to accommodate more participants at one time, the instructor can provide four additional supplementary exercises, requiring no equipment. This will increase the amount of time required to complete the circuit.

e. Leadership. Close control of all Marines is necessary to ensure that a minimum amount of time is spent in moving them to their initial stations and in moving them between stations. One instructor can control the activity on the strength circuit. A stop watch or wrist watch with a second hand is required. When Marines arrive at the strength circuit, the instructor will form them for exercise and conduct the warm up drill. The group is then reassembled and formed into a number of files equal to the number of stations being used in the circuit. Each file is then directed to a station. As soon as all participants have reached an exercise position at a station, the instructor gives the command READY, GO. After 45 seconds of exercise, the instructor gives the command STOP, CHANGE OVER. The instructor allows Marines 45 seconds for moving to the next station and for preparing for the next exercise before the command EXERCISE is again given. In lieu of verbal commands, a whistle may be used to stop and start the exercises. For large groups, a megaphone or loudspeaker is useful.

f. Individual Conditioning Program. For the Marine working alone on the strength circuit, it would be impractical to time the exercise periods. For individual exercise, the Marine should select a number of repetitions of each exercise to complete, then rotate to the next stations after completing these repetitions. The number of repetitions selected should be at or near the maximum that the Marine is capable of doing without halting for rest.

g. Fixed Circuit Stations

(1) Primary Stations (See fig. 3-16.)

(a) Pullups/Chinups. A horizontal bar placed 8 feet above the ground is necessary for each Marine at this station. The Marine will also need a space on the bar that is 45 inches wide. On the command**EXERCISE**,the Marine grasps the bar with both palms facing either forward or to the rear, arms fully extended, and feet free of the ground and executes the exercise as described in MCO 6100.3_.The exercise is repeated as many times as possible until the command **STOP** is given. Then the Marine moves to the next exercise. If a Marine has done his maximum number of pullups/chinups prior to the command **STOP**, he will remain in the "dead" handing position until the command **STOP** and move to the next station.

(b) Twist Grip. The apparatus is a horizontal bar, free to turn, held between uprights placed 30 inches apart. The bar is 52 inches above the ground. A weight of 20 pounds is attached to the center of the bar by a light rope long enough to permit the weight to rest on the ground. The Marine stands at arm's length from the bar and grasps it with his hands on either side of the rope, palms down, thumbs under the bar. On the command **EXERCISE,** the hands are rotated so that the backs of the hands are rotated away from the body, thus winding the rope on the bar. The elbows are kept straight to ensure that the exercise is performed by the hand and forearm. When the weight is drawn up to the bar, the bar is then rotated in the opposite direction to lower the weight to the ground. This exercise is continued until the command **STOP** is given. The Marine then moves to the next station.

(c) Pulley Weights. The apparatus is a T frame with a system of pulleys that suspends a weight of about 90 pounds. The weight is attached to a light steel cable which has a drawbar attached to the other end. The Marine grasps the drawbar and sits down directly under the bar, legs extended to the front and arms extended overhead. The exercise is executed by pulling the drawbar down behind the head, then extending the arms slowly again until they are fully extended overhead. The exercise is repeated as many times as possible until the command STOP is given. The Marine then will move to the next exercise. Upon completion of the exercise, the weight is lowered slowly to the ground.

(d) Barbell Curls. A barbell is necessary for each Marine at this station. The barbell is constructed of 1 1/4-inch pipe 5 feet long, and two concrete-filled No. 10 cans. Each barbell should weigh about 40 pounds. Variance in the weight of the barbells, up to about55 pounds, will allow appropriate overload to be applied to Marines who are above average in strength or weight. The Marine grasps the bar with the palms forward and assumes a standing position with the barbell held in front of the hips, hands approximately shoulder's width apart. On the command **EXERCISE,** the Marine flexes the elbows and draws the barbell up until it touches the upper chest. The elbows remain at the sides. Breath is inhaled with the upward movement and exhaled as the barbell is lowered to the starting position. The exercise is repeated as many times as possible until the command **STOP** is given. The Marine moves to the next apparatus.

(e) Step-Up. The apparatus is a platform or ledge 18 inches high and of such size. At accommodate 10 Marines. The Marine faces the platform and on the command **EXERCISE,** steps up onto the platform, bringing his trailing foot up beside the leading foot. He then steps back down to the original position, stepping down first with the same foot he initially used in stepping up. After 10 repetitions of the exercise, he changes the order of moving the feet to use the opposite leg for stepping up. Repeat this exercise until the command **STOP** is given. The Marine then moves to the next station.

(f) Leg Lift. The apparatus is a horizontal bar constructed as described in (a). To prevent the body from swaying, a horizontal back support is added 40 inches below the horizontal bar. The arms are kept fully extended. On the command **EXERCISE,** the Marine jumps up, grasps the bar with the palms forward and the back support behind him. The exercise is executed by raising the legs to a horizontal position then slowly lowering them to the vertical position. The Marine does not flex his knees. He does not swing the legs to the rear of a vertical position to gain momentum for raising them in the next repetition of the exercise. The movement is repeated until the command **STOP** is given. The Marine then moves to the next exercise. If unable to raise his legs to a horizontal position without flexing his knees, the Marine flexes his knees and draws them up to the chest, then lowers his legs to the vertical position.

(g) Rope Climb. The rope climb is 20 to 30 feet high with five ropes suspended from a horizontal bar which forms the uppermost part of the framework. To prevent the horizontal bar from sagging and to provide safety, only five ropes are attached to it. There are two frameworks per station. The ropes are 6 feet apart. Any method may be used to climb the rope, and the Marines climb as high as possible. Marines who are proficient should climb therope several times during the time allotted. Inexperienced Marines should be cautioned to take care during descent to avoid rope burns on their hands.

(2) Supplementary Stations. The following exercises are designed to expand the basic circuit by being inserted in specific places within the system. For each supplementary station used, there should be adequate room for 10 Marines to exercise.

(a) Bent Knee Situp or Bottoms-Up. These calisthenics are designed to strengthen the abdominal muscles. These exercises will be inserted between the pullup/chinup and twist grip stations. The primary stomach exercise is the situp. In case of inclement weather or other conditions that make ground contact undesirable, the bottoms up exercise is used. In situps on the command **EXERCISE,** the Marine lies on his back (supine position) with knees flexed and both feet flat on the ground and executes the exercise (minus an assistant) as described in MCO 6100.3_.The Marine then returns to the starting position, repeating the exercise until the command **STOP** is given. The Marine then moves to the next station. In bottoms up on the commandEXERCISE,the Marine assumes the front leaning rest and executes the bottoms-up exercise as described in Conditioning Drill 3. He continues this exercise at a moderate cadence until the command **STOP** is given. The Marine then moves to the next station.

(b) Pushup. This exercise is designed to strengthen the arm- and shoulder-girdle muscles. It should be included between the twist grip and pulley weight stations. Upon the command **EXERCISE,** the Marine executes the pushup as described in Conditioning Drill 1. He continues this exercise at a moderate cadence until the command **STOP** is given. The Marine then moves to the next station.

(c) Knee Bender. This exercise is designed to build leg muscles and is included between the pulley weight and barbell curl stations. On the command **EXERCISE,** the Marine executes the knee bender as described in Conditioning Drill 3.He continues this exercise at a moderate cadence until the command **STOP** is given. The Marine then moves to the next station.

(d) Trunk Twister. This exercise strengthens the major muscles of the trunk and is included between the step-up and pullup/chinup stations. On the command EXERCISE, the Marine executes the trunk twister as described in Conditioning Drill 1. He continues this exercise at a moderate cadence until the command **STOP** is given. The Marine then moves to the next station.

3703. MOVEABLE STRENGTH CIRCUIT

a. Description and Objective. The exercises in this circuit are progressive and the course is planned to gain and hold the interest of the participating groups. The circuit consists of a series of stations, with each station designed to develop a particular group of muscles. Along with muscular development, correct posture and deep rhythmic breathing should be stressed on this circuit at all times.(See figs. 3-17 and 3-18.)

Figure 3-17 is available in the illustration section at the end of the book.

b. Equipment. The equipment is set up in files. Six files of 8 stations will accommodate a platoon

c. Formation. The platoon marches to the area where the equipment is positioned and forms a file within each lane of stations, covering down on a piece of equipment. Movements are made on the double, the important factor being that no time is wasted in getting to work.

d. Progression. Initially 40 to 45 seconds per station is adequate. As individuals become stronger the time should be increased in 5-second increments until a minute to a minute and a half is reached.

e. Leadership

(1) The leader stands in front of the barbell station and controls the rotation from this position. The leader supervises the entire group, with the assistance of several instructors who move about in the platoon correcting and encouraging the Marines.

(2) The leader starts each group but does not count cadence nor lead them through the exercises. Each individual exercises rapidly but individually.

(3) As each Marine finishes his repetitions with the barbell, he places the barbells on the ground. The leader calls, READY, followed by the command, FALL OUT ONE. All will double-time to the station directly in front of them, while the individuals on the barbell stations will do an about face to the rear station in their lane.

f. Movable Circuit Stations. The Marine can obtain best results on the movable circuit if the exercises on the various stations are given in the following manner:

(1) Station 1—The Barbell. (See fig. 3-17, 1.)The exercises at this station stress proper posture; deep, rhythmic breathing; and development of the muscles of the arms, shoulders, and upper body. An instructor teaches the proper methods of lifting before the exercises begin. The methods are to lift with the legs, to keep the back straight, and to merely grip with the hands. Two recommended exercises are given below. The instructor will use only one exercise per period .He may use either exercise. At successive periods, the other exercise is used.

(a) Exercise 1--Two-Hands Military Press. (See fig. 3-18, 1.)Grasp the barbell with both hands, knuckles up at shoulder width, and lift to the chest. Steadily press to arm's length overhead; lower to the chest resisting weight all the way. Inhale as the weight is pressed up and exhale as the weight is brought down.

(b) Exercise 2--Two-Hands Regular Curl. (See fig. 3-18, 1.) Lift the weight to the waist, with the palms of the hands out, heels together, stomach in, chest lifted and arched, shoulders back, elbows in close to the sides; inhale deeply and curl the weight to the shoulders, using the arms only, at the same time keeping the elbows close to the sides; exhale Emphasize posture and the use of the arms only. This is a very valuable exercise for the development of the biceps and the grip and should be repeated from 8 to 16 times, depending on the ability of the Marine.

(2) Station 2--The Jump Rope. (See fig. 3-18, 2.) This exercise develops strength and agility in the legs and stamina of the whole body. It makes the Marine agile on his feet and increases his footwork efficiency and timing. Each Marine should progress until able to jump rope at least 3 minutes at top speed.

(3) Station 3--The Twist Grip. (See fig. 3-18, 3.) The twist grip is an excellent exercise for the hands and forearms, and adds greatly to the Marine's ability in hand-to-hand combat. The handle is gripped and twisted, winding the rope until the weighted can is level with the height of the hands, which are held horizontal. The weight is lowered in the same manner; the Marine resists the weight all the way, occasionally stopping the twisting motion and alternately removing first one hand, then the other, from the handle. A variation of the above exercise is to wind the handle with the palms up and the arms bent and with the elbows held close in to the sides. Each Marine should maintain a good posture and keep the stomach muscles taut throughout this exercise.

(4) Station 4--The Incline Plane. (See fig. 3-18, 4,)The use of the incline plane is a very strenuous exercise and well designed for the development of the abdomen. Although 6 to 10 repetitions are sufficient for the beginners, more can be added as ability increases. Marines with hernias or recent operations will be excused from participation at this station.

(5) Station 5--The War Club. (See fig. 3-18, 5.) The war club is a simple and effective means of exercising the principal muscle groups of the body, especially those of the trunk, back, and shoulders. To gain the maximum benefit from this exercise, the Marine must keep both feet flat on the ground at all times. Throughout the exercise period, the weight is swung from arm's length as follows:

(a) As in chopping wood, first on one side, then on the other.

(b) As a batter warming up with a number of bats.

(c) In large circles, first with one hand and then with the other.

(6) Station 6--The Bicycle Ride. (See fig. 3-18, 6.) The bicycle ride is well suited to exercising many of the muscle groups of the body, particularly those of the abdomen. Vary the speed of the exercise, but keep everyone "riding" the entire period. A variation exercise may be performed by placing the legs together, raising them slowly to a height about 2 feet from the ground, and then lowering them slowly to the ground.

(7) Station 7--The Step-up. (See fig. 3-18, 7.) The step-up exercises the legs. The step-up is performed by initially stepping up with the left foot, followed by the right, then stepping down with the left foot followed by the right. Continue for 20 seconds, then change to the right foot as the lead foot for 20 seconds.

(8) Station 8--The Isometric Pull. (See fig. 3-18, 8.) Two trainees work at this station with a cable pull and perform the following exercises:

(a) Initially start with one Marine in the supine position and one Marine sitting. The sitting Marine lowers the upper body to the ground and pulls the partner up to the sitting position. The partner then performs this same action and this is continued for 20 seconds at a rapid rate.

(b) During the last 20 seconds, the same action takes place but, in this case, the Marine in the supine position resists the pull of the partner for approximately 5 seconds before allowing to be pulled up into the sitting position.

3704. CIRCUIT-INTERVAL TABLE

a. Description and Objective. The circuit-interval table is designed to develop strength and endurance within a short period of time, with no equipment requirement, through a rapid and vigorous routine of exercise. Fifteen minutes is an adequate period to execute all exercises and to secure a vigorous workout with the circuit-interval principle.

Figure 3-18 is available in the illustration section at the end of the book.

b. Formation and Starting Positions

(1) A leader forms platoons or smaller groups in an oval or circular formation with 3- to 5-yard intervals between Marines. The Marines face to the right and move forward at quick time and then double time. (See fig. 3-19, A.)After running several platoon circle laps, the leader calls out the name of an arm and shoulder exercise from the list below, orders quick time and commands, for example, PUSHUPS. On this command, all Marines immediately hit the ground and individually and rapidly begin doing pushups. No cadence is counted. (See fig. 3-19, B.) After 30 seconds of exercise, the leader commands. **QN YOUR FEET, FORWARD, MARCH** The platoon resumes the quick time cadence and the leader, when ready, gives the necessary commands for double time. The double time is continued for one or more laps and the leader calls out the name of the next exercise and the process is repeated. This continues, with running between each exercise, until every body part has been exercised.

(2) The instructor controls the running and quick time to observe the effects of the exercise upon the Marines. Cadence, step, and precision are not important to the objective and the instructor should not use them. What is important is speed and the instructor should stress this. After the exercise period is started, the Marines do not stop. This circuit method emphasizes stress and recovery, the recovery occurring during the quick time periods.

c. Activities. The leader can use the following exercises. These exercises can be repeated if necessary, during a second round. The leader can use other calisthenic exercises in the circuit-interval table.

(1) Arms and shoulders-push-ups.
(2) Stomach situps.
(3) Back-squat thrusts.
(4) Legs-bicycle (on back).

d. Progression. lhe progress is controlled by the leader, who must pace the running, quick-time movement, and exercise in such a way that Marines will receive a vigorous workout yet be able to participate throughout the 15-minute period. Marines who are in the initial stages of physical condition will not be able to double time or exercise as long as those who are better conditioned. The idea is to set a pace which can be increased during each workout, thus progressing gradually to a higher level of physical fitness.

e. Leadership. The platoon leader, platoon sergeant, or section leader can lead the group. The leader must execute the exercise with the unit to feel the effects and thereby adjust the pace.

f. Place in the Program. This activity may be scheduled whenever a short period of time is available. The only requirement is that enough space, indoors or out, be available to form the circle.

Figure 3-19 is available in the illustration section at the end of the book.

Section VIII. Basic Physical Skills and Obstacle Courses

3801. GENERAL

The purpose of this section is to list the basic military skills and the methods for their development. Many of these skills are best developed by obstacle courses but other drills are also discussed. The section explains types of obstacle courses, details of construction, and methods of negotiating the various obstacles.

3802. BASIC PHYSICAL SKILLS

a. Objective. The objective is to develop proficiency in the various military physical skills which are essential to personal safety and effective combat operations. In travel by foot over rugged terrain and in the execution of combat duties, Marines must be trained to perform basic skills, such as running, jumping, climbing, and carrying. During training, Marines will develop agility and coordination in these skills .Fast and skillful execution of these skills may mean the difference between success and failure on the battlefield.

b. Place in the Program. These skills are practiced throughout the entire program in many different activities. Many of these skills are best practiced on obstacle courses.(See par. 3803.)

c. Description. These basic skills are the minimum skills required by all Marines. The basic skills are as follows:

(1) Running. Running is used to strengthen the legs and develop cardiovascular endurance. Marines should be exposed to running in various situations: on roads, over rough ground, up and down hills, cross-country, and running over low obstacles.

(2) Jumping. In broad jumping, the take-off foot is planted firmly and the spring comes from the extension of this leg as the other leg reaches for the far side of a ditch or similar obstacle. (See fig. 3-20.) The arms are forcibly raised forward and upward to assist in propelling the body up and forward. Landing may be on one or both feet depending upon the length of the jump. In vertical jumping downward from a height, the jumper should aim his feet at the desired landing spot and he should also jump with the knees slightly bent and feet together, with the trunk inclined slightly forward. As the feet touch the ground, the shock is absorbed by bending the knees into a full squatting position. If the height is too great or the ground too hard to absorb the shock, then the jumper should forward roll or side roll thus eliminating some of the momentum.

Figure 3-20 is available in the illustration section at the end of the book.

(3) Dodging In combat situations, it is often times necessary to change directions quickly. To execute this movement while running, a lead foot is firmly planted, left foot if the direction is to the right and right foot if the direction is to the left. The opposite foot is moved toward the new direction. The knees are slightly flexed during the movement and the center of gravity is low and balanced. (See fig. 3-21.) At the time of the change of direction, the head and trunk are turned quickly in the new direction.

Figure 3-21 is available in the illustration section at the end of the book.

(4) Climbing and Surmounting. All Marines should know how to climb and surmount various types of obstacles. (See fig. 3-22.)

Figure 3-22 is available in the illustration section at the end of the book.

(a) Vertical Climbing, as in Climbing a Rope or Pole. The hands grasp the rope or pole overhead with the palms toward the face. Gripping the object, the body is pulled upward with the arms and shoulders, assisted by the feet which grip the object and assist by pushing downward. If shoulder girdle strength and body coordination are not adequate to permit alternating the hands, the arms act together in pulling upward.

(b) Climbing as in Surmounting a Wall. Ingoing over a wall, the body should be kept as close to the top as possible, since in combat operations it is important to offer as small a target as possible to the enemy. If an individual climbs a wall while carrying a rifle, both hands should be freed by slinging the rifle over the back. There are two methods commonly used for surmounting a wall of moderate height, but only one for dropping from it.The methods are as follows:

1 Run, Jump, and Vault. Approach the wall at a run, jump forward and upward at the wall and place one foot against it as high up as possible. Use the foot in contact with the wall to help push the body upward while grasping the top of the wall with the hands. Pull the body up with the arms, assisted by pressure of the foot against the wall and swing the legs over, propelling the body weight over the wall.

2 Hook and Swing. Approach the wall at a run and jump forward and upward. Hook one elbow over the wall, locking the arm in place by pulling up until the top of the wall is underneath the armpit. Grasp the top of the wall with the other hand. Draw the leg which is closer to the wall up as far toward the top as possible. Then swing the other leg over the top of the wall.The body is then carried over with a rolling motion. A variation of this leg action can be used by Marines who are unable to draw up the leg as described. While hanging with both legs fully extended, start a swinging motion with the legs together. When the legs have enough momentum, swing the outside leg over the top of the wall with a vigorous kick, then follow with the body.

3 **Dropping.** All drops from a height are executed in the same manner regardless of the method used to gain the top. One hand is placed against the far side of the wall while the other hand grasps the top. From this position, the body is rolled over the wall and vaulted away from it with the legs swinging clear. As the body passes over the wall and drops, it should at all times face the wall. This will keep the rifle and other equipment clear. Break the fall by retaining a grasp on the top of the wall as long as possible.

(c) Climbing Ladders and Cargo Nets. Rope ladders, stationary vertical ladders, and cargo nets employ the same general technique. The important element is to grasp the side supports firmly in the hands about shoulder height and place the feet on a rung which would cause the body to be fully extended. In movement upward, one hand is moved upward and a new grasp is secured and, at the same time, the opposite leg moves up a rung. As the knee straightens, the body is elevated. This process is repeated using the opposite arm and leg. Alternation continues in this manner until the climber reaches the objective.

(5) Traversing Horizontal Objects. The traversing of horizontal objects puts heavy stress on the arms and shoulder girdle area as the feet are usually suspended in the air with all of the body weight on the arms and shoulders.(See fig. 3-23.)

(a) Traversing Horizontal Ropes or Pipes. The hands grasp the horizontal support overhead with the palms facing. To propel the body forward, one hand is released and moved forward to secure a new grasp. At the same time, the opposite side of the body is swung forward (some people are able to "walk" in the air, keeping the body to the front and moving the legs in time with the arms as in walking on the ground). The other hand is then released and moved forward; this alternation is continued until the objective is reached.

Figure 3-23 is available in the illustration section at the end of the book.

(b) Traversing Horizontal Ladders. In this situation, the movement is the same as used in traversing a rope or pipe. The hands, however, are placed on the rung with the palms away from the face. Other than this difference the technique is the same.

(6) Crawling. Crawling in combat situations is an often used skill. Crawling may be high or low.(See fig. 3-24.)

Figure 3-24 is available in the illustration section at the end of the book.

(a) High Crawl.

In the high crawl, the Marine moves on hands and knees, moving one hand and the opposite knee and then continuing to move the hands in alternation with the opposite knee following the companion hand.

(b) Low Crawl. The Marine is in the prone position usually with the forearms and palms of the hand on the ground. He propels forward by bending the knee of one leg and pushing with the inside edge of the shoe. At the same time, the opposite arm moves forward and pulls to the rear. The body remains low and movement is continued by bending the opposite knee and pushing, and at the same time sliding the opposite arm forward and pulling. Alternation of hands and legs continues until the objective is reached.

(7) Throwing. Throwing may be executed from the kneeling or standing positions. The object to be thrown is held in the hand, and the throwing arm is bent at the elbow; the hand is then moved to the rear until the hand is behind the ear. The body is turned so that the lead foot and balance arm on the side toward the target are pointing at the target. (See fig. 3-25.) The balance arm is used to sight over and align the throwing hand and the target. When properly aligned, the elbow is moved rapidly forward until it is at a pointjust in front of the body where the arm is straightened and the wrist snapped. This whip motion propels the object to the target. Underhand throws secure momentum by the thrower bending his knees and swinging the throwing arm to the rear. As the knees are straightened, the arm is forcefully swung forward from the shoulder and the object released.

Figure 3-25 is available in the illustration section at the end of the book.

(8) Vaulting. Vaulting is employed to overcome low barriers or fences. (See fig. 3-26.)The object to be surmounted is approached at an angle. The hand on the side next to the obstacle is placed on the top of the obstacle and, with a straight arm, the body weight is pushed upward. At the same time, the leg on the side next to the obstacle is thrown upward and over the top followed by the other leg (side approach). In landing, the weight comes down on the leading leg first followed by regaining the balance on both legs. The free arm serves as a balance. A direct (front) approach can be used at which time both legs go over the object together.

(9) Man Carrying. There are three basic individual means of carrying personnel in combat situations and one of these methods may be used in carrying objects.

Figure 3-26 is available in the illustration section at the end of the book.

(a) Fireman's Carry."A" stands sideways in front of "B", "A" bends his knees and leans forward, placing one arm through "B's" crotch, grasps the wrist of "B's" arm, which is hanging over the shoulder, and then "A" runs forward.

(b) Saddle-Back Carry. With his back toward "B", "A" stands in front of "B". "B" mounts "A's" hips and clasps his arms in front of "A's" chest. "A" grasps "B's" thighs.

(c) Single-Shoulder Carry. "A" stands facing "B"."A" assumes a semi-squatting position. "B" leans forward until "B" lies across "A's" right shoulder. "A" clasps his arms around "B's" legs and straightens up, lifting "B" from the ground. "A" then runs forward. This method may also be used to carry heavy objects.

(10) Balancing. Balancing the body while walking or running on a narrow object when crossing obstacles is a skill which requires practice and confidence. Balance is required in negotiating a log placed across a stream, in crossing a narrow beam or rail, and in similar situations.(See fig. 3-27.) To perform this skill, place the feet on the object to be crossed, hold the arms to the sides at shoulder level, and fix the eyes on the object approximately 5 yards in front of the feet. Generally, it is not a good practice to look down at the feet. Walk the beam by placing first one foot and then the other in the center of the beam, thereby moving forward, using the arms to aid in maintaining balance.

Figure 3-27 is available in the illustration section at the end of the book.

(11) Falling. Injury can be avoided if Marines are taught to fall properly. They should know how to use the body momentum to their advantage during a fall rather than to try resisting that force. (See fig. 3-28.) If enough force is present, such as occurs during a fall while running or in jumping downward from a height, individuals can extend their hands to catch the weight. At the same time, duck the head and roll forward onto the feet. The key to falling without injury from the standing position is relaxation and rolling on the outside of the leg, hip, and buttocks to take the brunt of the fall.

Figure 3-28 is available in the illustration section at the end of the book.

3803. OBSTACLE COURSES

a. General. Obstacle courses are a valuable part of physical readiness training. The challenge presented by the obstacles assists in developing and testing the basic physical skills. In many combat situations, success will depend upon the Marine's ability to perform one or more of these skills, often while carrying field equipment and when fatigued. In this section, two types of obstacle courses will be discussed-Conditioning Obstacle Course and Confidence Obstacle Course.

b. Course Safety Precautions. Commanders and course instructors should take certain precautions to prevent injury on obstacle courses. A few of the precautions are:
(1) Inspect the course for faulty construction of obstacles, protruding nails, rotten logs, condition of the landing pits, and other hazards to safety.
(2) Conduct warm up exercises before the unit runs the course.
(3) Explain and demonstrate the correct techniques for negotiating all the obstacles before allowing the Marines to try them.
(4) Give Marines at least two weeks of conditioning exercises before scheduling the obstacle and/or confidence courses.
(5) Ensure that negotiation of the higher and more difficult obstacles are under the supervision of an instructor.
(6) Do not permit individuals who have neither practiced the basic skills nor run the conditioning obstacle course to participate in the confidence obstacle course.
(7) Weather conditions may cause footing or handhold surfaces to be slippery. If such is the case, postpone training on the course.

3804.CONDITIONING OBSTACLE COURSES

a. Description and Objective. The Conditioning Obstacle Course is commonly known as the Obstacle Course. This course consists of fairly low obstacles which are designed to be negotiated quickly. The obstacles serve to test various basic skills, and running the course is a test of the Marine's physical condition. After receiving instruction and an opportunity to practice the skills, Marines run the course against time.

b. Area and Equipment
(1) Complete standardization of obstacle courses should not be attempted since topographical conditions always vary. Commanders should use ingenuity in constructing a course, making good use of streams, hills, trees, rocks, and other natural obstacles. Since the course is eventually run at high speed, it should not be dangerous.

(2) The course should be wide enough for six or eight men to run simultaneously, encouraging competition. The lanes for the first several obstacles should be wider and the obstacles themselves easier than those that follow. This avoids congestion until the contestants scatter out over the course. The last two or three obstacles should not be too difficult and should not involve high climbing. This prevents injuries and falls resulting from fatigue.

(3) The total distance of the course should range from 300 to 450 yards and include from 15 to 25 obstacles. Normally the obstacles should be 20 to 30 yards apart and arranged so that those which exercise the same groups of muscles are separated.

(4) The obstacles should be substantially built. Peeled logs, 6 to 8 inches in diameter, are ideal for many of the obstacles. Sharp points and corners should be eliminated. Landing pits for jumps or vaults should be filled with sand or sawdust to prevent injuries.

(5) The course should be constructed and marked so that it is not possible to sidestep or detour obstacles. However, it is desirable to provide alternate obstacles of varying degrees of difficulty.

(6) The course should be in the shape of a horseshoe or figure eight so that the finish is close to the start and signs should be placed to indicate the course route.

c. Leadership. Before Marines run an obstacle course, they should be instructed in the proper technique of negotiating each obstacle. In each case this technique should be explained and demonstrated in detail, with emphasis on avoiding injury. Every individual should be given an opportunity to practice on each obstacle until he becomes reasonably proficient at negotiating it. Before the course is run against time, it is advisable for individuals to make several runs at a slower pace. During such practice or trial runs, the instructor should observe the performances and make appropriate corrections. Marines should never be permitted to run the course for time until they have practiced on all obstacles. The best method of timing the runners is to have the timer stand at the finish and call out the minutes and seconds as each individual finishes. If several watches are available, each wave may be timed separately. If only one watch is available, the different waves should be started at regular intervals, such as every 30 seconds. If an individual fails to negotiate an obstacle, a previously determined penalty should be exacted.

d. Types of Obstacles

(1) Jumping-Type Obstacles. These obstacles may be ditches which are cleared with one leap, trenches which the individuals can jump into, heights which require jumping downward, or hurdles which an individual can jump over. (See fig. 3-29.)

Figure 3-29 is available in the illustration section at the end of the book.

(2) Dodging-Type Obstacles. Obstacles of this type are usually mazes consisting of posts set in the ground at irregular intervals. (See fig. 3-30.)The intervals between posts should be rather narrow so that the Marines must pick their way carefully through and around them. Lane guides may be constructed to guide the Marines to dodge and change direction. Obstacles may be put into a maze pattern to cause the Marines to change direction.

Figure 3-30 is available in the illustration section at the end of the book.

(3) Vertical Climbing and Surmounting Type Obstacles. These obstacles may be climbing ropes, either plain or knotted and 1-5 inches in diameter; cargo nets or walls 7 or 8 feet high; or vertical poles 6 to 8 inches in diameter and 38 feet high.(Seefig. 3-31.)

Figure 3-31 is available in the illustration section at the end of the book.

(4) Horizontal Traversing-Type Obstacles. Horizontal obstacles may be pipes, beams, ladders, or ropes. (See fig. 3-32.)

(5) Crawling Type Obstacles. Obstacles which require crawling may be constructed of large pipe sections, low rails, and wire. (See fig. 3-33.)

Figure 3-32 is available in the illustration section at the end of the book.

(6) Vaulting-Type Obstacles. Obstacles of 3 to 3.5 feet in height such as low walls or fences may be used as a vaulting obstacle.(See fig. 3-34.)

(7)Balancing-Type Obstacles. Beams, logs, and planks may be used as balancing-type obstacles.(See fig. 3-35.)These items may be used to span water obstacles and dry ditches, or raised off the ground somewhat to simulate natural depressions.

Figure 3-33 is available in the illustration section at the end of the book.

3805.CONFIDENCE OBSTACLE COURSES

a. Description and Objective. The Confidence Obstacle Course is commonly known as the Confidence Course. This course is composed of higher and more difficult obstacles than those used in the conditioning course. The confidence obstacle course is designed to give Marines confidence in their mental and physical capacities and to cultivate their spirit of daring. They are encouraged but not compelled to negotiate this course and the course is not run against time. The negotiation of a confidence course, however, is strenuous enough to be an excellent physical conditioner. Marines should NEVER attempt to take the obstacles at high speed and should not compete for speed. The obstacles vary from fairly easy to extremely difficult. Some are of considerable height to accustom Marines to climbing such heights without fear. Considerable emphasis is placed on obstacles that train and test an individual's balance.

Figure 3-34 is available in the illustration section at the end of the book.

b. Area and Equipment

(1) The confidence course accommodates four platoons, one platoon at each group of six obstacles. The course should be made up of about 24 obstacles, numbered and marked as follows:! to 6, white numbers on red background; 7 to 12, black numbers on a white background;

13 to 18, white numbers on a blue background; and 19 to 24, white numbers on a black background.

(2)A few simple pieces of equipment should be provided for individuals who do not have the strength, courage, or ability to negotiate the obstacles.

Figure 3-35 is available in the illustration section at the end of the book.

c. Formation. The obstacles should be divided into groups of six, and each group is designated by a different color. Each platoon starts at a different color. Individuals are separated into groups of 8 to 12 at each obstacle. At the starting signal from the company commander, they proceed numerically through. Anyone may skip an obstacle who is afraid to try. Individuals proceed from obstacle to obstacle until time is called, then assemble as ordered.

d. Leadership. If the Marines are new to the confidence course, an instructor will demonstrate or will give a brief orientation at each obstacle, including an explanation and demonstration of a method of negotiating it. Marines are encouraged to try the various obstacles, but they are not compelled to do so. No compulsion is to be used. The manner of negotiating any obstacle is left to the discretion of the Marine. However, the instructor assists anyone who experiences difficulty. Instructors must supervise closely at all times to prevent injuries, as some of the obstacles are quite high. Also, some of the obstacles should not be used when slippery or wet. The example of instructors and especially selected demonstrators will serve to inspire the individuals to greater effort.

e. Negotiating the Obstacles. Although personnel need not conform to any one method of negotiating the obstacles, there should be some uniformity in the approach to them. A general method of negotiating the obstacles is indicated below.

(1) Red Group This group contains the first six obstacles. (See fig. 3-36.)

(a) The Belly Buster Individuals may vault, jump, or climb over. Warn them that the log is not stationary.

(b) Reverse Climb. Climb the reverse incline and go down the other side to the ground.

(c) The Weaver. Move from one end of the obstacle to the other by weaving the body under one bar and over the next.

(d) Hip-Hip. Step over each bar, either alternating legs or using same lead leg each time.

(e) Balancing Logs. Step up on log, retaining the balance, walk or run along log.

(f) Island Hopper. Jump from one log to another until the obstacle is negotiated.

(2)White Group. This group is composed of the second six obstacles. (See fig. 3-37.)

(a) Tough Nut. Step over each X in the lane.

(b) Slide for Life. Climb the tower, grasp the rope firmly, and swing the legs upward. Hold the rope with the legs to distribute the weight between them and the arms. Braking the slide with the feet and legs, proceed down the rope. Warn Marines that there is danger of getting rope burns on their hands. When the rope is slippery or wet, this can be a dangerous obstacle.

(c) Low Belly Over. .Mount the low log and jump onto the high log, both arms grasping over the top of the log, the stomach area in contact with it. Swing the legs over the log and lower the body to the ground.

(d) Belly Crawl Move forward under the wire, belly down, to the end of the obstacle.

(e) The Dirty Name Mount the low log and jump to or reach the higher logs in succession, then jump or drop to the ground. Warn the Marines about the height of the final log.

(f) The Tarzan Mount the lower log and walk the length of it and each successive, higher log until reaching the horizontal ladder. Grasp two rungs of the ladder and swing the body into the air. Negotiate the length of the ladder by releasing one hand at a time and swing forward, grasping a more distant rung.

(3) Blue Group. This group is formed by the third group of six obstacles. (See fig. 3-38.)

(a) High Step over. Step over each log, alternating the lead foot or using the same lead foot.

(b) Swinger. Climb onto the swinging log and over to the ground on the opposite side.

(c) Low Wire. Move under the wire on the back, using the hands to raise the wire to clear the body.

(d) Swing, Stop and Jump. Gain momentum with a short run, grasp the rope, and swing the body forward to the top of the wall. Release the rope while standing on the wall and jump to the ground.

(e) Six Vaults. Vault over the logs, using one or both hands.

(f) Easy Balancer. Walk up one inclined log and down the one on the other side to the ground.

(4) Black Group. The last group is formed by the final six obstacles. (See fig. 3- 39.)

(a) Incline Wall. Approach the underside of the wall, jump up and grasp the top and pull the body up and over. Slide or jump down the incline to the ground.

(b) Skyscraper. Jump or climb to the first floor, climb up the corner posts or assist each other to any desired floor. Descend to the ground in any desired manner.

(c) Jump and Land. Climb up the ladder to the platform and jump to the ground.

(d) Confidence Climb. Climb the inclined ladder to the vertical ladder. Go to the top of the vertical ladder, then down the other side to the ground.

(e) Belly Robber. Step on the lower log and assume the prone position on the horizontal logs. Crawl over the logs to the opposite end of the obstacle.

(f) The Tough One Climb the rope or pole on the higher end of the obstacle, then go down the ladder and across the log platform. Climb over or between the logs at the end and go down the rope or pole to the ground. Vault over the final log.

Figure 3-36 is available in the illustration section at the end of the book.

Figure 3-37 is available in the illustration section at the end of the book.

Figure 3-38 is available in the illustration section at the end of the book.

Figure 3-39 is available in the illustration section at the end of the book.

Section IX Individual Exercise Programs

3901.GENERAL

Marines at times will **be** stationed on independent duty and consequently will be responsible for their own physical fitness program. This section will assist them in understanding the need for exercise and will aid in the planning and execution of an individual exercise program. Exercise activities included are the Bench Conditioner, 6-12 Plan, Weight Training, and Isometric Contraction. Keeping physically fit is a problem that faces every Marine. Even though we are frequently engaged in training that requires some physical effort, in many cases, it is not enough to prepare us to meet the intense physical demands of combat. Attaining a satisfactory level of physical readiness is not an insurmountable objective. Available time appears to be the most difficult obstacle to the development of physical readiness. In most cases, regular physical training programs are centralized, requiring the individual to temporarily leave the work area. The problems involved in setting an hour aside two or three times each are numerous. However, most of us can devote a few minutes each day to physical fitness with little, if any, impact on our daily work schedule, especially if it does not require us to leave our work area.

a. Type of Program. There are many good physical fitness programs available to the individual or group. Regardless of the type or duration, to be effective, the program must contain exercises that are strenuous and are challenging to the individual. Space will not permit the inclusion of all available means of individual exercise. The programs selected for this chapter have met the requirement of minimum space and minimum time.

b. Need to Augment Program. These programs are quite strenuous and will develop a satisfactory level of physical readiness. However, if the Marine desires additional development of endurance, it is recommended that he supplement these programs with a 15- to 30-minute period of wind sprints and double timing on an alternating daily basis.

c. Progressive Training. If Marines are performing duties which require little or no physical activity, they must plan a physical conditioning program that assures a moderate beginning, moderate but steady progression, and sufficient warmup before starting the vigorous exercise. To avoid organic of bodily harm, a Marine activity without adequate warmup. He should conduct conditioning programs on a daily basis over an extended period of time, never on an unduly accelerated or crash basis.

3902.THE BENCH CONDITIONER

a. Description and Objective. The bench conditioning program uses a modified bench to employ both isotonic (moving) and isometric (stationary) exercises are as the nucleus of the program. The exercises are designed to develop strength and endurance in all the major muscle groups of the body. The principles of progression, overload, and balance are employed when the exercises are performed properly.

b. Area and Equipment. The conditioning apparatus can be constructed in any unit motor pool with welding equipment found in most salvage yards.(See figs. 3-40, 3-41.)Additionally, there are available any number of commercially produced apparatus available. The important thing is that in utilizing this equipment the following exercise routine be adhered to.

c. Starting Level and Progression.. The program consists of two tables, each with 10 exercises. The Marine controls progression by required repetitions or, in some cases, by application of maximum effort. Each table can be completed within 15 minutes.

d. Starting Position. To start the program, the Marine begins with the Table 1 and executes each exercise for the required number of repetitions as indicated. The Marine controls the starting level and progression. When the Marine executes the maximum repetitions for Table I within a 15-minute exercise period, he progresses to Table II. To maintain this level of development, the Marine should also execute the maximum repetitions for Table II within a 15-minute exercise period. The Marine should keep substitution of exercises to a minimum. However, it he completes a full 15 minutes of strenuous exercise and exercises all muscles, then there should be no appreciable difference in the overall development.

e. Bench Conditioning, Table I

(1) Exercise 1: Side-Straddle Hop. This is a two-count warmup exercise done at moderate cadence. The starting position is the position of attention. On count ONE jump slightly into the air, swinging the arms out to the sides and up to a vertical position, hands touching.(See fig. 3-40, A.)At the same time, spread the feet wider than shoulder-width apart. On count TWO, using a slight flexing of the knees and ankles, jump slightly into the air and return to the starting position by swinging the arms back down to the sides. Twenty repetitions of this exercise is the standard dosage throughout the program.

(2) Exercise 2: Hand Walk. Remove the lower horizontal bar. Adjust the upper horizontal bar so that it is high enough to permit a "dead hanging" position, with the feet off the ground.(See fig. 3-40, B.j From the "dead hanging" position, release one hand and drop the arm to the side of the body. Then raise that arm and regrasp the horizontal bar. Release the bar with the other hand and drop that arm to the side. Repeat this as many times as possible.

(3) Exercise 3: Situps. Lie down with the fingers interlocked and placed behind the head. Hook the toes under the foot braces. Raise the trunk and upper body to an upright sitting position, twisting it to the left and then forward and downward until the right elbow touches the left knee.(See fig. 3-40, C.)Lower the body to the starting position .Sit up again but twist the body to the opposite direction as before, touching the left elbow to the right knee. Again lower the body to the starting position. The starting dosage is 20 situps. The Marine should continue the progression until he has attained 40 situps.

Figure 3-40 is available in the illustration section at the end of the book.

(4) Exercise 4: Double Step-upStarting at one end of the bench, step up onto the bench, and walk across it. Step down from the other end; turn around and repeat the process to return to the starting point.(See fig. 3-40, D.)Each return to the starting point constitutes a repetition. The starting dosage is 20 repetitions. Maximum dosage is 35 repetitions. This exercise should be done at a rapid cadence.

(5) Exercise 5: Isometric Bar Lift. Adjust the lower bar so that it is slightly higher than the beltline. Placing the feet on the footplates at the base of the frame, grasp the lower bar so that the hands are spread shoulder-width apart. Assume a crouched position and lift with maximum effort using the arms, back and legs.(See fig. 3-40, E.)Starting dosage is 4 repetitions of a stress time of 5 seconds followed by a 5-second rest prior to the next repetition. The Marine obtains progression by lengthening stress periods to 6 and later 7 seconds. Do not increase the number of repetitions.

(6) Exercise 6: Knee Lift. Adjust the upper bar to the same height used in Exercise 2.Adjust the lower bar so that is stops rearward movement of the hips when the "dead hanging" position is assumed.(See fig. 3-40, F.) Keeping the arms extended, flex the legs and raise the knees as high as possible. Hold this position for 5 seconds, then return to the starting position. After 2 seconds in the starting position, raise the knees again. Each return to the starting position constitutes one repetition. The dosage is five repetitions. Progression is the same as in Exercise 5.

(7) Exercise 7: Isometric Pull. Adjust the lower horizontal bar to a position where it is slightly higher than the beltline. Grasp the handles and pull outward.(See fig. 3-40, G.)Apply maximum effort and hold for approximately 5 seconds. Relax for 5 seconds between repetitions; perform four repetitions. Moving the body closer to or farther away from the bar will change the stress from the upper arms to the forearms. Progression is the same as in Exercise 5.

(8) Exercise 8: Isometric Compression. Maintain the position as in exercise 7.(See fig. 3-40, H.)Grasping the handles in the same manner, press in **with** maximum effort and hold for approximately 5 seconds. Relax for 5 seconds between repetitions. Progression is the same as in Exercise 5.

(9) Exercise 9: Isometric Press. Remove the lower horizontal bar. Adjust the upper horizontal bar until it is about 6 inches lower than the extended arms can reach. Stepping on the footplates at the bottom of the frame, grasp the bar with both hands and push up.(See fig. 3-40,1.)Keep both the legs and arms slightly flexed and the back straight. Apply maximum effort for 5 seconds then relax for 5 seconds. Complete four repetitions. Progression is the same as in Exercise 5.

(10) Exercise 10: Pushups. Grasping the foot braces with both hands, assume the front leaning rest position.(See fig. 3-40, J.)Keeping the back and legs straight, lower the body until the chest is lower than the **hands,** then return to the starting position. The Marine should complete the maximum possible number of repetitions.

f. Bench Conditioning, Table II. There is no limit on the maximum number of repetitions attainable in Exercises 3, 4, and 6 of Table II. The only limit imposed is that the entire program of 10 exercises outlinted in either table should not exceed 15 minutes.

(1) Exercise 1: Side-Straddle Hop. This is a two-count warm up exercise done at a moderate cadence. The starting position is the position of attention. On count ONE, jump slightly into the air, swinging the arms out to the sides and up to a vertical position, hands touching.(See fig. 3-41, A.)At the same time, spread the feet wider than shoulder-width apart. On count TWO, using a slight flexing of the knees and ankles, jump slightly into the air and return to the starting position by swinging the arms back down to the sides. Twenty repetitions of this exercise is the standard dosage throughout the program.

(2) Exercise 2: Pull-up. Adjust the horizontal bar so that is is high enough to permit a "dead hanging" position with the feet off the ground. Grasp the bar with both hands, palms facing forward. By flexing the arms, raise the body to a position where the chin is higher than the bar.(See fig. 3-41, B.)Then lower the body to the "dead hanging" position. Repeat as many times as possible.

(3) Exercise 3: Bench situps. Sit on the bench and hook the feet under the foot braces. With the fingers interlocked behind the head, lean back until the head touches the floor.(See fig. 3-41, C.)Return to the starting position. The starting dosage is 15 situps.

(4) Exercise 4: Step-Up. Face the bench and step up on it with one foot, bringing the trailing foot up next to the leading foot. Step back down again, leading with the same foot used first in stepping up.(See fig. 3-41, D.)Perlorm half of the total repetitions, then change the sequence of moving the feet to use the other leg in stepping up, and repeat the same amount of exercise. The starting level is a total of 40 step-up. **This exercise should be done at a rapid cadence.**

(5) Exercise 5: Isometric Bar Lift\djust the lower bar so that it is slightly higher than the beltline. Placing the feet on the footplates at the base of the frame, grasp the lower bar so that the hands are spread shoulder-width apart. Assume a crouched position and lift with maximum effort using the arms, back, and legs.(See fig. 3-41, E.)Starting dosage is four repetitions of a stress time of 8 seconds followed by a 5-second rest prior to the next repetition. Progression is obtained by lengthening stress periods to 10 seconds. Do not increase the number of repetitions.

Figure 3-41 is available in the illustration section at the end of the book.

(6)Exercise 6: Leg Lift. Adjust the bars and assume the starting position as shown in figure 3-41, F. Keeping arms and legs extended, raise the legs to a horizontal position and hold in that position for 2 seconds. Then lower the legs slowly to the starting position. Five repetitions is the starting level.

(7) Exercise 7: Isometric Pull. Adjust the lower horizontal bar so that it is slightly higher than the beltline. Grasp the handles and pull outward.(See fig. 3-41, G.)Apply maximum effort and hold for approximately 8 seconds. Relax for 5 seconds between repetitions; perform four repetitions. Moving the body closer to or farther away from the bar will change the stress from the upper arms to the forearms. The Marine obtains progression by lengthening the stress period to 10 seconds.

(8) Exercise 8: Isometric Compression. (See fig. 3-41, H.)Maintain the position as in exercise 7.Grasping the handles in the same manner, press in with maximum effort and hold for approximately 8 seconds. Relax for 5 seconds between repetitions; perform four repetitions. The Marine obtains progression by lengthening the stress period to 10 seconds.

(9) Exercise 9: Isometric Pres. Remove the lower horizontal bar. Adjust the upper horizontal bar until it is about 6 inches lower than the extended arms can reach. Stepping on the footplates at the bottom of the frame, grasp the bar with both hands and push up.(See fig. 3-41,1.)Keep both the legs and arms slightly flexed and the back straight. Apply maximum effort for 8 seconds, then relax for 5 seconds. Complete four repetitions. The Marine obtains progression by lengthening the stress period to 10 seconds.

(10) Exercise 10: Inclined Pushup. Assume the front leaning rest position with the feet on the bench.(See fig. 3-41, J.)Keeping the back and legs straight, lower the body until the nose touches the ground. By extending the arms, raise the body to the starting position. Repeat as many times as possible.

3903. THE 6-12 PLAN

a. Description and objective. The 6-12 plan of physical fitness has been developed to assist in regulating quantity and progression and to provide a convenient set of exercises. This is a basic program and will take 18 weeks to complete if you follow the moderate progression as to the time prescribed for each level of achievement. The time can be shortened as explained below. This plan consists of six basic exercises a day which can be completed in 12 minutes. There are six tables of six exercises each, thus allowing you to progress from table to table. The plan is progressive, fits any age group, contains balance and variety, and applies the principle of overload in a safe and gradual manner. Begin at Table I, Progression Guide, with the number of repetitions as indicated by age.

b. Formation. If just starting an exercise program, do not rush through the first table. Remember, individuals should remain at each level for about a week before moving upward. The time allotment stated for each exercise at the bottom of the tables is a guide; some people may take more and some less time on the individual exercises. At the end of a one-week period, when the individual can comfortable perform the six exercises in 12 minutes, move on to the next level. To a certain degree, the individual must be the judge of his ability to progress form level to level and table to table. If starting with a certain degree of fitness, some of the beginning tables may present little challenge.

c. Starting Level **and Progression.** There are three levels of achievement for each age group, indicated as A, B, and C. Start at the C level for the appropriate age group. At the end of a one-week period, or when the individual can do all exercises at that level within 12 minutes, progress to the B level. At the end of the second week, or when the individual can accomplish that level within 12 minutes, progress to the A level. At the conclusion of the third week or when the individual can achieve the A level within the time limitation, move on to table II.(See Tables I through VI, Progression Guide.)

d. Maintenance Level. Attempt to work through all six tables. If this proves to be too difficult, then maintain exercise at the—

(1) A-level on Table IV, Progression Guide it in the 45 to 49, 50 to 59, or over 60 age group.

(2) A-level on Table V, Progression Guide if in the 17 to 29, 30 to 39, or 40 to 44 age group.

e. Precautions. To achieve the maximum benefit, perform each exercise exactly as specified. Read the descriptions and study the illustrations. Do not slight the movements. Use a sensible approach and follow these four points as they apply before starting or during your exercise program.

(1) If you have the slightest doubt about your ability to participate in this exercise program, consult a physician.

(2) Stop immediately if you notice unusual breathlessness or chest pain while taking part in these exercises. If these conditions persist, consult a physician.

The following Tables of Progression Guides and Exercises are the progression of the 6-12 Plan program.

Table I. Progression Guide is available in the illustration section at the end of the book.

Table I. Exercises.is available in the illustration section at the end of the book.

Table II. Progression Guide.is available in the illustration section at the end of the book.

Table II. Exercises.is available in the illustration section at the end of the book.

Table III. Progression Guide.is available in the illustration section at the end of the book.

Table III. Exercises.is available in the illustration section at the end of the book.

Table IV. Progression Guide.is available in the illustration section at the end of the book.

Table IV. Exercises.is available in the illustration section at the end of the book.

Table V. Progression Guide.is available in the illustration section at the end of the book.

Table V. Exercises.is available in the illustration section at the end of the book.

Table VI. Progression Guide.is available in the illustration section at the end of the book.

Table VI. Exercises.is available in the illustration section at the end of the book.

(3) Unless you have exercised regularly and know yourself to be in good physical condition, start at table I with the C level appropriate to your age.
(4) If you are out of shape, admit that fact to yourself. Hide your pride; after all, you are in the privacy of your own quarters. Set your goal for the longer, steadier pull toward fitness. Resist the urge to pass over the lower numbered tables to find a table that will test your fitness. You are not trying to test, but rather to develop.

3904.WEIGHT TRAINING

Weight or barbell training should not be confused with the more common types of weight lifting used as a competitive sport. Weight lifting is designed to develop specific muscle groups so that the individual is capable of lifting a large amount during a single lift. In contrast, weight training is the systematic development of all the major muscle groups by the use of calisthenics reinforced with weight to provide resistance.

3905.WEIGHT LIFTING

a. Description and Objective. The weight lifting program is progressive and applies the principle of overload in a safe, gradual manner. The exercises of the table can be completed in 15 minutes. The objective of the exercises is to develop strength and muscular endurance, and muscle tone of the five major muscle groups:legs, arms, back, trunk, and shoulder girdle.(See fig. 3-42.)

b. Warmup. A warm up exercise is important to prepare the body for the more vigorous exercises that are to follow. Ten repetitions of the high jumper exercise are excellent for a warm up period. Also, an exercise which will require fast body movement is needed.To provide such exercise, 3 to 5 minutes of rope skipping is recommended to increase the individual's development potential.

c. Formation. Marines should take care in the completion of these weight lifting exercises. They should ensure that their back is straight during the lifting phase of all exercises. When exercises require assuming the standing position with the weight, they should always grasp the weight while in a squatting position and then rise to a standing position.

d. Progression. Each exercise has a starting number of repetitions and specified pounds of weight. After each 4th or 5th day of exercise, the Marine should increase the repetitions by one until he has reached the maximum of 10 repetitions. At this time, the Marine should increase the weight by 5 pounds. Then he should repeat the process again with the initial number or repetitions.

e. Weight Training

(1) Exercise 1:Squat(See fig. 3-42, A.)Starting level—six repetitions, 50 pounds (commonly called the flatfoot deep knee bend). Place the bar upon the shoulders. Stand with feet about 18 inches apart. Keeping the feet flat, lower the body into the low squat position. Stand erect and repeat. Exhale while lowering into the squat position and inhale while standing. This constitutes one repetition.

(2) Exercise 2:Waist Bender(.See fig. 3-42, B.)Starting level-6 repetitions, 40 pounds. Assume the standing position with the bar across the shoulders and the feet a shoulder-width apart .Bend forward at the waist until the upper body is parallel to the ground; return to the starting position. Each return to the upright position constitutes one repetition.

(3) Exercise 3:Curl (See fig. 3-42, C.)Starting level—6 repetitions, 40 pounds. Grasp the barbell with the palms facing to the rear and assume the standing position with the feet a should-width apart. With the barbell held in front of the hips, flex the elbows and lift the weight until the bar touches the upper chest. Lower the barbell back to the hip level position. Inhale deeply with the upward movement and exhale on the downward movement. Each time that the bar touches the chest will constitute one repetition.

(4) Exercise 4:Side Bender(.See fig. 3-42, D.)Starting level—6 repetitions per side, 40 pounds. Assume the standing position, with the bar across the shoulders, with feet a shoulder-width apart. Bend to the left as far as possible and return to the starting position. Repeat six times and then execute the same procedure to the right for six repetitions.

(5) Exercise 5:Standing Press(See fig. 3-42, E.)Starting level—6 repetitions, 45 pounds. Grasp the bar with the palms facing forward and assume the starting position. Curl the weight to the upper chest position. Inhale deeply and press the bar upward to an overhead position. Exhale while lowering the bar to the chest position. Each time that the bar is pressed upward constitutes one repetition.

(6) Exercise 6:Upward Row. .see fig. 3-42, F.)Starting level—6 repetitions, 40 pounds. Grasp the bar, hands close together, palms to the rear, and assume the standing position. Starting with the bar held in front of the hips, flexing the elbows and the shoulder girdle muscles, lift the bar straight up to an overhead position. Inhale deeply while lifting the bar. Exhale while lowering the bar to the hip position. Each time that the bar returns to the hips constitutes one repetition.

(7) Exercise 7:Shoulder Curl. See fig. 3-42, G.)Starting level—6 repetitions, 25 pounds. Grasp the bar, palms down, and assume the standing position. Keeping the elbows locked, curl the bar, pivoting the arms at the shoulders until the bar is in an overhead position and as far to the rear as possible. Return the bar in the same manner to the hip position. Each time that the bar returns to the hip position constitutes one repetition.

Figure 3-42 is available in the illustration section at the end of the book.

3906.ISOMETRIC EXERCISING

a. Description and Objective. Isometric exercising is the application of maximum effort during an exercise period. The isometric principle is to apply force gradually over a 5- to 10- second period until the maximum effort is applied. Relaxation follows for approximately 5 seconds and then force is applied again. The Marine continues this process at the prescribed level for each exercise for a period of 15 minutes or less. The objective of isometric exercises is to create muscle growth. Isometric exercises are the fastest means of creating muscle growth. They are founded on the fact that a muscle will grow only so fast regardless of the type or duration of the activity. The principle of overload—that the muscles develop commensurate with demand—reinforces that fact. However, isometrics will not develop cardiovascular or muscular endurance. Consequently, Marines who choose to use isometrics in their exercise routine must also include running or some other type of aerobic activity.

b. **Area and Equipment** isometric exercises may be designed to be performed with or without equipment.

c. Use With Other Programs. The isometric exercises presented in this section will provide a variety from which to choose. In addition to the exercises contained in this section, the application of isometric force is used in some of the exercises in section II.

d.Door Frame Exercises. The following exercises are designed for use with a standard door frame found in all offices or barracks.(See fig. 3-43.)

(1) Exercise l:Arm Press. Stand in the doorway with the legs straight, knees locked. Using the arm muscles, press hard upward against the top of the door frame. Repeat for three repetitions applying gradual effort to maximum contraction.

(2) Exercise 2:Leg Press. Stand in the doorway with the hands on the top of the door frame, elbows locked .With knees bent, press hard with the leg muscles .Repeat for three repetitions beginning with a gradual effort and increasing to maximum contraction. A low platform may be necessary to reach the top of the door frame and still maintain a bent knee position.

(3) Exercise 3:Side Press. Extend both arms to the side of the doorway. Palms are shoulder high, facing outward. With both arms, press hard against the sides of the door frame. Repeat for three repetitions. Begin gradually and increase to maximum contraction.

(4) Exercise 4:Lateral Raise. Extend both arms to the sides of the doorway, arms down, palms facing inward. With the back of the hands, press hard against the sides of the door frame .Repeat for three repetitions. Begin with a gradual effort and increase to maximum contraction.

Figure 3-43 is available in the illustration section at the end of the book.

(5) Exercise 5:Neck Press?lace the forehead against the door frame, hands clasped behind the back Using the neck muscles, press hard against the door frame. Repeat for three repetitions, then reverse position so that the back of the head is resting on the door frame. Again do three repetitions .Begin gradually with both exercises and increase to maximum contraction.

(6) Exercise 6:Door Pull Stand facing the edge of the open door and grasp the doorknobs. Pull outward with both arms (if doorknobs are not available grasp the edge of the door).Applying outward pressure, move the body toward and away from the door .Repeat the three repetitions. Begin with gradual effort and increase to maximum contraction.

Chapter 4
COMBAT WATER SURVIVAL
Section I. Marine Corps Water Survival Program

4101.MILITARY SWIMMING

Military swimming emphasizes strokes that result in staying power rather than speed. Marines need to swim easily, quietly, and with adequate vision. Strokes should be used that allow the Marine to carry basic combat equipment and to tow or push a wounded buddy while keeping the face out of the water to allow breathing. The best strokes to use are the side stroke and the breast stroke. Marines must be proficient in floating, drown-proofing, and in using their basic combat equipment to make flotation devices. They must be drilled on using the Kapok and Mae West buoyancy compensators and on emergency egress from helicopters, landing craft, and amphibious assault vehicles.

4102.PROGRAM DEVELOPMENT

Commanders of the Landing Force Training Commands (Atlantic/Pacific) are tasked with developing and maintaining the individual water survival and swimming training program. They are responsible for training all Marine Corps water safety and survival instructors and integrating this training in to their combat readiness training program. All Marines are deployable and as soldiers of the sea must constantly practice the skills necessary to achieve mission accomplishment. Commanders must be imaginative in program development, integrating basic skills development which enhances the total mission accomplishment. An example is teaching waterproofing of the Marine's backpack so that it can be used as a flotation device in streams or rivers. Once learned, this skill would be practiced first in a pool and then in a stream or a river. Routine inspections during deployments would ensure that backpacks could be used as flotation devices.

4103.QUAL STDS AND TRAINING GUIDELINES

The inherent nature of Marine Corps operations and training requires that Marines achieve an ability to survive in water. Water survival and swimming training is designed to reduce fear of water, instill self-confidence, and develop a Marine's ability to survive in water. As a minimum, every Marine should be qualified as swimmer, third-class (S3).This qualification should be met during recruit training or Officer Candidate School. Once qualified, a Marine need not be requalified unless requalification at a specific level (S3 or higher) is required by unit mission, military occupational specialty or duty assignment, or for other reasons determined by the commander. The following classifications and standards apply to Marine Corps water survival and swimming training. Abbreviations will be used for service, medical, and training record entries citing MCO 1510.29A, Individual Water Survival and Swimming Training, as authority.

a. Qualification standards and test procedures are described in paragraph 4104.

b. Swimmer classification and abbreviations are:

(1) Unqualified - UQ.

(2) Swimmer, third-class (S3).Minimum aquatic skill level for all Marines.

(3) Swimmer, second-class (S2)-

(a) Minimum standard for all naval aviators, naval flight officers, and aircrewmen with written waiver from the commanding officer.

(b) Requirements for military occupational specialty 1803 (assault amphibian vehicle officer), 1833 (assault amphibian vehicle crewman), 1302 (engineer officer), 1371 (combat engineer), 1381 (shore party specialist), and 1379 (engineer operations chief).

(4) Swimmer, first-class (S1)-

(a) Naval aviator, naval flight officer, and aircrewman qualification requirement.

(b) Requirements for military occupational specialty 0321 (reconnaissance man), 8652 (reconnaissance man, parachute jump qualified), 8653 (reconnaissance man, self-contained underwater breathing apparatus qualified), 8654 (reconnaissance man, parachute/self-contained underwater breathing apparatus qualified), 9952 (self-contained underwater breathing apparatus Marine (officer/enlisted)), and 9953 (parachutist/self-contained underwater breathing apparatus Marine (officer/enlisted)).

(5) Water Survival Qualified (WSQ).Ultimate water survival/swimming goal for Marines.

c.Instructor and assistant instructor classifications are-

(1) **American Red Cross Advanced Lifesaver.** Individuals who are currently certified as American Red Cross advanced lifesavers may assist qualified instructors during Marine Corps water survival/swimming training and testing. Certification is valid for 3 years.

(2) American Red Cross Water Safety Instruction. Individuals who are currently certified as American Red Cross water safety instructors may conduct swimming training and qualify individuals as 3d, 2d, and 1st class swimmers. Certification is valid for 2 years.

(3) United States Marine Corps Water Safety and Survival Instruction. Those personnel certified as water safety and survival instructors by one of the landing force training commands may conduct Marine Corps water survival and swimming training and American Red Cross swimming and water safety training. These instructors may also participate as assistant instructors during the conduct of landing force training command-sponsored water safety and survival instructor courses. Certification is valid for2 years.

(4) American Red Cross Water Safety Instructor Traine. Individuals qualified who have been certified as instructor trainers by the American Red Cross may train American Red Cross water safety and basic swimming instructors. Certification is valid for 2 years.

(5) United States Marine Corps Water Safety and Survival Instructor Traine. Individuals qualified who are certified as instructor trainers through special training courses conducted jointly by the American Red Cross and the Marine Corps, under the direction of the Commandant of the Marine Corps, may train water safety and survival instructors and American Red Cross water safety and basic swimming instructors. Certification is valid for 2 years.

d. Training Guidelines Instructors engaged in training Marines as swimmers will use the current edition of the American Red Cross swimming and water safety courses instructors' manuals as the authoritative textbooks for swimming procedures, strokes, breaks, and holds.

4104.WATER SURV/SWIM QUAL STD AND TEST PROCS

a. The following qualification standards and test procedures are applicable to Marine Corps water survival and swimming training. The utility uniform without cover, boots/oxfords or socks will be worn for all swimming tests through swimmer, first class. The utility uniform with boots/oxfords will be worn as prescribed for the water survival qualification.

(1) Swimmer, Third-Class (S3) To qualify as a swimmer, third-class, a Marine must enter the water feet first from a minimum height of 5 feet and remain afloat for 5 minutes. During this time the Marine must swim 50 yards using any stroke or combination of strokes. This test should be taken by all Marines as early as possible in their initial training. Those unable to pass this test will be classified as UQ and should be given instruction in fundamental swimming skills. Those who barely meet requirements are swimmers who need help and should receive additional instruction in fundamental swimming skills, whenever possible.

(2) Swimmer, Second-Class (S2.) To qualify, a Marine must enter the water feet first from a minimum height of 10 feet and remain afloat for 10 minutes. The Marine must swim 100 yards and use the three basic survival swimming strokes (side stroke, backstroke, and breast stroke) for a minimum distance of 25 yards each.

(3) Swimmer, First-Class (S1)A prerequisite to qualification as swimmer, first-class, is successful completion of the test for swimmer, second-class. To qualify as a swimmer, first-class, a Marine must be able to do each of the following:

(a) Approach a person of approximately the same size while in the water, demonstrate a release, get that person in a carry position, and tow the "distressed swimmer" 25 yards.

(b) Enter water feet first and immediately swim underwater for 25 yards. Swimmer is to break the surface twice for breathing during this distance at intervals of approximately 25 feet.

(c) Remove trousers in water, inflate for support, and remain motionless for a minimum of 1 minute.

(d) Swim 220 yards using any survival swimming stroke or any combination of survival strokes.

(4) Water Survival Qualification (WSQ)A prerequisite to water survival qualification is successful completion of the test for swimmer, first-class. To be water survival-qualified, a Mannemust:

(a) Enter water by jumping from a height of 10 feet.

(b) Stay afloat in full utility uniform (less cover) for 1 hour. Boots/oxfords will be removed after 5 minutes in the water but will be retained by the swimmer.

(c) Traverse 75 yards in deep water in full utility uniform (less cover) with boots/oxfords.

b. All water survival and swimming training will include instruction in the following:

(1) Employment of safety lines or other comparable expedients to assist in swift stream/river crossings.

(2) Adverse physiological effects caused by cold water (hypothermia) and the precautionary measures to be taken prior to exposure to such an environment.

(3) Employment of a standard and expedient flotation devices in a water survival situation.

(4) Removal of individual field equipment after unscheduled water entry. Entry will be made from a minimum height of 10 feet with a complete set of field equipment, properly packed as an expedient flotation device and waterproofed. The ALICE pack will then be removed, the rifle laid across the top of the pack, the helmet remaining or put back on the individual's head, and the individual will traverse 25 yards to the side of the pool or water's edge.

c. The additional water survival qualification (flight physiology training) and requalification required of personnel assigned to flight status will be conducted in accordance with OPNAVINST 3710.7_,Promulgation of NATOPS General Flight and Operating Instructions. Swimming qualification (first-class, SI) is a prerequisite to water survival (flight physiology) training and is a one-time requirement.

Section II. Water Sports and Competitive Activities

4201.GENERAL

The commander, when developing the combat readiness program, needs to look carefully at the mission. If the mission requires Marines who must perform mission-essential water-related activities of long duration, he may want to look to water sports/competitive activities as a method of developing endurance. Endurance and confidence can only be developed by extended time in the water. While defined training can accomplish this goal, the goal can be enhanced through training which is integrated with the physical conditioning program. This approach will provide variety and enthusiasm. Two activities which will accomplish this goal are water relays and water polo. The important consideration is that the unit achieve total participation by all members of the unit.

4202.WATER RELAYS

Development of water relays will generally conform to the guidelines in chapter 5, section IV. When contemplating relays, the commander should review chapter 5 and adapt the principles to the size of the unit and the facility to be used.

4203.WATER POLO

The official rules provided in this paragraph will need to be adapted to the unit's unique situation, its size, and the size of the facility. Commanders should make every effort to have all members of the unit participate.

a. NCAA Rules This style of play is based on two popular sports: swimming and basketball. Playing is done on the surface of the water by teams of seven players each. The size of the playing area is 25 yards by 14 yards, which is the size of the typical high school or college indoor pool. The deeper the water, the better. If the entire playing area is deep-6 feet or more—this is ideal, although most indoor pools have a shallow end. At each end of the pool is a goal. In deep water, the goal is 10 feet across and 3 feet high. In shallow water, where the goalie can stand on and jump from the bottom, the goal is 10 feet across and 5 feet high. One of the seven players on each team is the goalie. He is the only player on the team who can stand on or jump from the bottom (if it is shallow enough to permit this), catch and pass the ball with both hands at the same time, or hit the ball with a clenched fist. However, the goalie may not swim across or pass the ball across the middle of the pool.

b.Equipment. The goalie on the visiting team must wear a white cap with the number 1 clearly marked on it; his teammates, who are designated as guards and forwards, must wear white caps numbered 2 through 7, with the substitutes wearing white caps numbered 8 through 21.The goalie on the home team must wear a dark-colored cap with the number 22 clearly marked on it; his teammates designated as guards and forwards, must wear similar dark-colored caps numbered 29 through 42.The guards and forwards on each team may swim freely up and down the pool as they see fit, interchanging positions as often as they wish. They may not stand on or jump from the bottom, touch the ball with more than one hand at a time when catching or passing or shooting it, or enter inside the opposing team's 2-yard line unless preceded by the ball. The ball is similar to a soccer ball, except that it is yellow and covered with a waterproofing substance that makes it easier to handle with one hand.

c. Playing Time. In intercollegiate competition the game consists of four 7-minule quarters, the teams changing ends after every period of play. In high school competition the game consists of four-,five-, or six-minute quarters, depending on the ages of the participants. Here again the teams change ends after every period. There is a very brief interval between each quarter. At the start of each period the teams line up at their respective ends, and with ablast of the whistle, the head referee throws the ball into the middle of the pool. The fastest swimmers on each team then race to gain possession.

d. Officials. The head referee walks along one side of the pool watching for infractions. He is aided by the assistant referee, who patrols the opposite side of the pool. The referees are both equipped with whistles and two official's flags. These flags should be 12 inches square and should correspond in color to the caps worn by the respective teams. Whenever a referee sees an infraction, he blows his whistle and signals with his flag indicating that the player nearest to the ball with the appropriate colored flag will be given possession of the ball. Whenever the whistle is blown, all players should react by looking immediately to see which color flag is being waived which will indicate which team should be moving into the defense. Other officials are a timer and a scorer. As in all sports, it is important that the officials be skilled and competent.

e. Technical Fouls. The following are some of the common technical fouls: starting before the referee blows his whistle to open play; holding onto or pushing off from the side of the pool during play; taking hold of the ball underwater when tackled by an opposing player; swimming inside an opponent's 2-yard line unless preceded by the ball; touching the ball with both hands at the same time (goalie excepted); standing, walking on or jumping from the bottom when taking an active part in the game. When a technical foul occurs, thereferee blows the whistle and with his flag awards possession. The player on the team awarded possession nearest the point of infraction then has 5 seconds to put the ball into play by passing to a teammate or by dropping the ball into the water and swimming after it.

f. Personal Fouls. The following are some of the more common personal fouls: committing any of the technical fouls mentioned above for the purpose of scoring or preventing a goal; holding, ducking, pulling, pushing off from, swimming over, or impeding the arm or leg movement of any opponent who is not touching the ball; and splashing water in the face of an opponent. When a personal foul occurs, the referee will follow the same procedure as with a technical foul. While doing this, the referee will also call out clearly and loudly the number of the player who was guilty of the infraction and a personal foul will be marked against that player by the official scorer. When a player accumulates five personal fouls, he fouls out of the game and must be replaced by a substitute.

g. Penalty Shot. A penalty shot can be awarded by either of the officials or the scorer when:

o An offensive player inside the opponents' 4-yard line not touching the ball is held, ducked, pulled-back, kicked, or struck.
o A team has accumulated a total of ten personal fouls. When the former occurs, the head or assistant referee should immediately blow the whistle and by holding a two-flagged stick in a vertical position above his head, signal that a penalty shot has been awarded. When the latter occurs, the scorer should use a buzzer located at the scorer's tabic to signal that one team has accumulated ten personal fouls, thereby entitling the other team to a penalty shot. A penalty shot is taken from the 4-yard line in front of the goal. All players except the defending goalie must leave the 4-yard line until the shot is taken. No player can be within 1 yard of the shooter. After ascertaining that the shooting player is on the 4-yard line and the goalie is onThegoal line, the referee will ask the shooter to lift up the ball. When he docs, the referee will give a sharp blast from the whistle. At the whistle, the shooter must shoot without delay and without any faking at the goal. The goalie may try to block the shot. If he does, or the shot is otherwise missed, it is immediately in play and action continues.

h. When a Goal is Scored. When a goal is scored, either from a shot taken by a player out in the field or by penalty shot, the defending goalie must then pass the ball to the nearest referee. The referee will then pass the ball back to the goalie, who puts it into play by a pass to one of his teammates as soon as possible.

Section III. The Battle Swimming Test

4301.GENERAL

The battle fitness test is the evaluation contained in the Guidance for Basic Warrior Training (BWT) as part of the Basic Warrior Training Concept Plan.

4302.THE BATTLE SWIMMING TEST

The battle swim test consist of the following:

a. Be able to climb a three meter high board, while wearing a field uniform, carrying a M-16, web gear, and four canteens. Jump off blindfolded, recover and swim 100 meters with the M-16.

b. Repeat the same process, without the blindfold, and upon entering the water a second time, recover and retrieve all submerged gear.

c. Remain afloat and tread water for five minutes.

d. Demonstrate a knowledge of how to cross a stream using field expedient techniques and a single strand rope bridge.

e. Successfully complete helicopter egress training.

Chapter 5
COMPETITIVE CONDITIONING ACTIVITIES
Section I. Organization of Competitive Activities

5101.LEADERSHIP OF COMPETITIVE ACTIVITIES

a. Place in the Program. Competitive conditioning activities consist of dual combatives, relays, team contests, and team sports in which individuals or teams compete against an opponent to win. Competition is one of the best ways of maintaining interest in the physical training program during the sustaining stage. Organized competition provides enjoyable, vigorous physical activity that has proved to be one of the best supplements to conditioning drill activities. The benefits of competition are the development of aggressiveness, unit pride and identity, teamwork, and the will to win. These activities help develop cardiovascular and muscular endurance, strength, and coordination. Scheduling competitive activities in an orderly and progressive manner is desirable. The progression should be from relays to dual combatives, to team contests, and finally to team athletics. These activities should be conducted as part of the program after a basic level of conditioning has been developed, usually during the slow-improvement stage of conditioning. Muscles and joints should be strengthened by preconditioning to withstand the strain placed upon them by sudden stops and turns, body contact, bearing of weight, and falls. Competitive activities, however, should not be allowed to dominate the physical training program.

b. Area and Equipment. Some of the competitive activities included in this chapter require specific types of areas and equipment; others do not. The area requirement can usually be satisfied on available training fields. When items of equipment are required or specific courts or field layouts are to be marked off, such information is included.

c. Leadership. The principal factor for success as a leader of competitive activity is an energetic, dynamic, enthusiastic approach. The leader's attitude is reflected by the group, so the commander must carry on the activity in a snappy and vigorous manner. Confidence on the part of the commander will create an impression of decisiveness and certainty. Confidence grows out of experience and a thorough knowledge of the activity. Mastery of subject is the first step in developing confidence, assurance, and poise.

(1) The following suggestions are offered for leaders of competitive combatives, contests, and sports:

(a) Get the activity underway quickly by selecting and teaching only the essentials.

(b) Use rules to add to the enjoyment of the activity and do not allow them to interfere with the spirit of competition.

(c) Stop the activity before interest begins to lag.

(d) In team contests, clearly distinguish sides by attempting to maintain unit identity.

(e) Always insist on fair play, enforcing the rules impartially.

(2) The following procedure is recommended for presenting a competitive activity.

(a) Name the activity.

(b) Briefly explain the objective of the activity and give only the pertinent rules.
(c) Have a demonstration at slow speed and answer questions.
(d) Organize groups into teams and appoint captains.
(e) Arrange teams in the proper starting positions.
(f) Conduct the activity.

d. Competitive Units. Units for competition should be the same organizations in which Marines train: battalions, companies, platoons. In most situations, the unit is the squad.

e. Provisions for Instruction. One of the most effective methods of maintaining interest and participation in competitive activities is to provide instruction in activities with which most Marines are unacquainted. Such instruction can be conducted during regular physical training periods. Careful planning is required to keep all individuals continuously engaged in vigorous activity. The materials in the following sections should be used as a guide for instructional purposes.

f. Officiating and Control. Every effort should be made to provide good officials for all competitive activities. Poor officials quickly cause dissatisfaction among participants in team activities and create a situation which can turn an organized activity into a brawl. Each company should have or develop several competent and qualified officials available for games on company and platoon levels.

5102.MILITARY FIELD MEETS

When units reach the latter part of the slow-improvement and sustaining stage of training, interest in the program may lag. A change in course content can arouse the desire to participate. An event which does not require a high degree of skill yet demands strenuous activity is ideal for the military field meet. This is a series of team contests conducted on a station-to-station basis during a given period of time. Team contests carried on simultaneously provide essential training, vigorous exercise, and stimulating competition in an atmosphere that is enjoyable for all. Because of the healthy rivalry that a field meet arouses, it is an excellent form of inter-unit competition.

a. Objective and Advantages. The objective of the military field meet is to provide activity for everyone in the participating units. Activities are chosen that will develop aggressiveness, teamwork, a will to win, and competitive spirit, and that will stimulate interest and build esprit de corps. The military field meet can **be** included as part of the physical training program or as part of the off-duty recreational program. It is a form of contest that can be conducted in nearly all circumstances because it can be easily modified, requires a minimum amount of equipment, and can be readily organized.

b. Level of Competition. The military field meet is flexible. It can be adjusted to large or small groups. A company-size unit is the most desirable, but it may also be administered within a larger unit. If it is conducted within a company, the participating units will be squads. If it is held within a larger unit, platoons compose the teams. A larger-size unit requires more extensive organization and administration.

c. Selection of Events. In organizing a military field meet, select events that are simple and easy to administer. All rules and regulations should be clearly understood by everyone. No event should require previous practice. In selecting the events, consider the interest and capabilities of the Marines and available equipment and facilities. Select events which will require that all members of the units participate. Events should not be dominated by the athletes; rather, MOS-related events which have significance in producing skills and combat readiness should be used.

d.Equipment and Facilities. The site must be large enough to permit events to be grouped about a central control point. Use available facilities such as parade decks, Softball fields, or tracks. A public address system is desirable at the control point for the initial orientation of teams and for subsequent announcements of time lapses, cumulative scores, and final standings. A tally board is necessary so that all units can constantly monitor how the events are progressing.

e.Administrators. Efficient administration of the military field meet depends on the referees, judges, and scorers. For these positions, choose individuals within the units who have had athletic or officiating experience. Prior to the day of the event, all administrators should be briefed and assigned a specific task in order that they may become familiar with the rules and organization of the contests they will conduct. The following officials should be available:

(1) A primary instructor or supervisor who is in charge of the control center and who is responsible for the successful operation of the athletic carnival. The supervisor must have an assistant to act as a timer and scorer.

(2) One assistant instructor in charge of each event. This person should be-

(a) Familiar with the rules of the event.

(b) Effective as a leader to ensure proper supervision and control over the participating teams.

(c) Enthusiastic to provide proper motivation.

(d) Self-confident of the ability to judge infractions of the rules. The assistant instructor must be fair in judgment and penalize without hesitation when infractions occur.

(3) Runners between event stations and the control point. These individuals collect and deliver scores.

f. Team Organization for Competition. The size of the teams is determined by the level on which the field meet is organized. Maintaining the integrity of the unit promotes esprit de corps, but this does not preclude grouping two squads into one team.

g.Conduct of Events

(1) The assistant instructor at each station takes charge of the group and gives a brief explanation of the major rules of the event. The assistant instructor speaks clearly and distinctly from a position to be seen and heard by all. A short demonstration is desirable if it will help clarify the event.

(2) The assistant instructor should make certain that teams can be clearly distinguished by the use of, for example, T-shirts and fatigue jackets, or caps and no cap, or colored jerseys. The assistant instructor should teach a whistle response (teams stop play immediately upon hearing whistle), get the event started as quickly as possible, and make any necessary corrections as the contest progresses. The rale of good officiating is to use a minimum of calls, yet maintain control of the contest. Penalize when necessary, but refrain from disqualifying contestants or teams.

(3) Keep the activity moving as rapidly as possible. When the central control point sounds the whistle to stop the play, all competition ends immediately. The assistant instructor then assembles the group, forwards the team scores to the central point, and, upon a signal from the central control point, rotates the teams to their next station. It is essential that rotation and orientation of teams be quick and orderly to reduce time spent between contests.

(4) Upon completion of the final event, assistant instructors move their teams to the central control point for the announcement of winning teams, presentation of awards (if any), and final critique.

h. Scoring System. The system for determining the winner of the military field meet should be simple and efficient. At the completion of competition, the scorer totals the points that each team has scored in all contests. The scorer then subtracts the number of points scored against a team (penalty points for nonparticipation,etc.) from the total. The resulting scores are placed in a column with the highest score at the top and the lowest at the bottom. The team with the highest total is the winner. This type of scoring system encourages a team to ensure total unit participation.

Section II.Combatives

5201.DESCRIPTTON AND OBJECTIVE

Combatives are strenuous, short competitive contests in which two individuals attempt to overcome each other in a bout of skill and strength. These contests help to develop Marines' resourcefulness, confidence, strength, agility, coordination, and will to win. Any level ground area can be used. However, extremely hard ground should be avoided as some of the combatives require ground contact. A whistle is needed to control the bouts during the competition since voice commands may go unheeded. The extended rectangular formation is used for dual combatives. Combatives are conducted on an informal basis. Marines are allowed to remain at ease between activities and are allowed to brush themselves off after being on the ground. To get the most out of combatives, individuals must be urged to overcome their opponents as quickly as possible as would be required in combat.

a. Benefits for Marine Training. The possibility of close contact with an enemy in combat faces Marines at all times. They must be trained to react aggressively and violently. Combatives may be used as an introduction to such hand-to-hand contact and should be followed by hand-to-hand combat training. Marines enjoy competition, and this type of activity is a welcome change from formal conditioning activities.

b. Instructor Responsibility

(1) The instructor tells the individuals that all combatives begin and end on the whistle signal. The instructor demonstrates each activity before having it performed, explaining it in simple terms.

(2) After stopping one activity, the instructor gets everyone in place for the next.

(3) The instructor must closely supervise combatives to insure that contestants do not use unfair or unsportsmanlike tactics. To avoid unnecessary injury, instructors must see that the bouts are closely controlled and opponents equally paired. Adjustments should be made in apparent cases of mismatched abilities.

5202.COMBATIVES TABLES

There are three tables of combatives, each of which can be completed in 15 minutes. These tables become progressively more difficult from lower to higher numbers.

a.Combatives Table I

(1) Open-Hand Slap Boxing(See fig. 5-1, A. individuals assume a boxer's stance, palms open, fingers extended and joined. Each contestant tries to slap the opponent about the head and upper body with the open hand. This is a good warm- up activity.

(2) Wrist Tug-of-War(See fig. 5-1, B.)Two individuals sit on the ground with the soles of their feet in contact. Each grasps the opponent's wrists so that the hands are directly over the feet. At the whistle, each individual tries to pull the opponent from sitting to standing position.

(3) Arm-Lock Wrestling.(See fig. 5-1, C.)Two individuals sit back-to-back with legs spread and arms locked at the elbows. Each right arm is inside the opponent's left. Each tries to force the opponent's left arm or shoulder to the ground. The individual who first wins three bouts is the winner.

(4) Bulling(See fig. 5-1, D.)Two individuals assume the Westmoreland wrestling hold, each grasping the opponent's neck with the right hand and the opponent's right elbow with the left hand. Each tries to force the other to move one foot by pushing, pulling, or otherwise manipulating. The individual who first wins two boutsis the winner.

(5) Indian Wrestling(See fig. 5-1, E.)Two individuals lie on the ground, side by side, with their heads in opposite directions. They link right elbows. On the instructor's signal or by mutual agreement, each individual raises the right leg approximately straight and far enough to engage the opponent's heel. To start the contest, each person usually raises the leg three times rhythmically and on the third time engages the opponent's heel. Each tries to roll the other over backward. The right leg is used for three bouts, then the left leg for three bouts.

Figure 5-1 is available in the illustration section at the end of the book.

b. Combatives Table II

(1) Wrist Bending.(See fig. 5-2, A.) Opponents pair off, face each other, raise their arms forward, and, with palms forward, interlock their fingers. At the starting signal, each individual attempts to bend the opponent's wrist. The arms are kept up and forward and are not allowed to swing around and down to the sides. The individual who first wins two bouts is the winner.

(2) Back-to-Back Push.(See fig. 5-2, B.)Two individuals stand back-to-back with elbows locked. Each right arm is inside the opponent's left arm. At the starting signal, each pushes backward, trying to move the opponent. Opponents are not allowed to lift and carry each other; only pushing is permitted. The one who pushes the opponent the farthest wins the bout. The individual who first wins two bouts is the winner.

(3) Hop and Pull Hand.(See fig. 5-2, C.) The individuals are matched in pairs. Each grasps the opponent's right hand and, hopping on the right foot, attempts to pull the opponent off balance. Contestants automatically lose if they touch their free hand or their lifted foot to the ground. For successive bouts, they alternate hands and feet.

(4) Westmoreland Wrestling. (See fig. 5-2, D.)Each contestant grasps the back of the opponent's neck with the right hand and the opponent's right elbow with the left hand. In this position, each attempts to pull, push, or force the opponent to touch the ground with any part of the body other than the feet. The individual who first wins two bouts is the winner.

(5) Crab Fight.(See fig. 5-2, E.)Two individuals sit on the ground facing in opposite directions with their hands on the ground behind them. At the whistle, they raise their hips and push with their shoulders and bodies, each trying to make the other's hips touch the ground. The individual who first wins two bouts is the winner.

Figure 5-2 is available in the illustration section at the end of the book.

c. Combatives Table III

(1) Hand Wrestling.(See fig. 5-3, A.) Two individuals stand facing each other. Their right feet are forward and braced side by side. They grasp right hands for the first bout, left hands for the second bout. Each pulls, pushes, makes sideward movements, and otherwise maneuvers to force the opponent to move one or both feet from the original position. The contestant who first wins two bouts is the winner.

(2) Back-to-Back Tug.(See fig, 5-3, B.) Two individuals stand back-to-back with both arms linked at the elbows. Each has the right arm inside the opponent's left arm. At the starting signal, each attempts to pull the opponent. Lifting and carrying are permitted. The contestants must maintain their original direction and keep their arms linked. After a predetermined time, the player pulled or carried the farthest is the loser.

(3) Wrestling to Lift off Feet.(See fig. 5-3, C.) Contestants face each other. Each places the right arm under the left arm of the opponent and around the body. The left arm is over the opponent's right shoulder. Each tries to lift the other off the ground. The individual who first wins two bouts is the winner.

(4) Arm Pull Between Legs.(See fig. 5-3, D.) Two individuals are paired off, back-to-back. Each bends forward and, extending the right arm between the legs, grasps the opponent's right wrist. At the starting signal, each person attempts to pull the opponent After a predetermined time, the player who has pulled the opponent the farthest is the winner of the bout. The one who first wins two bouts is the winner. Repeat with the left hand and then both hands.

(5) Rooster Fight.(See fig.5-3, E.) Each contestant grasps the left foot with the right hand from behind, and the right arm with left hand. Each hops on the right foot, and, by shoulder-butting the opponent or by feinting and sudden evasions, forces the opponent to let go of the foot or arm. The contestant who first wins two bouts is winner.

Figure 5-3 is available in the illustration section at the end of the book.

Section III .Relays

5301. DESCRIPTION AND OBJECTIVE

Relays are races in which each member of a team runs one leg of the race. The team effort decides the winner. Relays provide stimulating competition and contribute to the conditioning of personnel. They also develop aggressiveness, team spirit, and the will to win. Relays should be dispersed throughout the program for short periods of time to provide a change of activity.

a. Team Organization. Relays are conducted most efficiently in platoon-size groups. Teams of equal size must be organized. Competitive spirit is encouraged and teams are organized better by basing teams on units such as squads, crews, or sections. Team captains should be designated. Extra Marines may be used as officials. The number of individuals on a relay team should be limited to squad size. If larger teams are used, the runners will spend too much time awaiting their turns and too little time actually participating. Two to six teams are ideal for relay competition. It is difficult to keep track of winners when more teams compete.

b. Administration of Relays

(1) The time spent on any one relay should be relatively short. If one team achieves a substantial lead in a long relay, the competitive spirit and enthusiastic participation of the other teams may decrease. Several short relays are generally better than one long relay.

(2) To maintain competitive spirit throughout a number of relays, determine the teams that win, place, and show in each relay and their total points for all relays. This can be done by awarding points to all teams on the basis of position at the finish of each relay. The team with the greatest number of points is the winner of the entire set of relays.

(3) Difficulties commonly encountered in conducting relays may be avoided by the following procedures:

(a) The last player in a relay race should be conspicuously identified. For example, the last runner can tie a handkerchief around the head or arm, take off the shirt, put on a hat or take it off, or use some other means.

(b) Another way to keep track of the progress of the race is to have each player sit or squat as soon as each is finished.

(c) Judges at the starting line can keep the runners from starting too soon.

(d) To prevent contestants from turning before they run the full distance, they should be required to run around a peg, pole, or assistant instructor.

(e) Batons, handkerchiefs, tent pegs, or other objects should be passed from one runner to the next when relays are run on a circular track.

(4) Before a unit's first participation in a relay, inform participants of the rules and scoring system. Violation of the rules should not result in disqualification. Instead, impose point penalties. A point penalty is imposed by subtracting a point from the team total at the conclusion of the relay.

(5) Careful administration will prevent most violations. For efficient conduct of relays, follow this procedure:

(a) Announce the name of the relay.

(b) Form the individuals in relay position.

(c) Briefly explain the relay and the rules for running it.

(d) Demonstrate.

(e) Have a definite finish line, and ensure that all know where it is.

(f) Answer questions.

(g) Conduct the relay.

(h) Determine winner and award points.

5302. RELAY TABLES

a. Events. The following relays are grouped into a table of activities. Each relay table can be completed in 15 minutes. Thus, relays can be used as a sole activity or as a part of a longer period. Each table is planned for a platoon-size group (30-60Marines). Adequate warmup for participants is provided by conducting several repetitions of exercises 1 and 2 of a conditioning drill. The recommended relay tables require an area 40 by 60 meters in size. Each table provides a variety of activity. The tables are progressive in the overload applied and should be scheduled in numerical order although not necessarily on successive days.

b. Relay Table I

(1) **60-Meter Lane Relay** (See fig. 5-4, A.) Each team is assembled in single file behind the starting line. On signal, the first individual of each team runs to the turn-around line 30 meters away, then runs back and touches the next teammate waiting at the starting line. The winning team is the first team to get its last member across the finish line. If an individual starts before being touched by the preceding runner, the team may be penalized.

(2) Wheelbarrow Race.(See fig. 5-4, B.) The players of each team pair off and line up in single file behind the starting line. The first individual walks on the hands while the partner grasps the ankles. They advance to the distance line (25 meters) behind which they exchange positions and return to the starting line. The rear individual must always hold the partner's ankles. After the first pair returns across the starting line, the next pair starts.

Figure 5-4 is available in the illustration section at the end of the book.

(3) Squad Front Relay .(See fig. 5-4, C.) The teams form in a line along the starting line with a 10-foot interval between teams. The members of each team lock elbows so that they are linked together. At the starting signal, the teams run to the distance line (20meters) where the left flank individual acts as a pivot. The team swings around on the pivot and returns to the base line. If a team breaks its links, it must reform before continuing. The first team to completely cross the base line intact is the winner.

(4) Crab-Walk Race.(See fig. 5-4, D.) The players of each team line up in single file. The first person of each team assumes the crab-walk position with feet forward on the starting line, hands on the ground behind it. At the starting signal, contestants move forward to the distance line (10 meters). They touch the line with their feet and then return to the starting line in the reverse position with the head and hands leading. The second person may not start until the first individual touches the finish line.

c.Relay Table II

(1) 100-Meter Lane Relay.(See fig. 5-5, A.)This relay is conducted exactly as the first relay of table 1 except that the start and turn-around line are 50 meters apart. This relay provides progression in sprinting.

(2) Frog-Jump Relay. (See fig. 5-5, B.) Each team lines up in single file. The first individual assumes a squatting position on the starting line. At the starting signal, the contestant progresses to the distance line (15 meters) and back by leaping forward, catching the weight on the hands, and bringing up the legs to the squat position for the next leap forward.

(3) Simple Relay. (See fig. 5-5, C.)Each team lines up in single file. Place a marker on the distance line (20 meters) in front of each team. Each team member, behind the first one, grasps the belt of the individual ahead. At the starting signal, each team runs as a unit to the marker, circles it, and returns to the starting line. The first team to completely cross the starting line intact is thewinner.

(4) Fireman's Carry Relay.(See fig. 5-5, D.)The players of each team line up in pairs, one behind the other. One individual in each pair carries the partner to the distance line (30 meters), using the fireman's carry. At the distance line, individuals exchange places and return to the starting line. As a variation, the person to be carried lies on the ground. The one carrying lifts the partner to the proper position. This relay may be performed with the other carries described in guerrilla exercises in chapter 3.

d.Relay Table III

(1) 200-Meter Circle Relay .(See fig. 5-6, A.)A course is laid out in a circular, rectangular,or oval pattern that is 200 meters around. Each team provides one runner on the starting line. On signal, the runner races around the 200-meter track and touches the next teammate, waiting at the starting line, who runs the same course. Each team member runs one lap of the course.

Figure 5-5 is available in the illustration section at the end of the book.

(2) Bear-and-Crab Race.(See fig. 5-6, B.) Each team lines up in single file. At the signal to start, the first contestant in each column assumes the bear-walk position (hands on ground in front), walks to the distance line (15 meters), and then runs back to the starting line. The returning contestant touches off the second individual and goes to the rear of the line. The second contestant assumes the crab-walk position (hands on ground behind) and crab-walks with feet leading to the distance line. This contestant also runs back to the starting line and touches off the next individual, who walks bear fashion. The rest of the members of each team alternate in this manner. The relay ends when the first person is back at the head of the line.

(3) Pilot Relay.(See fig. 5-6, C.) The players are grouped in threes, arms interlocked at the elbows, and outside players facing backwards. The middle person runs forward; the two outside individuals run backward. They run to the turning point (15meters), where they start back, this time with the middle person running backward and the two outside individuals running forward. The next set of three players starts when the first set crosses the starting line.

(4) Saddle-Back Relay.(See fig. 5-6, D.) Mark two parallel lines 15 meters apart. Each team selects a rider. The remaining members of each team count off. The even-numbered players from each team form in single files behind one line and the odd-numbered players from each team form in single files behind the other line directly across from their teammates. At the starting signal, the rider mounts the back of the first player of the team who carries the rider across the other line. There the rider changes mounts to the first player in the second line without touching the ground. This person carries the rider to the next player waiting in the first line. The relay continues until all of the mounts have carried the rider. A rider who falls off must mount again at the point of the fall. A rider who falls in changing mounts must get back on the original mount before making the change.

e. Relay Table IV

(1) 100-Meter Circle Relay (See fig. 5-7, A.)A course is laid out in either a circular, rectangular, or oval pattern that is 200 meters around. Each team is divided in half with each half positioned at starting lines on opposite sides of the track. Each runner races halfway around the track and touches a teammate who completes the lap. Each runner then waits in file at the halfway point until touched by the runner and then completes the second half of the lap. The first team to return all runners to their original starting line is the winner.

Figure 5-6 is available in the illustration section at the end of the book.

(2) In-and-Out Relay.(See fig. 5-7, B.) Each team lines up in a file with players 2 meters apart. At the starting signal, the first player runs back through the column in a zigzag fashion. He alternates, going to the right of one teammate and to the left of the next. Upon completing the run, the player lines up 2 meters behind the last one in the column. As soon as the first runner has passed the second person, the latter starts to run. This continues until all the players have realigned their original order. The team that finishes first is the winner. It may be desirable to have this relay continue until all individuals have run through their entire team two or three times in succession.

(3) Circle Race.(See fig. 5-7, **C.)** Each team forms a circle and holds hands with all individuals facing out except one who faces in and is the "driver."At the starting signal, the teams race to the distance line (20 meters) and back, keeping the circle intact. All the individuals in the circle must completely cross the distance line. The "driver" gives directions and orders. When the circle breaks, it must be reformed before it can continue. The first team completely over the starting line is the winner.

(4) Horse-and-Rider Relay.(See fig. 5-7, D.)Each team lines up in a single file. At the signal to start, the second player in each column leaps upon the back of the first who carries the rider across the distance line (30 meters). At the distance line, the rider dismounts and runs back to the starting line. There the rider picks up the third individual in the column, and carries this player to the distance line where the first player has remained. This continues until the last person is carried across the distance line.

Figure 5-7 is available in the illustration section at the end of the book.

Section IV. Team Contests and Athletics

5401. DESCRIPTION AND OBJECTIVE

Team contests are competitive activities in which Marines as a team compete with another team to win. They are guided by simple rules and organization. The function of team contests is to provide competition and an opportunity for body contact and to contribute to the development of physical readiness. In competing and working together as a team, individuals develop aggressiveness, the will to win, and teamwork. Team athletics deserve a prominent place in the physical training program because they contribute to increased combat efficiency. Because of the competitive nature of athletics and their natural appeal, individuals take part with enthusiasm. Athletic teams formed at intramural and higher levels are a strong unifying influence and provide one of the best means of developing esprit de corps.

a. Area and Equipment A level training field is sometimes the only area required. Many contests need no equipment. In contests requiring equipment, the need is for standard items such as logs, balls, nets, goals, and similar types of equipment. Specific requirements for area and equipment are listed with each contest description.

b. Progression. Team contests of a strenuous nature should be introduced after a basic period of conditioning has been completed and individuals are in the slow improvement stage of conditioning. Progression can take place from the less active to the more vigorous contests, and then from the non-contact to the contact or combative-type contests.

c. Necessity for Preconditioning Individuals must undergo conditioning prior to participation in athletics. Muscles, organs, joints, and ligaments not accustomed to stress and strain from sudden stops and starts, falls, body contact, rapid turns, prolonged running, and other rigors of athletic competition are subject to injury. Although athletics should not be introduced until players are physically prepared, there is still opportunity to engage in competition through lead-up contests. Individuals learn many of the skills required for athletics while participating in team contests.

d. Benefits. Athletics are beneficial primarily to sustain interest in the program and to maintain an achieved level of physical fitness. Athletics are a supplement, not a substitute for other types of conditioning activities which should continue. All of the desirable traits of physical fitness cannot be developed through athletics, yet their contribution is significant. For athletics to make a proper contribution to physical conditioning, the selected sports must bevigorous. The team contests recommended here are carefully selected for their simplicity, aggressiveness and applicability to squad- teaching teamwork. They are generally strenuous and involve the whole team throughout the entire duration of the contest.

5402. TEAM CONTESTS

a. Pushball

(1) **Players.** 10 to 50 players on a side. (See fig. 5-8.)

(2) Equipment A large pushball 5 to 6 feet in diameter.

(3) Area. A field 240 to 300 feet long, 120 to 150 feet wide. Mark a center line in the middle of the field parallel to the end lines. Mark a line 45 feet on either side of this center line and parallel to it, extending it across the width of the field. Mark another parallel line 15 feet from each end line, extending it across the width of the field.

(4) Game. Four 10-minute quarters are played. Give 2-minute rests between quarters and 5-minute rests between halves. The object of the game is to propel the ball over the opponent's goal line by pushing, rolling, passing, carrying, or any other way except kicking.

(a) The ball is placed on the center line with the opposing captains 3 feet from the ball. The rest of the players are 45 feet from the ball, on their half of the field. On the referee's starting whistle, the captains immediately play the ball with their respective teams coming to their assistance.

(b) At quarter time, the ball remains dead for 2 minutes at the spot where it was when the quarter ended. At half time, the teams exchange goals. The play is then started as it was in the beginning.

Figure 5-8 is available in the illustration section at the end of the book.

(c) Players may use any means of interfering with an opponent's progress except striking and clipping. Clipping is throwing the body across the back of an opponent's leg, while the opponent is running or standing. Legal use of force may be applied to all opponents whether or not they are playing the ball. For striking an opponent, the offender is removed from the game. The team penalty is half the distance to their goal. The penalty for clipping is the same.

(d) When any part of the ball goes out of bounds, it is a dead ball. The teams line up at right angles to the side lines and 3 feet apart at the point where the ball went out. The referee then tosses the ball between the teams.

(e) When, for any reason, the ball remains in one spot for more than 10 seconds, the referee declares the ball dead. The ball is then put into play as it is for an out-of-bounds situation.

(5) Scoring. A goal is scored when the ball, or any part of it, is propelled across the opponent's end line. A goal counts 5 points. The team scoring a goal has the privilege of trying for a point after the goal. To try for this extra point, the ball is placed on the opponent's 5-yard (or 15-foot) line. The teams line up on either side of this line separated by the width of the ball. Before the whistle blows, one player on each team may place hands on the ball. On the referee's signal, the ball is put into play for 1 minute. If any part of the ball is driven across the goal line in this 1-minute period, the offensive team scores 1 point. The defending team may not score during the opponent's try for the extra point.

b. Line Rush

(1) Players. Any number up to 50 on each side. (See fig. 5-9.)

(2) Equipment. None.

(3) Area. A field, 75 by 100 feet.

(4) Game. One team lines up behind one goal line and the other in midfield. On the starting signal, the team standing behind the goal line seeks to cross the goal line at the opposite end of the field within 30 seconds. The team in the center seeks to prevent it by catching and holding the runners. At the end of 30 seconds, the teams change.

(5) Scoring. Count the number of individuals who have crossed the far goal at the end of 30 seconds. After each team has had from three to five tries, the scores are added and the winner declared. A player scores 1 point when any part of the body is across the goal line.

Figure 5-9 is available in the illustration section at the end of the book.

c. Human Tug-of-War

(1) Players. 10 to 20 on a team.(See fig. 5-10.)

(2) Equipment. None.

(3) Area.40 to 60 feet.

(4) Game. Draw a line in the center of the area. Divide the players into two equal teams. Place them in single file on opposite sides of the center line facing each other. Each individual places the arms around the waist of the teammate in front. The two leaders of the opposing teams grasp each other around the waist. On signal, the teams try to pull each other over the center line within 30 seconds.

(5) Scoring. The team pulled across the center line loses. If neither team is pulled over the center line, but one team breaks its file, that team loses the match.

(6) Variation. Use a 3/4- or 1-inch rope. Space the leading players on each team 10 feet apart, each holding one end of the rope. The team pulled across the center line loses.

Figure 5-10 is available in the illustration section at the end of the book.

d. Master of the Ring

(1) Players. Any number. (See fig. 5-11.)

(2) Equipment. None.

(3) Area. A clearly marked circle large enough to contain all the players.

(4) Game. All the players stand inside the circle. At the signal, all players attempt to throw each other out of the circle. All tactics are fair except unnecessary roughness. When any part of the body touches across the line, the player is out and leaves the circle at once. Several officials are needed to spot the players who cross the line.

(5) Scoring The player who remains in the circle when all the others are out is the master of the ring.

(6) Variation 1. The players are divided into two equal teams. Each team is clearly marked. On signal, each team tries to throw the opponents out of the circle. The winning team is the team that eliminates all opponents from the circle.

(7) Variation 2. The players are divided into two equal teams. Each team sends only one individual into the circle. When one person has been forced out of the circle, the losing side sends in only one player. The team which eliminates all opponents is the winner.

(8) Variation 3.A pit, approximately 4-feet deep, is used rather than a circle on level ground. The contest may then be an individual or team activity.

Figure 5-11 is available in the illustration section at the end of the book.

e. Log Pivot Circle. Space teams far enough apart so each team can pivot in a circle without colliding with any other team. Each team holds a log in the bend of the arms in front of the chest. (See fig.5-12.) At the command **CIRCLE RIGHT, MOVE**,the left-flank individual holds the pivot while the log is carried around 360 degrees in a counter- clockwise motion, back to the original position. This movement may also be performed to the left by facing in the other direction and pivoting clockwise(**CIRCLE RIGHT, MOVE**).Other commands may be used such as: **CIRCLE HALF-RIGHT, CIRCLE HALF- LEFT** and so on. The first team to complete the prescribed movement is the winner.

Figure 5-12 is available in the illustration section at the end of the book.

f. Log-Rolling Race. Each team tries to roll its log a measured distance by pushing it with the hands and driving the body forward with the legs. (See fig. 5-13.)The first team to get the entire length of the log across the finish line wins.

Figure 5-13 is available in the illustration section at the end of the book.

g. Prone Pushing Contest.. Two teams lie prone, facing each other with a log between them. (See fig. 5-14.)Both teams place their hands against the log, keeping their arms straight. Then, by driving with the legs, each team attempts to push the other a measured distance to the rear.

Figure 5-14 is available in the illustration section at the end of the book.

h. Shuttle Relay Race. Each team in this race divides into two parts, A and B. Members of part A run 50 yards with a log held under their right arms. (See fig 5-15.)At the distance line, they give the log to their teammates in part B who bring it back to the starting line. The team pair finishing first is the winner.

Figure 5-15 is available in the illustration section at the end of the book.

5403.CROSS-COUNTRY AND DISTANCE RUNNING

a. Value of Running. Long-distance running gives some benefits that cannot be obtained in the same degree from any other sport. It builds powerful leg muscles, increases lung capacity, and develops endurance. For these reasons, cross-country and distance running should be included in the physical training program. These sports require only a few miles of open space which is generally available. Short cross-country runs and middle-distance runs can be used to supplement other activities, particularly team sports or the sports that develop precision or agility rather than endurance. Short cross-country runs can be scheduled once a week, gradually increasing the distance as the physical condition of the runners improves.

b. Cross-Country Runs. A cross-country run is a distance run held on a course laid out along roads, across fields, over hills, through woods, and on any irregular ground. A flat cinder or dirt track is not a suitable surface for cross-country running. Some runs are as long as 6 miles while others are as short as 3 or 4 miles. If cross-country running is being used to supplement other activities, the 3-mile course is long enough for most. Cross-country runs should be used only after individuals reach the sustaining stage of conditioning. This type of running should then be scheduled occasionally to provide variety. Cross-country running has the advantage of allowing mass participation. Interest can be stimulated by putting the runs on a competitive basis.

c. Practice Methods. Conditioning is essential to distance and cross-country running. Championship distance running depends on stamina, which can be developed only through constant training. An individual of only average ability can become an outstanding distance runner by steady and careful training. Hiking is an excellent method for getting into condition before the season opens. Long walks build up leg muscles. During the first month of the season, training should be gradual, starting with short distances, and increasing day by day. At first the legs will become stiff, but the stiffness gradually disappears if running is practiced for a while every day. To prevent strain, it is essential to limber up thoroughly each day before running.

(1) In training a large group, leaders should be stationed at the head and the rear of the column. They should make every effort to keep the runners together. After determining the abilities of the unit in cross-country running, it is advisable to divide them into three groups. The poorest-conditioned group is started first, the best-conditioned group, last. The starting time of the groups should be staggered so that all of them come in at about the same time.

(2) In preliminary training, running is similar to ordinary road work in that it begins with rather slow jogging, alternating with walking. The speed and distance of the run is gradually increased. As conditioning improves, occasional sprints may be introduced. Atfirst, the distance run is from one-half to 1 mile. It gradually increases to 2 or 3 miles. On completing runs, require runners to walk for 3 or 4 minutes before stopping to permit a gradual cooling off.

d. Facilities and Equipment. There should be at least one timer with a stopwatch (preferably three) for timing the runners. A course 3 or 5 miles long should be measured and marked by one of the three following methods:

(1) Fasten directional arrows to the top of tall posts and place them at every point where the course turns. Such signs should also be placed at every other point where there may be doubt as to the direction of travel.

(2) Place a lime line on the ground over the entire course.

(3) Place flags, clearly visible to the runners.

- **(a)** A red flag indicates a left turn.
- **(b)** A white flag indicates a right turn.
- **(c)** A blue flag indicates the course is straight ahead.

e. Rules/Scoring. A cross-country team consists of seven individuals, unless otherwise agreed. In dual meets, a maximum of 12 on one team may be entered, but 7 or fewer are scored. First place scores 1 point, second place 2, third place 3, and so on. All who finish the course are ranked and tallied in this manner. The team score is then determined by totaling the points scored by the first five of each team to finish. The team scoring the least number of points is the winner. If fewer than five (or the number determined prior to the race) finish, the places of all members of that team are disregarded. If two or more teams score the same number of points, the event is called a tie.

5404.TEAM ATHLETICS

Team athletics add variety to the combat readiness program. At no time should they be allowed to replace intramural activities or to provide practice time for intramural or base teams. If team athletics are integrated into the program, the goal should be total participation and competition. Therefore, the games selected can be played by the skilled or the unskilled. They require limited equipment, yet will teach competitiveness, aggressiveness, and teamwork while they improve conditioning.

a.Soccer. Soccer is one of the best athletic activities for developing endurance, agility, leg strength, and great skill fulness in using the legs. The most popular sport in the world, soccer is the national game of many European, Asian, and Central and South American countries. In recent years, it has become popular in the United States. A soccer ball is the only equipment needed for the game, and Marines can learn to play it easily. Players do not need much skill to participate, but the amount of skill they can develop is almost unlimited.

(1) Place in the Program. Soccer should be introduced during the latter part of the slow improvement stage of physical training. It can be used as a competitive activity in the sustaining stage. It is primarily a spring or fall sport. Any level field is suitable for competition. The boundaries for the soccer field are similar to the dimensions for a football field. (See fig. 5-16.) Goal posts are essential, but they are easily constructed and usually temporary so that they may be removed when not in use.

Figure 5-16 is available in the illustration section at the end of the book.

(2) Basic Skills

(a) Passing. Passing with the feet is the primary means of moving the ball. Short passes are easier to control. Emphasis should be continually placed on the skill of passing.

(b) Dribbling. The ball is dribbled by a series of kicks with the inside or outside of the foot. Do not kick with the toe. Keep the head over the ball when kicking and propel it only a short distance at a time. Keep it close to the feet. If a player lets a ball move far from the feet, an opposing player can easily take it away.

(c) Instep Kicking. The instep kick is the basic soccer kick. The toe does not come in contact with the ball. The toe is pointed downward, and the instep (the area under the shoe laces) is applied to the ball with a vigorous snap from the knee.

(d) Inside-of-the-Foot Kicking. The ball is kicked with the inside of the foot, and the leg is swung from the hip. The toe is turned outward, and the sole of the foot is parallel with the ground as the foot strikes the ball. This kick is used for short passes and for dribbling.

(e) Foot Trapping. The foot trap is the method of stopping the ball by trapping it between the ground and the foot. Place the sole of the foot on top of the ball at the instant it touches the ground, keeping the foot relaxed. This is an effective way to stop a fast-moving ball.

(f) Shin Trapping. .The shin trap is a method of stopping the ball with the shins. Stand just forward of the spot where the ball should strike the ground and allow it to strike the shins in flight or on the bounce. Use either one or both legs from the knee down so as to absorb the bounce.

(g) Body Trapping. .Intercept the ball with any part of the upper body except the arms and hands. Keep the body relaxed and inclined toward the ball. To keep the ball from bouncing, move backwards from it as it strikes the body. This drops the ball at the feet in position for dribbling or passing.

(h) Heading. Heading is a technique for changing the direction of the flight of a ball by butting it with the head. Tense the neck muscles and jump up to meet the ball. Butt the ball with the forehead at about the hairline to reverse its direction. Use the side of the head to deflect it to the side.

(3) Offensive and Defensive Positions. Forwards usually play on the offensive half of the field. Fullbacks usually play on the defensive half of the field. Halfbacks are the backbone of the team. They move forward on the offense and back on defense. The goalkeeper almost always remains within a few feet of the goal.

(4) Abridged Rules

(a) A soccer team is composed of 11 players. (See fig. 5-16.)

(b) The player propels the ball by kicking it with the feet or any part of the legs, by butting it with the head, and by hitting it with any portion of the body except the arms or hands.

(c) The goalkeeper is the only individual allowed to place hands on the ball, but only in the goalkeeper's area. The term "hands" includes the whole arm from the point of the shoulder down.

(d) A goal is made by causing the ball to cross completely the section of the goal line lying between the uprights and under the cross bar.

(e) Each goal scores 1 point for the team scoring the goal.

(f) The penalty for a foul committed anywhere on the playing field (except by the defensive team in its penalty area) is a free kick awarded to the opposing team.

(g) All opponents must be at least 10 yards from the ball when a free kick is taken.

(h) The penalty for a foul committed by the defensive team in its penalty area is a penalty kick.

(i) A penalty kick is a free kick at the goal from the spot 12 yards directly in front of the goal. The only players allowed within the penalty area at the time of the kick are the kicker and the defending goalkeeper.

(j) An official game consists of two 30-minute halves,

(k) Teams change goals at the end of every quarter.

(l) In the event of a tie, an extra quarter may played. This may be followed by a sudden death period. If the tie still exists, the game can be decided by penalty kick.

(m) After a team has propelled the ball across a side line and out of play, the ball is put back into play by a throw-in from the side line by a member of the opposing team. The ball is thrown in from the point where it crossed the side line as it went out of bounds. The ball must be thrown in with both hands using an over-the-head motion, keeping both feet on the ground.

(n) When the offensive team propels the ball across the defensive team's goal line but not in the goal, the defensive team is awarded a goal kick-a free kick taken from within its own goal area. The ball must come out of the penalty area to be in play.

(o) When the defensive team causes the ball to go behind its own goal line, excluding the portion between the goal posts, the offensive team is awarded a corner kick—a free kick taken by a member of the offensive team at the quarter circle, at the cornerflag post nearest to where the ball went behind the goal line. The flag post must not be removed.

(p) The game is started and, after a goal has been scored, is resumed by placing the ball in the center of the midfield line. Players must be on their side of the line until the ball is kicked. The ball must be kicked forward and must move at least 2 feet to be legal. The first kicker may not touch the ball twice in succession at the kick off. The opposing team must be 10 yards from the ball until it moves.

b.Speedball. Speedball offers vigorous and varied action with plenty of scoring opportunities. It is easy to learn and provides spontaneous fun. Little equipment is needed-a ball is all that is absolutely necessary. Speedball combines the kicking, trapping, and intercepting elements of soccer; the passing game of basketball; and the punting, drop kicking, and scoring pass of football.

(1) Place in the Program.Speedball, like soccer, should be introduced into the physical training program during the latter part of the toughening stage and used as a competitive activity in the sustaining stage. Speedball may be played any time weather permits but is primarily a spring or fall activity.

(2) Method of Play.The game is played as follows:

(a) Two teams of 11 each play the game under official rules, but any number of players may successfully constitute a team.An inflated leather ball, usually a soccer ball, is used. The playing field is a football field with a football goal post at each end. (See fig. 5-17.)

(b) The game starts with a soccer-type kickoff. The kicking team tries to retain possession of the ball and advance it toward the opposite goal by passing or kicking it. Running with the ball is not allowed, with the result that there is no tackling or interference. After the ball touches the ground, it cannot be picked up with the hands or caught on the bounce but must be played as in soccer until it is raised into the air directly from a kick. Then the hands may be used.

(c) When the ball goes out of bounds over the sidelines, it is given to a player of the opposing team and is put into play with a basketball throw- in. When it goes over the end line without a score, it is given to a player of the opposing team who may either pass or kick it onto the field.

(d) When two opposing players are contesting possession of a held ball, the official tosses the ball up between them as in basketball.

(e) Points are scored by kicking the ball under the crossbar of the goal post, drop kicking the ball over the crossbar, or completing a forward pass into the end zone for a touchdown.

Figure 5-17 is available in the illustration section at the end of the book.

(3) Basic Skills

(a) Shared Skills. Skills include kicking, passing, heading, and trapping (from soccer); punting, dropkicking and forward passing (from football); and passing, receiving, and pivoting (from basketball).

(b) Kickups. The kickup is a play in which a player lifts the ball into the air with the feet so as to legally play the ball with the hands. The kickup generally makes the transition from ground play to aerial play. The technique of making the play depends upon whether the ball is rolling or stationary. If the ball is rolling or bouncing toward the player, the foot is held on the ground with the toe down until the ball rolls onto the foot. Then the foot is raised, projecting the ball upward. If the ball is stationary, the player rolls it backward with one foot. Then the player places a foot where the ball will roll onto it. The player then raises the foot, projecting the ball as before. If a ball is rolling away, the player should stop it with a foot and play it as a stationary ball. The player can also raise the ball by standing over it with a foot on either side. The player presses the feet against the ball and jumps into the air, propelling the ball into the hands.

(4) Offensive and Defensive Play. The positions of players in speedball are much the same as in soccer. However, some of the positions have different names. Each team has 11 players. The forward line is composed of five players, the right end, right forward, center, left forward, and left end. The second line consists of right halfback, fullback, and left halfback. In the next line is the right guard and left guard. The player who defends the goal is the goal guard. The strategy employed in speedball during offensive play is very similar to that of soccer. There are two types of defensive formations in speedball: man-for-man and position defense. Man-for-man defense is recommended for beginning players.

(5) Abridged Rules

(a) Field. A speedball field is 100 yards long with additional 10-yard end zones on each end and 55 yards wide, the same as a regulation football field.

(b) Players. Eleven players are on a team. The goal guard has no special privileges.

(c) Time. Ten-minute quarters. Two minutes between quarters. Ten minutes between halves. Five minutes for extra overtime periods.(Begin the first overtime by a jump ball at center, same goals; change goals in the event of a second overtime period.)

(d) Winner of Toss. The winner of the toss has the choice of kicking, receiving, or defending a specific goal.

(e) Kickoff. A kick off starts the first and second half of a game and starts play after a score. The kick off is made from any place on the 50-yard line. The kicking team must be behind the ball when it is kicked. The receiving team must stay back of its restraining line (10 yards distance) until the ball is kicked. The ball must go toward the opposing team's goal before it can be played. Kick off out of bounds goes to the opponent at that spot. A kick off touched by the receiving team before going out of bounds, with no impetus added, still belongs to the receiving team. A kick off, in possession and control of the receiving team and then fumbled out of bounds, belongs to the kicking team at that spot.

(f) Starting Second and Fourth Quarters. The ball is given to the team that had possession at the end of the previous quarter, from out of bounds, as in basketball.

(g) Second Half. The first-half receiving team kicks off at the beginning of the second half.

(h) Fly and Ground Ball. The most characteristic playing rule of speedball is the difference between a fly ball (or aerial ball) and a ground ball. A player is not permitted to touch a ground ball with the hands and must play it as in soccer. A fly ball is one that has risen into the air directly from the foot of a player (for example, a punt, drop kick, place kick, or kickup).Such a ball may be caught with the hands provided the catch is made before the ball strikes the ground again.

(i) Kickup and Overhead Dribble. A kickup is a ball that is so kicked by a player that he can catch it himself. A bounce from the ground may not be touched with the hand because it has touched the ground since being kicked. This rule prohibits the ordinary basketball dribble, but one overhead dribble (throwing the ball into the air and advancing to catch it before it hits the ground) is permitted.

(j) Out of Bounds. If a team causes the ball to go out of bounds over the sidelines, a free throw-in (any style) is given to the opposing team. When the ball goes over the end line without scoring, it is given to the opponents who may pass or kick from out of bounds at that point.

(k) Tie Ball. In case two players are contesting the possession of a held ball, even in the end zone, a tie ball is declared. The ball is tossed up between them.

(6) Scoring

(a) Field Goal(3 points).A soccer-type kick, in which a ground ball is kicked under the crossbar and between the goal post from the field of play or end zone, is a field goal. (A punt going straight through is not a field goal for it is not a ground ball. The ball must hit the ground first.) A drop kick from the field of play that goes under a crossbar does not count as a field goal. A drop kick from the end zone that goes under the crossbar counts as a field goal;if it goes over the crossbar, it is ruled as a touch back, and the other team takes possession of the ball.

(b) Dropkick(2 points). A scoring drop kick must be made from the field of play and go over the crossbar and between the uprights. The ball must hit the ground before it is kicked (usually with the instep).

(c) End goal(1 point). This is a ground ball which receives its impetus (kicked or legally propelled by the body) from any player, offensive or defensive, in the end zone and which passes over the end line but not between the goal posts.

(d) Penalty Kick(1 point).A ball kicked from the penalty mark that goes between the goal posts and under the crossbar is a penalty kick. The penalty mark is placed directly in front of the goal at the center of the goal line.

(e) Touchdown(1 point). A touchdown is a forward pass from the field of play completed in the end zone. The player must be entirely in the end zone. If the player is on the goal line or has one foot in the field of play and the other in the end zone, the ball is declared out of bounds. If a forward pass is missed, the ball continues in play, but it must be returned to the field of play before another forward pass or drop kick may be made.

(7) Substitutions. Substitutions may be made any time when the ball is not in play. A player who is withdrawn may not return during that same period.

(8) Timeout. Three legal timeouts of 2 minutes each are permitted each team during the game.

(9) Fouls

(a) Personal Foul. Kicking, tripping, charging, pushing, holding, blocking, and unnecessary roughness of any kind, such as running into an opponent from behind, are personal fouls. Kicking at a fly ball and thereby kicking an opponent is a personal foul. Four personal fouls disqualify a player from the game.

(b) Technical Foul. Illegal substitution, more than three timeouts in a game, unsportsmanlike conduct, and unnecessarily delaying the game are technical fouls.

(c) **Violation.** Traveling with the ball, touching a ground ball with the hands or arms, double overhead dribble, violating tie ball, and kicking or kneeing a fly ball before catching it are violations.

(d) **Penalties.** The offended player shall attempt the kick.

(e) **Officiation of Fouls**

1 Fouls in the field of play allow no follow-up while fouls in the end zone always allow follow-up.

2 On penalty kicks, with no follow-up, only the kicker and goalie are involved.

3 On penalty kicks, with a follow-up, the kicking side is behind the ball and the defending side behind the end line or in the field of play. No one is allowed in the end zone or between the goal posts except the goal guard. The kicker must make an actual attempt at goal and cannot play the ball again until after another player plays it.

c. Volleyball. Volleyball is a popular sport. The game entails much physical activity, yet it is not strenuous. It is, therefore, a game for young and old alike, for beginners and for skilled players. It may be played indoors or outdoors on any type of terrain. As an informal activity, volleyball can be played by any number; as an organized activity, it provides, as few other sports do, a game for 12 to play in a limited area. While volleyball requires no great skill to play, it permits a high degree of proficiency. Players naturally get more enjoyment when they know the game and play it well. For this reason, instruction in the basic skills should be provided.

(1) Place in the Program Volleyball may be used occasionally as a competitive activity during the sustaining stage. Though it is a year- round sport, it should be included in the physical training program only when a more strenuous activity is impractical. Volleyball is an excellent off-duty activity.

(2) Instruction. Usually during the first three or four classes, a 10- to 15- minute period of instruction, followed by scrimmage, is enough to teach the basic skills, rules, and techniques of volleyball. More time, if available, can be given to teaching basic skills, but emphasize competitive play rather than formal instruction. Divide the unit into 6-person teams, organized on the basis of ability .Teams should be as nearly equal as possible.

(3) Basic Skills

(a) Passing the High Ball. The chest pass is the most effective method of playing the ball. To receive the ball, the knees are fixed, and the body is tilted forward. The elbows are raised sideward to a point in line with the shoulders. The wrists are extended in line with the forearm. The arms, wrists, and hands are rotated inward. To pass the ball, the hands are chest-high, thumbs pointing inward. The fingers are flexed, forming a cup, allowing them to contact the ball. On contact with the ball, the wrists are snapped while the fingers and elbows are pushed upward, sending the ball upward. A high ball is much easier to handle than a low one.

(b) Passing the Low Ball. A ball that is lower than the waist is one of the easiest to hit, but it is also a frequent cause of fouls (holding or carrying the ball). The best position for handling a low ball is to have the knees flexed and arms flexed at the elbows and rotated so the thumbs are pointing outward, the palms up. When the fingers contact the ball, the entire body reacts in a lifting motion. The arms and hands swing upward in a scooping action. It is important that the fingers, not the palms, contact the ball, and that the ball is batted, not thrown.

(c) Underhand Serve. Take a position behind the back line facing the net, left foot forward, holding the ball in the palm of the left hand. The left knee is flexed; the right knee is straight. Swing the right arm back and at the same time move the left hand (holding the ball) across the body in line with the right hip. Then swing the right arm forward hitting the ball off of the left hand with the palm of the right hand, raising the hips and arching the back in the same motion. Be certain to swing the right arm in a straight line, or the ball will be difficult to control. When the opposition is in formation, the server should try to place the ball in the right or left back area, and not near the net.

(d) Setup. A setup is a ball hit into the air near the net by one player, so that a teammate may hit or "spike" it sharply downward into the opponent's court. The chest pass is the best pass to use. The ball is sent approximately 10 feet into the air toward the spiker so it will descend from 4 to 20 inches from the net.

(e) Spiking. The spike is a leap into the air and a sharp downward hitting of the ball into the opponent's court. A spiker must be able to spring easily from the floor, judge the movement of the ball, and strike it with a downward movement of the arm. To jump from the floor, step off with one foot and jump with the other. Stand with the right or left side to the net, facing the setup player. Much depends upon the setup player to place the ball in the proper position. The spiker jumps into the air and strikes the ball above its center so as to drive it downward. A snapping movement of the arm and wrist will drive the ball forward and downward with power and control. Aim for a weak spot in the opponent's defense.

(f) Blocking. The block is a defense technique used to prevent a spiker from driving the ball across the net. It is an attempt by one or more defensive players at the net to block a hard-hit shot by using the force of the ball to send it immediately back into the opponent's court. In an effective block, forwards on the defensive team spring into the air at the time of the spike, placing both hands and arms in the expected path of the ball. An effective block tends to upset the offense and presents another element for the spiker to worry about. To be effective, the blocker must anticipate the path of the ball and time the block with the spike.

(4) Abridged Rules

(a) The volleyball court is 30 feet wide by 60 feet long. (See fig. 5-18.)

(b) The top of the net is 8 feet high.

(c) A volleyball team consists of 6 players.

(d) A match consists of the best two out of three games.

(e) The first team scoring 15 points wins the game, provided that they have 2 points more than their opponents.

(f) A deuce game is a game in which both teams score 14 points. The game is continued until one team obtains a 2-point advantage over the other.

(g) Only the serving team can score. If the serving team commits a fault, it loses the serve to the opposing team.

(h) The team members who receive the served ball rotate one position in a clockwise direction.

(i) The ball is put into play by serving from behind the back line.

(j) A served ball touching the net results in the loss of the serve. At any other time during play, a ball touching the net is still in play.

Figure 5-18 is available in the illustration section at the end of the book.

(k) The ball is out of play when it touches the ground or goes outside one of the boundary lines.

(l) All line balls are good.

(m) The players must hit or bat the ball; they may not throw, lift, or grasp it.

(n) A player may not touch the ball with any part of the body below the knees.

(o) A player may not play the ball twice in succession. In receiving a hard-driven spike, a defensive player may make several contacts with the ball even if they are not simultaneous. All such contacts, however, must constitute one continuous play, and all must be above the knees.

(p) The ball may be touched no more than three times on one side of the net before being returned across the net to the opposing team.

(q) A player must not touch or reach across the net.

(r) A player may touch the line under the net, but may not cross it.

Chapter 6
EVALUATION OF PERFORMANCE DURING TRAINING

6001.OBJECTIVES OF COMBAT READINESS

The objectives of the Marine Corps physical fitness program are-

a. To contribute to the health and well-being of every Marine through regular exercise and health education.

b. To develop Marines who are physically capable of performing their duties in garrison and in combat.

c. To develop Marines with a reserve level of physical fitness that will help them win in combat.

d. To provide a medium for developing the self-confidence of the individual Marine and thereby enhancing overall discipline, morale, esprit de corps, unit efficiency, and the desire to excel within the Marine Corps.

6002. THE PHYSICAL FITNESS TEST

The physical fitness test (PFT) has distracted attention from the proper goal of the physical fitness program. Simply stated, the goal of the physical fitness program is the success of Marines in combat. Unit commanders must not allow the PFT to become an end in itself. The PFT should be used within the commander's program of combat readiness training as a standard, an index of the physical fitness of individual Marines at a given time and place. The PFT is not difficult, and it must be kept in perspective. All athletes should be first class, and first class is attainable for all Marines. The conduct of the PFT is described in Marine Corps Order 6100.3_.

6003. OTHER METHODS OF EVALUATION

The commander should use other ways of evaluation to provide variety and imagination. Intangible objectives, such as confidence and aggressiveness, are hard to measure. The use of either inspection or observation is necessary to evaluate objectives of this nature. The commander has several other methods available when evaluating the physical condition of the command. The additional methods of inspection, observation, medical examination, and testing are summarized as follows:

a. Formal Inspections. Formal inspections, using inspection officers and standardized rating criteria, may help evaluate unit physical fitness.

b. Daily Observation. Routine observation of physical performance and appearance can serve as an indicator of an individual's or unit's physical readiness. However, mere observation is not a totally reliable or accurate means of evaluation.

c. Physical Examination. Medical examination may detect any individual disability or detrimental physical condition. It may guide in application of remedial, therapeutic, or limited exercise programs.

d. Other Physical Fitness Tests. Some other tests are described in detail in this manual for information purposes. Commanders are cautioned, however, not to place too great an emphasis on testing.

6004. RESPONSIBILITY

The commander is responsible for the physical fitness of the command and for the measurement and evaluation of its physical readiness. The goal of combat readiness training remains to prepare Marines and units for the physical demands of combat. The bulk of the commander's combat readiness training program should be such activities as obstacle and confidence courses, endurance courses, and progressive load-bearing marches which will prepare Marines for combat. Group activities which are competitive and combative in nature are advantageous and add imagination and enthusiasm to the program. Activities such as orienteering, which is not only physically demanding but also competitive and which teaches essential subject skills, are excellent and motivate all concerned. A commander's goal should be to provide an integrated training program that is so aggressive and demanding of Marines and leaders alike that a PFT administered at any time will reflect that each and every Marine is both physically and mentally prepared for the demands of combat.

6005. COMMANDER'S PHYSICAL FITNESS PROGRAM

The Commandant of the Marine Corps has directed that the physical fitness test shall be the universal measure of adequate individual physical fitness. The focus of a commander's physical fitness training program, however, should not be to prepare Marines to pass the physical fitness test, but to prepare them for the physical rigors of combat.

a. Administration. The physical fitness program requires each Marine to spend a minimum of 3 hours per week in physical fitness training, to be tested semiannually, and to obtain a minimum standard of third class. The attainment of a higher score is a laudable individual goal which should be encouraged but should not become a unit objective. Overemphasis could be detrimental to training required to develop the complete Marine. All commanders should exercise sound judgment as well as positive, aggressive leadership in striving toward the training objective.

b. Variety of Activities. The program must include the following elements:
(1) Physical Readiness Training. physical readiness training is a complete physical training program which develops and maintains the strength, endurance, and physical skills needed to sustain the individual during combat.
(2) Remedial Physical Conditioning. Remedial physical conditioning is a process by which physically substandard Marines are conditioned to meet prescribed standards.
(3) Competitive Conditioning Activities. Competitive conditioning activities consist of teams or individuals competing against an opponent to win. This involves a combination of sports and military skills designed to foster competitive attitudes and develop unit pride and esprit de corps.
(4) Occupational Conditioning. Marching long distances, dry net training, etc., should not be ignored as a means to fulfill the time requirement.

c.Time. In order to develop the desired level of physical fitness, unit commanders are advised that a minimum program of 3 hours per week is mandatory. The program must encompass at least three exercise periods weekly. Each period should include calisthenics, running, or other forms of vigorous activity .The minimum desirable length for each exercise period is 60 minutes. Shorter periods may be authorized by the unit commander if dictated by the local training situation. Unit commanders are encouraged to use the normal working day to satisfy this training requirement. However, commanders are authorized to conduct required physical fitness training during off-duty hours when the mission, workload, personnel status or other significant factors preclude fulfilling it during duty hours.

d.Action. Marine Corps Order 6100.3_ requires that commanders-

o Establish and maintain an effective physical fitness program which ensures that all Marines maintain an acceptable level of physical fitness.

o Ensure that all individuals are medically qualified prior to participating in the physical fitness program.

o Establish a minimum physical fitness program of at least 3 hours per week. Physical training may be authorized on an individual basis at the discretion of the unit commander.

o Ensure that all Marines participate in physical fitness conditioning activities commensurate with their medical qualifications and limitations.

o Conduct physical fitness testing for all Marines.

o Place all Marines who fail to pass the PFT on a daily command-supervised remedial physical conditioning program until they pass the test.

o Ensure that results of physical fitness testing are entered on section A of the fitness report. Comments are placed in section C in connection with attainment of superior physical performance or a medical excuse from the PFT.

6006. OTHER FORMAL TESTS

Commanders must realize that the best index of combat readiness is personal observation of performance on conditioning marches of 15 miles while under load. This manual presents three additional indicators of unit fitness. These indices are the physical readiness test, which was formerly the Marine Corps standard and is now routinely administered at Officer Candidate School; the airborne trainee physical fitness test which many Marines who attend U.S. Army formal schools are required to take; and the battle fitness test, the ultimate test of the combat readiness of Marines. Commanders may wish to include these tests or portions thereof within their combat readiness programs.

6007. PHYSICAL READINESS TEST

a. Purpose. The purpose of this test is to measure the Marine's ability to meet the minimum standard of physical readiness for duty in the field. This test is currently administered to all male officer candidates at Officer Candidate School, MCCDC, Quantico, VA. Female candidates do not take this test.

b. Test Area. Administration of this test requires an area suitable for crawling, jumping, and running 160 yards to include a trench that is 8 feet wide.

c. Equipment. The equipment required for this test includes—
o A platform, rail on a ledge, or a box 18 inches high,
o A 20-foot climbing rope,
o A stopwatch.

d. Conduct of Test. Adequate timers and other supervisory personnel must be provided by the company staff. All events should be conducted in a single session of one morning or afternoon. The events may be run in any sequence.

e. Uniform. The uniform for the physical readiness test is as follows: full utilities, cartridge belt, belt suspender straps, two canteens (full of water), rifle, and helmet. Depending on the heat condition, the PT shirt may be worn in place of the utility jacket. All equipment will remain secure throughout the test. Except for emergencies, canteens are not to be emptied.

f. Events

(1) Event 1: Climbing Uphill (Step-ups)This event simulates marching uphill at a rapid and steady pace. The Marine stands in front of the step. On the command GO, the Marine places either foot on the step and steps up. Hand pressure on the knee may be used if desired. The Marine then stands erect and steps backward and down, on one foot at a time. The same leg may be used for each repetition or the legs may be alternated. For 100 points, Marines will complete 60 repetitions in the maximum allotted time of 1 minute and 20 seconds (80 seconds).

*(2) Event 2: Rope Climb. T*his event resembles entering and leaving a hovering helicopter, using ropes in house-to-house fighting, and ascending and descending landing nets. The Marine assumes a sitting position on the ground at the bottom of the rope with hands at the highest point that can be reached. On the command GO, the Marine jumps up and climbs to the top. The Marine must touch the 20-foot mark with one hand, then descend without sliding or dropping. Marines must reach the 20-foot mark in the maximum allotted time of 30 seconds or less.

(3) Event 3: Evacuation. This event simulates reaching and evacuating a wounded Marine under fire. The Marine assumes the prone position at a distance of 50 yards from a casualty who is of approximately equal weight. On the command GO, the Marine springs up and covers the distance in a sprint. The Marine lifts the casualty to a fireman's carry and returns to the starting point. Assistance by the casualty is permitted. The tested individual must carry all equipment; the casualty has none. Marines must negotiate the course in a maximum time of 47 seconds or less.

(4) Event 4.'Advance by Fire and Maneuven. This event simulates advancing and assaulting during an attack. The Marine assumes the prone position at the starting point. On the command GO, the Marine then alternately crawls, runs in a zigzag manner, rolls, jumps, and assumes different firing positions for the next 130yards. Before reaching the objective, the Marine must "hit the deck," roll over, and assume the firing position three times. While assuming the firing position, the Marine must pull back the charging handle, take aim, and squeeze the trigger before rising from the ground. After rising from the ground for the third time,the Marine continues running and jumps to clear the 8-foot trench in one leap. The Marine then finishes the course in

the maximum allotted time of 2 minutes. The Marine is allowed multiple attempts only within the specified time limit.

(5) Event 5: Forced March. This event is a combat run of 3 miles without halts and against time. The Marine must double-time. Marines must complete the 3-mile course in a maximum allotted time of 30 minutes or less.

g. Event Standards

(1) To successfully pass the physical readiness test, the Marine must pass each event with a minimum of 80 points for each. Points are then assessed for each event in accordance with the scoring tables. (See fig. 6-1.)

(2) If a Marine fails one event, the highest possible score that the Marine may receive for the entire test is 79 percent. (See example 1 in fig. 6-2.) If two events are failed, the highest possible score for the entire test is 69 percent, regardless of the total points accumulated. (See example 2 in fig. 6-2.) If three events are failed, the highest possible score is 59 percent. Failure in four or more events will result in a score of zero for the entire test.

(3) To convert the total points for all five satisfactorily passed events to a final percentage score, total the accumulated points for all five events and divide by five. (See example 3 in fig. 6-2.)

Figure 6-1 is available in the illustration section at the end of the book.

Figure 6-2 is available in the illustration section at the end of the book.

6008. AIRBORNE TRAINEE PHYSICAL FITNESS TEST

a. Use and Composition of the Test.

(1) Use. The airborne trainee physical fitness test is a means of determining the physical ability of the applicant for acceptance to and retention in the airborne course of instruction.

(2) Test Events The test battery consists of five events as follows: chinups, knee bender, pushups, situps, and an endurance run.

b. Method of Scoring and Standards

(1) Scoring. The Marine will be scored by a trained scorer who is thoroughly familiar with the minimum standards for the test events. The Marine will be scored on a pass or fail basis. The performance on each event may be recorded on the reverse side of DA Form 705 (Physical Fitness Testing Record).The examinee records the personal information on the face of the card as specified on the first, second, and third lines. The face of the card is identified by writing diagonally across the lower half of the card "Airborne Trainee PPT—See reverse side."Using line 7 on the reverse side of the scorecard, the scorer enters the test title in the block entitled "(Other (Specify))" and completes the required information on the remainder of that line. The "Remarks" section can be used to record the test event titles and the applicant's performance on each test event.

(2) Standards. To successfully pass the test, the examinee must reach the standard in each test event. The standards follow:

(a) Chinups

(b) Knee bender 80(2-minute period)

(c) Pushups 45(2-minute period)

(d) Situps45(2-minute period)

(e) Endurance runs:2 miles in 15:54 minutes or less in athletic gear and 4 miles in 32 minutes in utilities and running shoes.

c. Uniform for Testing

(1) Examinees. The prescribed uniform for test participation is boots and the work uniform of the season. No headdress is worn. When climatic conditions permit, jackets or outer shirts may be removed.

(2) Officials. Scorers and other test officials should be uniformly and distinctively dressed for contrast with Marines being tested.

d. Test Administration Procedure

(1) Preparation for Test. The administration of the test to a large group makes it mandatory that the test effort be organized and efficiently operated. All testing is not completed with large test groups; on certain occasions, individuals and small groups are tested. Care must be exercised to administer the test uniformly and to standardize the conduct of all elements of the test. Regardless of the size of the test group, the following elements of sound test administration should be included:

(a) An orientation to include the purpose, method of administration, scoring of the test, preparation of the score card, and required standards.

(b) A correct demonstration of each event to ensure that there is no misunderstanding of the proper form and required standards.

(c) Completion of all five test events in one test period with all Marines taking the events in the same sequence.

(d) Adequate rest periods between the test events to allow for recovery before the next event.

(2) Method of Administration with a Large Group. With a minimum of 12 lanes per test event, 14 officials can administer the test battery to 150 or 200 Marines in 2 hours. The officials are designated as follows: one officer in charge, one demonstrator,and 12 scorers. If more or fewer Marines are to be tested, a greater or lesser number of officials will be required. The following procedure is recommended:

(a) Conduct an orientation and ensure the examinees have properly completed their score cards.

(b) Assign Marines to lanes and caution them to remain in the same lane order throughout the test.

(c) Explain and demonstrate the chinup event, administer it, and score it. Then proceed to the knee bender and pushup events, and administer them in the same manner.

(d) Grant a 5- to 10-minute rest period after the pushup event. Advise against excessive consumption of water during the break period.

(e) Explain and demonstrate the situp event, administer it, and score it as prescribed. Then move to the run area, explain the running event, and complete it.

(f) Retain the scorecards at the completion of the running event.

(3) Method of Administration with a Small Group. A similar procedure is followed for the testing of individuals and small groups. The informality usually associated with small groups must not conflict with sound test administration. With fewer examinees, fewer officials are required.

e. Description and Explanation of Test Events

(1) Test Event l:Chinups

(a) Purpose. This event is devised to test arm and shoulder flexor strength.

(b)Equipment There is one horizontal bar per lane, made of plumber's pipe or a gymnasium horizontal bar 1 1/2 inches in outside diameter. The bar must be rigidly supported at a height of 8 feet above the ground. The upright supports must be 5 feet apart. There must be a movable stand at each bar for short Marines to stand on to reach the bar.

(c) Officials There is one scorer per lane.

(d) Organization. The examinees, holding their scorecards, stand in order behind the restraining line in their respective lanes. The scorers take each scorecard when the Marine is called forward for the test.

(e) Starting Position The bar is grasped with the palms turned toward the face, the thumbs underneath the bar. The body is fully extended in a "dead" hanging position with the arms straight and the feet above the ground.

(f) Movement. Pull the body directly upward until the chin is placed over the bar. Lower the body until the elbows are completely straight and the body is again in the "dead" hanging position. Repeat as many times as required.

(g) Instructions. Explain and demonstrate the fully extended "dead" hanging position with the proper grasp. Show that the chin is placed over the bar at the top of the movement and that the arms are fully extended, the elbows completely straight, at the bottom of the movement (the hanging position). Explain that the body must be kept from swinging and that it is permissible to raise the legs and flex the hips when pulling up, but that any kicking, bicycling, or jerking motion with the trunk or legs is not acceptable. Inform the examinees that no penalty is exacted for hanging on the bar to rest in the bottom position but that this is not to their advantage. Tell them that half-completed chinups are not counted, and that the scorer will repeat the number of the last correct chinup when incorrect execution is detected.

(h) Administration and Scoring. Caution the examinees to assume the "dead" hanging position and wait for the scorer's command to begin. The scorer is at the examinee's left with a clear view of the bar. If the examinee begins to swing widely, the scorer should stop the swinging by extending the left arm across the front of the examinee's body, being sure not to hinder the execution of the chinup. The scorer counts aloud the number of chinups correctly executed. When a chinup is not correctly executed, the scorer repeats the number of the last correct one. The scorer records the number of correct chinups on the scorecard and returns the card to the examinee.

(2) Test Event 2:Knee Bender

(a) Purpose. This event measures the strength and endurance of the leg muscles.

(b) Equipment. None.

(c) Officials There is one scorer per lane.

(d) Organization. Marines stand in numerical order behind the restraining line in their respective lanes. The scorer takes each scorecard when the Marine is called forward for the test.

(e) Starting Position. The feet are spread less than shoulder width apart, hands on hips, thumbs in the small of the back, elbows back.

(f) Movement. Do a knee bend and at the same time bend slightly forward at the waist and thrust the arms between the legs until the extended fingers touch the ground. The hands are about 6 inches apart. The bend is approximately a three-quarters bend. From this knee bend position, recover to the starting position by moving the body upward, straightening the knees, and returning the hands to the hips. Repeat as many times as required.

(g) Instructions. Explain and demonstrate the correct starting position. Be certain examinees understand the correct knee bend and that only the tips of the fingers touch the ground. Tell them the scorer will repeat the number of the last correct knee bender when incorrect execution is detected. Some of the common errors are failure to correctly bend the knees, failure to touch the ground, and failure to assume the completely erect position after the bend has been executed.

(h) Administration and Scoring. The scorer stands to one side so as to see that the knees are properly bent and the fingers touch the ground as prescribed. From this position, the scorer can view the examinee to see that a properly erect position is assumed after each knee bend. The scorer counts aloud the number of correctly executed knee bends. When a knee bend is done incorrectly, the scorer repeats the number of the last correct one. The scorer records the number of correct knee bends on the scorecard and returns it to the examinee.

(3) Test Event 3:Pushups

(a) Purpose. Pushups measure arm and shoulder extensor strength.

(b) Equipment. None.

(c) Official. There is one scorer per lane.

(d) Organization. Marines stand behind the restraining line in their respective lanes until the scorer calls on them to perform. The scorer takes each scorecard when the Marine comes forward.

(e) Starting Position. The front-leaning rest position is the starting position. The body is straight from head to heels, palms are flat on the ground directly underneath the shoulders, and elbows are straight and locked. The body weight is supported on the hands and toes throughout the event.

(f) Movement. Bending only the elbows, lower the body in one straight plane until the chest touches the scorer's hand. Straightening and locking the elbows, raise the body in one straight plane, returning to the original front-leaning rest position. Repeat as many times as required, keeping the body in a straight line from head to heels.

(g) Instructions. Explain and demonstrate that the arms are straight at the beginning and completion of the movement and that the chest must touch the scorer's hand, but the stomach and thighs must not touch the ground. Also explain that the whole body must be maintained in a straight line as it is lowered and raised; that is, there is to be no breaking at the hips or shoulders so that any body part is lowered or raised in advance of the other or as a separate segment .Dipping or rolling through the shoulders is illegal, as is lowering or raising the body with one arm or shoulder at any time. Resting is not permitted during repetitions. Instruct

Marines that the scorer will repeat the number of the last correct pushup when incorrect execution is detected.

(h) Administration and Scoring. It is recommended that Marines assume a prone position while placing their feet and hands in the proper positions. This permits them to rest while the scorer gets into position and, at the same time, provides a feel of the body in a straight plane from head to heels. The scorer lies on the right hip and side to the right of the examinee. This gives the scorer a clear view of the examinee's body to see any errors. The palm of the scorer's right hand rests flat on the ground underneath the lowest part of the examinee's chest. By keeping the right forearm flat on the ground at an angle in front of the examinee's right arm, the scorer's position will not prevent the examinee from lowering the body completely. The scorer's left hand is free to test the straightening of the elbow at the completion of the movements and to point out body segments being lowered or raised separately. When in position and ready, the scorer has the examinee assume the starting position and begin doing pushups. The scorer counts aloud the repetitions done correctly and repeats the number of the last correct pushup if one is done incorrectly. There is no penalty if the contour of the examinee's body causes the hips to protrude slightly out of line, provided that the whole body is raised and lowered simultaneously. The scorer enters the number of repetitions on the scorecard and returns it to the examinee.

(4) Test Event 4: Situps

(a) Purpose. This event primarily measures abdominal strength.

(b) Equipment. None.

(c) Official. There is one scorer per lane.

(d) Organization Marines stand behind the restraining line in their respective lanes until the scorer calls on them to perform. The scorer takes each scorecard when the Marine comes forward.

(e) Starting Position. The examinee lies flat on the back with knees flexed, both feet flat on the ground. The correct angle of the thighs to the ground is 45 degrees. If the heels are too near the buttocks, the applicant will not be able to sit up. Legs are spread shoulder width apart. The examinee interlaces fingers and places them behind the head in contact with the ground. The feet are not held by another person.

(f) Movement. Bend forward at the waist and raise the upper body until the head is directly over the knees. Heels are not to leave the ground. Elbows remain in the same plane to the head and body throughout the event. The upper body is slowly lowered to the starting position until the head touches the ground. Repetitions are done at a slow cadence with no rest periods.

(g) Instructions. Explain and demonstrate the correct starting position and the proper execution of the situps to be sure that Marines understand the movement. Warn them that their knees must remain flexed during each situp, the heels cannot leave the ground at any time, and they may not roll up on one side and push up with one elbow. Tell them they must do the repetitions at a slow cadence with no rest periods. Instruct Marines that the scorer will repeat the number of the last correct situp when incorrect execution is detected.

(h) Administration and Scoring. When the performer is in position and ready, the scorer has the performer assume the starting position and begin doing situps. The scorer counts aloud the correct executions. When a situp is improperly done, the scorer repeats the number of the last correct one. No situp is credited if the hands are unclasped from behind the head, if the back is used to bounce up from the ground (which means the shoulders would not touch the ground), or if one shoulder or elbow is used to push up. The examinee is not penalized if the heels slide forward slightly so long as the knees remain flexed and the heels maintain contact with the ground. The scorer enters the number of repetitions on the scorecard and returns it to the examinee.

(5) Test Event 5:Endurance Run

(a) Purpose. This event measures cardiovascular endurance.

(b) Equipment One stopwatch or watch with a sweep second hand.

(c) Area. A large training field on which a quarter-mile track has been staked out or a level road over flat terrain may be used as a running surface. A 1-mile route is designated with wooden stakes marking the start point, finish point, and one-quarter mile intervals.

(d) Officials For large groups, a scorer times the event and controls the conduct of the run, and a guide runs with the group and sets the pace.

(e) Organization The run is conducted with groups of Marines in a column formation. Company-sized units may run at the same time with platoons serving as running groups. The scorer issues the command to assume double-time.

(f) Starting Position. Marines are assembled in proper column formation (column of twos, threes, or fours, as appropriate to the size of the group) with short men to the front. When all is ready, the column is moved forward a short distance before the running period is started.

(g) Movement. At the command DOUBLE-TIME, MARCH, the examinees retain their places in the column formation and execute the command. Length of steps is about 40 inches. The scorer has the group execute the run. The formation is maintained during therun.

(h) Instructions. Marines are instructed to maintain formation while running and are informed that the guide will set the proper pace. They will be instructed in the command used to control the column in the execution of the test. The scorer should announce the 4-minute, 2-minute, 1-minute, and 30-second remaining time intervals.

(i) Administration and Scoring. The event may be administered as previously prescribed to a large group, to several Marines, or to an individual. An individual examinee usually does not require a guide or pacer. If the event is administered on a training field, the scorer may stand in the center as the runner(s) circles about the field. This method of administration relieves the scorer of running with each group to be tested. Scoring is based on successful completion of the run as prescribed. The scorer should announce the remaining times as prescribed in (h).

6009. BATTLE FITNESS TEST

The battle fitness test is a physical fitness evaluation which is currently under development as part of the basic warrior training concept plan.

Chapter 7
THE HUMAN BODY

7001.GENERAL

To implement an effective physical conditioning program, a leader must understand how the human body functions and how exercise affects the body .This chapter provides only an introductory discussion of anatomy and functioning.

a. Body Functioning During the Stages of Conditioning. For more on the stages of conditioning, see chapter 1.

(1) During the toughening stage, the waste products of muscle activity (lactic acids) collect more rapidly than the blood can remove them. This acid waste builds up in the muscle tissue and irritates the nerves in the muscle fiber, causing pain and stiffness. As the exercise program continues, more blood is carried through the muscle, removing the waste materials more rapidly and eventually causing the soreness to disappear.

(2) During the slow improvement stage, the blood circulation in the muscles increases, and the body as a whole becomes a more efficient machine. The improvement is rapid in the first few weeks, but as a higher level of skill and conditioning is reached, the improvement becomes less noticeable. The body reaches its maximum level of performance after 6 to 10 weeks and should then be maintained at this peak.

(3) During the sustaining stage, it is possible to maintain this state of conditioning through 15 to 20 minutes of exercise a day, but the exercise must be quite strenuous.

b. Diet. Regular exercise increases the appetite. If the desire for greater amounts of food is satisfied by a balanced diet, the body benefits.

(1) There are two main types of foods: body-building and energy- producing.

(a) Body-building foods consist of proteins, which build up tissue and repair wear and tear.

(b) Energy-producing foods include carbohydrates and fats. Carbohydrates provide a quick source of energy, while fats act as a reserve store of energy.

(2) In addition, food contains vitamins, mineral salts, and water. During hot weather and strenuous training periods, the body requires greater fluid intake.

(3) Diet should be supplemented with proper rest to allow the digestive system to digest the food.

(4) Occasionally, especially during early stages of conditioning, strenuous exercise may cause vomiting. Although vomiting is not a frequent occurrence, it is not usually a cause for concern.

7002. SYSTEMS OF THE BODY

The systems of the body include the skeletal, muscular, circulatory, respiratory, endocrine, digestive, genitourinary, and nervous systems. Each has a different function, but all must work in cooperation with one another to insure a sound body. Of these systems, the first four are the most affected by exercise and are discussed in succeeding paragraphs.

7003. THE SKELETON

a. Bones. The skeleton (see fig. 7-1) is composed of about 206 bones, which-
o Provide structure for the body and support for the attachment of muscles,
o Protect vital organs, such as the brain, lungs, and heart,
o Manufacture red blood cells, which carry oxygen through the body.

Figure 7-1 is available in the illustration section at the end of the book.

b. Joints. A joint is a place of union between two or more bones. Joints can be-
o Immovable, such as in the face and head.
o Slightly movable, such as in the vertebrae or sternum.
o Freely movable, such as in the shoulder, hip, knee, ankle, and elbow. The bones in a freely movable joint are held in place by ligaments. Freely movable joints are of the greatest importance in physical training because exercise improves their mobility and stability.

c. Cartilage and Ligaments. The joints in the body, where bones connect, are supported by cartilage and ligaments.
o Cartilage is a tough, elastic, translucent tissue that acts as a shock absorber or buffer between bones. Examples are the discs between the vertebrae, the tissue attaching the ribs to the sternum, and buffers in the knee joints. Damaged cartilage does not heal.
o Ligaments are connective tissue that bind bones together. They contract and extend but are not elastic. Consequently, ligaments can be torn or strained. Damaged ligaments will heal to some extent.

7004. THE MUSCLES

a. General. Muscles are tissue, or an organ composed of tissue, which contract or extend to produce movement. Muscles are of three classifications: involuntary, voluntary, and cardiac. The voluntary and cardiac muscles are the most affected by physical training; physical training has little effect on involuntary muscles.(Seefigs. 7-2, 7-3, 7-4.)
o People have no control over the involuntary muscles, an example of which is the diaphragm,
o Voluntary muscles are the larger skeletal muscles which cause movement and which the individual can control.
o Cardiac muscle is found only in the heart and for all practical purposes is an involuntary muscle. Physical training has a significant effect on the fitness of cardiac muscles.

b. Muscle Structure. Muscle is composed of cells which are grouped into bundles called fibers. Groups of fibers form muscles. These bundles of fibers are held in place by sarcolemma, a thin, sheath-like material that surrounds the muscle bundles and secretes a fluid that lubricates the muscle tissue. The fused ends of the sarcolemma form the tendons which attach the muscles to the bones.

c. Attachment of Muscles. The arrangement of muscles on the skeleton provides the proper angle of pull to make movement possible. Voluntary muscles are usually attached to the skeleton in two places.
d.Action of Muscles. To produce motion and to do work a muscle usually shortens its fibers. The movement may be flexion such as bending the arm at the elbow, or extension, such as the straightening the arm. Some muscles raise the arms or legs, others depress the raised limb. Some muscles have the primary function of rotating the trunk from side to side. In addition, muscles act as stabilizers as well as prime movers.

Figure 7-2 is available in the illustration section at the end of the book.

Figure 7-3 is available in the illustration section at the end of the book.

Figure 7-4 is available in the illustration section at the end of the book.

7005. FUNCTIONING OF THE SKELETON AND MUSCLES

a. General. The skeleton and muscles function in close coordination. Muscles move the body, and bones support the muscles.

b. Effect of Exercise on Bones. Continuous exercise, particularly among younger people, usually strengthens the bones, allowing them to withstand greater stress and strain. Bones which are not used lose much of their strength, a factor which should be considered when individuals return to physical training after a prolonged period of inactivity. People in this category should refrain from activities which might result in bone breakage before the bone is exercised back to normal condition. The condition known as "march fracture" is usually a result of this condition.

c. Muscular Strength. When exercised vigorously to improve strength, muscle grows in size. In general, the larger the muscle, the stronger the muscle. Furthermore, trained muscle functions more efficiently than untrained muscle. Trained muscles contract more vigorously and with less effort. Regular and strenuous exercise also toughens muscle, making it firmer and able to stand more strain.

d. **Muscular Endurance.** Muscles naturally become fatigued during continuous, repeated work. Through specific exercises, such as chinups or situps, local muscle groups become exhausted long before an individual fatigues. Training to develop muscular endurance enables people to continue a relatively heavy load of exercise over a long period of time. Lengthened exercise periods improve muscular endurance.

e. Muscular Coordination. Improved speed and strength result in part from improved muscular coordination. An unskilled individual may use irrelevant muscles to perform a particular activity, increasing the amount of work without increasing the mechanical output. This increase in skill is highly desirable, but it should be offset by greater effort in duration or intensity to compensate for the loss in overload due to increased skill.

f. Muscular Fatigue. When the rate of work is increased, the energy required is proportionately much greater than the increase in rate. For example, if an individual doubles running speed, the amount of power demanded to do this is increased eight times.

g. Circulation in Muscles Regular, strenuous exercise causes the creation of new capillaries and the opening of inactive, latent capillaries, increasing blood circulation within the muscle by as much as 400 percent. This increases the supply of food materials and oxygen to the muscle, improving its endurance. This process takes about 8 to 12 weeks of regular conditioning in young adults and longer as age increases.

7006. THE CIRCULATORY AND RESPIRATORY SYSTEMS

a. The Circulatory System. The circulatory system transports blood to all parts of the body, removes waste products for disposal, and delivers protecting and repairing substances where needed. The circulatory system includes the heart and blood vessels.

(1) The Heart. The heart is a pump which forces blood through the blood vessels to the parts of the body. The heart is a little larger than a fist and is located in the left center of the chest between the lungs.

(2) Blood Vessels. The vessels carrying blood away from the heart are the arteries. They eventually divide into capillaries, the very small vessels through which diffusion and osmosis take place. The capillaries gradually increase in size, forming the veins which carry blood back to the heart.

b. The Respiratory System. The respiratory system performs the function of breathing, whereby oxygen is inhaled and carbon dioxide is exhaled. The respiratory system consists of the mouth, nose, trachea, lungs, and diaphragm.

o Trachea. The trachea, or windpipe, is a hollow, tube-like structure that carries air from the mouth to the lungs.

o Lungs. The lungs are elastic bags in the chest where the exchange of oxygen and carbon dioxide takes place.

o Diaphragm. The diaphragm is a thin, sheet-like muscle just below the lungs. During inhalation, the diaphragm flattens out and lowers, allowing the lungs to expand and fill with air. During exhalation, the diaphragm raises into a dome shape, helping to expel air from the lungs.

7007. CARDIOVASCULAR FUNCTIONING

a. General. Cardiovascular functioning is the combined functioning of the circulatory and respiratory systems. The chief organs of these systems, the heart and lungs, function together to provide oxygen to the body.(See fig. 7-5.)

Figure 7-5 is available in the illustration section at the end of the book.

b. Heart Action. The heart, the organ which propels blood through the blood vessels, is the chief organ of cardiovascular endurance. A person tires quickly if the heart lacks the capacity to circulate the blood through the body. The heart is a muscular organ and is developed by exercise just like any other muscle. The heart cannot be exercised alone; any exercise that puts a sufficient load on the heart, such as running or cycling, will also exercise other body parts. A conditioned heart tends to beat more slowly but pump more blood when at rest. This is known as an increase in "stroke volume,"a desirable condition because it enables the heart to pump more blood with a slower contraction rate.

c. Functions of the Lungs. Strenuous and regular exercise can improve the efficiency with which the lungs transmit oxygen to the blood by as much as 25 percent In a poorly conditioned person, some of the alveoli (air sacs) within the lungs are closed or collapsed. The forced breathing created by exercise over a period of weeks causes the air sacs to slowly expand, increasing the ability to absorb oxygen.

d. Relationship of Heart and Lungs. The heart pumps carbon dioxide-laden blood through the pulmonary artery to the lungs. In the lungs the carbon dioxide is exchanged for oxygen and the purified blood is returned to the heart by way of the pulmonary vein. The heart then pumps the blood through the aorta for circulation throughout the body. Blood moving into the muscles exchanges oxygen for carbon dioxide, after which time it is ready for the return trip to the heart.

e. Cardiovascular Functioning in High Altitudes. Marines to be employed in high altitudes should be acclimated in a similar area for 10 to 14 days prior to employment. Persons not accustomed to the rarefied air of higher altitudes tire more quickly and may collapse after rapid physical exertion. Air is much less dense at high altitudes than at sea level. Persons accustomed to sea level inhale only about 80 percent of the oxygen they are used to no matter how hard they breathe. Further, people accustomed to sea level or moderate altitude simply do not have enough red corpuscles in the blood to fulfill their needs at high altitudes. The red blood cells will increase over time.

f. Symptoms of Cardiovascular Malfunctioning. The brain is the first organ to react to a lack of oxygen. Unconsciousness results when the brain is denied sufficient oxygen. "Blacking out" is actually a defense mechanism to keep the body alive.

APPENDIX A
INSTRUCTOR HINTS AND INSTRUCTOR TRAINING

1. General. This appendix informs instructors on administration of exercise periods, commands, the extended rectangular and circle formations, methods of instruction and conduct of exercise activities, explanation of basic positions associated with the various activities, and instructor training.

2. Aim of the Physical Fitness Program. The aim of the physical fitness program is to prepare Marines physically for war. This can be conveniently split into two phases.

a. Preparation Phase

(1) Development of all-around physical fitness.

(2) Development of purposeful physical skills.

(3) Development of mental alertness.

(4) Development of character and leadership.

b. Maintenance Phase. Once the preparation is completed, the program must progress to a maintenance program. This phase revolves around the need to maintain the trained Marine's physical fitness for a particular role. This will include—

(1) Maintenance of a high standard of all-around physical fitness.

(2) Application of purposeful physical skills to combat skills.

(3) Development of leadership and self-confidence.

(4) Development of mental toughness and endurance.

3. Physical and Mental Development

a. Marines need to be developed physically and mentally. The physical training (PT) instructor must ensure that the program accomplishes this.

b. The system of the body functions best in response to progressive exercise. This means that Marines should master the easier exercises before they attempt the more difficult ones.

c. The body and mind are linked inseparably and must be considered as one in physical fitness training. Physical exercises assist in developing the mind.

d. The confidence in oneself and in one's comrades attained by physical achievement helps to promote the indomitable will to win. The Marine must be taught to realize this. It will help during arduous training, which at all times must be imaginative, realistic, and filled with enthusiasm. All physical fitness training must be associated with developing the qualities required of a Marine.

4. Planning and Preparation of Programs The following factors must be considered when compiling a program (see also ch. 2)~

a. Aim or desired skill level.

b. Duration of the complete training program and number of physical training lessons.

c. Purpose of included exercises,drills, or practical applications.

d. Beginning/inventory fitness levels.

e. Medical facilities, medical personnel, and Marine Corps or local command regulations governing their proximity to the training.
f. Number of Marines to be trained.
g. Availability of instructors, assistants, and facilities.
h. Apparatus and equipment available.
i. Alternative training areas for varying weather conditions,
j. Stages of progression and tests of achievement.

5. Five Basic Principles of Program Design
a. Regularity. Regularity of exercise is far more important than the amount of exercise performed on a given day. A Marine should exercise a minimum of three times per week. Human muscle begins to atrophy after 3 days without exercise.
b. Balance. The program must be balanced in terms of total physical fitness. Imbalance is created when, for example, an individual merely runs. While the cardiovascular system is improved, general upper body strength may be lacking. Conversely, a person who just does weight training may develop strength, but not cardiovascular endurance.
c. Overload. If total body fitness is to be developed, then the muscular and cardiovascular systems must be made to perform more work than previously. This is done by increasing the frequency, intensity, and duration of the exercise. The body adapts to the amount of stress placed upon it and becomes stronger and more efficient.
d. Progression. To be successful, any type of training program must be progressive. This is an area where untrained persons conducting PT can cause injury to Marines resulting in failure of the PT program. Close supervision must be made of untrained instructors. The result of progressing too fast is that the class begins to reject PT because it is no longer enjoyable or motivating. The program should start at a slow pace, and the pace should increase as proficiency improves. Average persons will encounter three stages of progressiom the toughening stage, the slow improvement stage, and the sustaining stage. (See ch. 1.)
e. Variety. Some programs fail because the training becomes boring. Variety is essential and perhaps presents the greatest challenge to the commander and staff.

6. Grouping. Providing for different levels of physical fitness is particularly recommended in the early stages of conditioning. Older Marines and those in poorer physical condition should be expected to attain a group level of fitness, but they should be given more time to do it.
a. Homogeneous Grouping One simple method of providing for the difference in levels is to group them according to their condition. A two-group classification would divide Marines into highly conditioned and average groups. A finer classification could be obtained by dividing them into three groups—a highly conditioned, a moderately conditioned, and an unconditioned group.
b. Performance or Age Grouping. The segregation of Marines into different exercise groups should be based on physical fitness test scores or on the level of fitness they demonstrate. They may also be grouped at first according to age. A common classification by ages is under 30, 30 to 34, and 35 and above.

7. Preparatory Commands and Commands of Execution. The prepreparatory command describes and specifies what is required. The command of execution calls into action what has been prescribed. All preparatory commands are given with a rising inflection. The interval between commands is long enough to permit the average Marine to understand the first one before the second one is given.

8. Extended Rectangular Formation. The traditional formation for carrying on many physical training activities is the extended rectangular formation (fig. A-1). This formation is more compact than any other. It is the best type to use for large numbers of Marines because it is simple and easy to assume.

a. For the formation of oneplatoon, the base man paces off five paces from the stand, faces left and moves five paces, halts, and again faces left. With the base man positioned facing the stand, the platoon leader then commands. **FALL OUT AND FALL IN ON THE BASE MAN**. At this command, all Marines run to the designated area and reform. This procedure is preferred to marching the unit into position. If more control is desired, the unit may march at double time to the vicinity of the base man and then be directed to fall out and fall in on him. Much valuable time is wasted in the field by needless maneuvering of Marines at quick time in an effort to position the platoon or unit on the exact spot for the exercises.

b. A company-size unit assumes the extended rectangular formation from a column of threes or fours at normal intervals between squads. This extension can also be executed from a company mass without interval between platoons. In extending either a platoon or company-size unit, take your place at the head of the column and command:
(1) TAKE INTERVAL TO THE LEFT, MARCH. At this command, the Marines in the right flank file stand fast with arms extended sideward. All others turn to the left and run forward at double time. After taking a sufficient number of steps, all Marines face the front with both arms extended sideward. The distance between finger-tips is about 12 inches, and dress is right.

Figure A-1 is available in the illustration section at the end of the book.

(2) ARMS DOWNWARD, MOVE. At this command, the arms are lowered smartly to the sides.
(3) LEFT, FACE
(4) EXTEND TO THE LEFT, MARCH. At this command, those in the right flank file stand fast with arms extended sideward. All others turn to the left and run forward at double time. Spacing is the same as in (1) above, and dress is right.
(5) ARMS DOWNWARD, MOVE. Same as in (2) above.
(6) RIGHT, FACE.
(7) FROM FRONT TO REAR, COUNT OFF. At this command, the leading Marine in each column turns the head to the right rear, calls of **ONE** and faces the front. Successive Marines in each column call off in turn **TWO, THREE, FOUR, FIVE** in the same manner.
(8) EVEN NUMBERS TO THE LEFT, UNCOVER. At this command, each even-numbered Marine stridejumps sideward to the left, squarely in the center of the interval. In doing this, each one swings the left leg sideward, jumps from the right foot to the left foot, and smartly brings the right into position against the left.

c. To assemble the unit, you commandASSEMBLE TO THE RIGHT, MARCHAt this command, all return to their original position in the column at double time and reform on the base man.

d.It is recommended that the area for grounding equipment and arms be at the edge of, or well away from, the area to be used for exercising. To conserve time and ensure proper position of the unit, the base man or, if the unit is composed of several platoon-sizegroups, the various base men may precede the unit and establish their positions in relation to the instructor's stand.

9. Circle Formation.The circle formation is effective for conducting various exercise activities (fig. A-2). This formation has an advantage in that supervision of all Marines is facilitated and a moving formation is available which provides control. Guerrilla exercises, grass drills, and some forms of running are examples of activities which are more easily conducted in the circle formation than in the extended rectangular formation.

a. It is not advisable to have more than 60 Marines in a circle. When more must be accommodated, separate circles should be used.Concentric circles may be employed to reduce the size of the circle or to accommodate more persons. If concentric circles are formed, the different circles are made by designating squads for each circle.Each additional circle requires more Marines than the one inside it.For example, one squad of a platoon may form the inner circle and theremaimng three squads the outer circle.

b. When a platoon is to form a circle, the commands ai€IRCLEFORMATION, MARCH, FOLLOW MEUpon this command, the right flank squad of the column moves forward at double time with the leader of the platoon group gradually forming a circle in a counterclockwise direction. Each succeeding file falls in behind that on the right. After the rough outline of the circle is formed, the leader commandsPICK UP A 5-YARD INTERVAL.

Figure A-2 is available in the illustration section at the end of the book.

c.The group may be halted and faced toward the center, or, if instruction is not necessary, the exercise activity may be executed without stopping the platoon.

lO.Leadership Techniques

a. Unless you experience all the exercises, you cannot appreciate how arduous they are, what movements are most strenuous and difficult, where the errors in performance are likely to occur, and what the proper cadence should be.

b. You must give everyone careful supervision and participate in the exercises to show that you can do them. When you participate, your assistant instructors should supervise because it is difficult for you to supervise and exercise at the same time.

c. Marines should never be kept too long in one position, especially a constrained one. They should never have to perform an exercise more times than they can do it without losing the proper form. You should insist upon proper form in the execution of all exercise activities.Even slight deviations from the proper form reduce the value of the exercise.

d. Avoid long explanations.As a rule, it should be necessary to give a full explanation of new exercises only. Explain the most essential features of an exercise first; add details later. Too many details at one time are more likely to confuse than to assist them.Minor corrections should be made to the entire class while the exercise is in progress (for exampHEADS **UP,** KNEES STRAIGHT). If necessary, follow this direction by the name of the person who is particularly at fault. If a Marine requires special attention, givethat person separate instruction by an assistant instructor to avoidwasting the time of the group.

e. The heavy demand on your voice can be lightened by training assistant instructors to assume some of the instruction. Using mass cadence is also an effective method of lessening the demand on your voice.

“Illustrations”

DAY	TIME	ACTIVITY
MONDAY	60 MIN	RUNNING ACTIVITY
TUESDAY	60 MIN	COMBATIVES
WEDNESDAY	60 MIN	CONDITIONING & LOG DRILLS
THURSDAY	60 MIN	CONDITIONING & RELAYS
FRIDAY	60 MIN	RUNNING & CONDITIONING
SATURDAY	60 MIN	RUNNING ACTIVITY

Figure 2-1. Sample A- -Single Activity.

DAY	TIME	ACTIVITY	REMARKS
MONDAY	20 MIN 20 MIN 20 MIN	RIFLE DRILL COMBATIVES RUNNING	PROGRESSIVE ACTIVITY SYSTEM
TUESDAY	30 MIN	GUERRILLA EXERCISES TABLE I	SINGLE ACTIVITY
WEDNESDAY	30 MIN	RIFLE DRILL	SINGLE ACTIVITY
THURSDAY	30 MIN	CIRCUIT INTERVAL TRAINING	SINGLE ACTIVITY
FRIDAY	20 MIN 20 MIN 20 MIN	STRENGTH CIRCUIT CONDITIONING DRILL CROSS-COUNTRY RUN	ROTATION ACTIVITY
SATURDAY	60 MIN	WARM UP RUN AND GRASS DRILL	SINGLE ACTIVITY

Figure 2-2. Sample B- -Various Time Periods, Activities, and Systems.

DAY	TIME	ACTIVITY
MONDAY	45 MIN	RUNNING – CROSS-COUNTRY
TUESDAY	75 MIN	TEAM ATHLETICS – TOUCH FOOTBALL
WEDNESDAY	45 MIN	RUNNING – CROSS-COUNTRY
THURSDAY	20 MIN 40 MIN	LOG DRILL DUAL COMBATIVES
FRIDAY	90 MIN	SPEED MARCH
SATURDAY	90 MIN	TEAM CONTESTS

Figure 2-3. Sample C--Sustaining Activities.

Route	Kilometers from SP	Recommended rate of march (kmph)	Remarks
SP: RJ 8th Div Rd—Superhighway	......	4	Bridge; hard surface; two guides.
RJ Jamestown—Hersey Rd	2,25	4	Traffic heavy; two guides; Jamestown Rd bears to right.
RJ Jamestown—Yankee Rd	7.10	4	Two guides
RJ Jamestown—Lightning Rd	10.50	4	Two guides
RJ Jamestown—Sunshine Rd	12.40	4	Light traffic; one guide Sunshine Rd; gravel, poor traction when wet.
RP; RJ Sunshine—Sedan Rd	14.80	4	Two guides

Figure 3-1.Example of Route Reconnaissance Report.

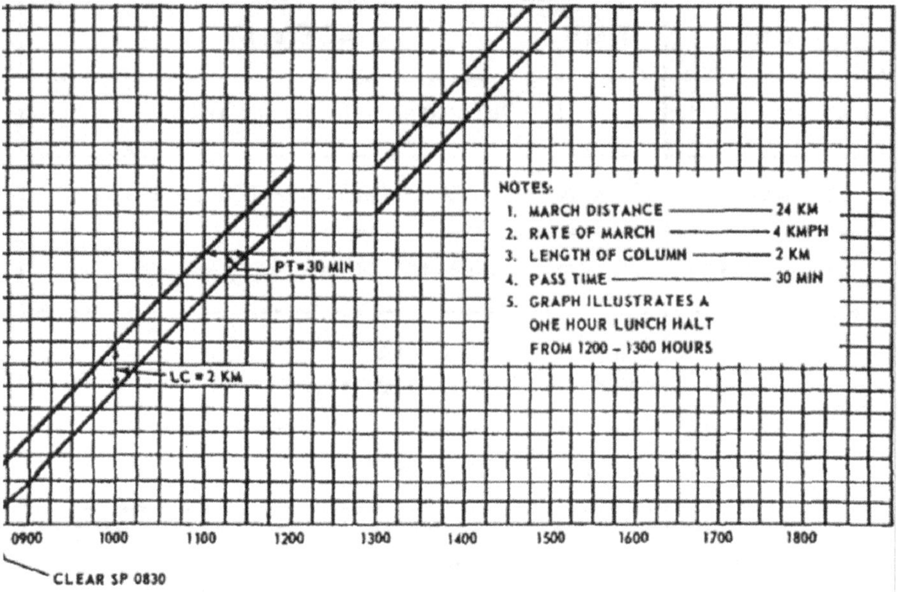

Figure 3-2.Road Movement Graph.

Figure 3-3.The Pace-Setter.

	SINGLE FILE	COLUMN OF TWO
2 m/MAN APART	24	12
5 m/MAN APART	54	27

GAP DISTANCES			
SERIAL:	(3-1) × 100 m	=	200 m
MARCH UNIT:	(12-1-2-) × 50 m	=	450 m
	TOTAL COLUMN GAP		650 m

Figure 3-4.Conditioning Drill One.

Figure 3-5.Conditioning Drill Two.

Figure 3-6.Conditioning Drill Three.

Figure 3-7. Rifle Drill.

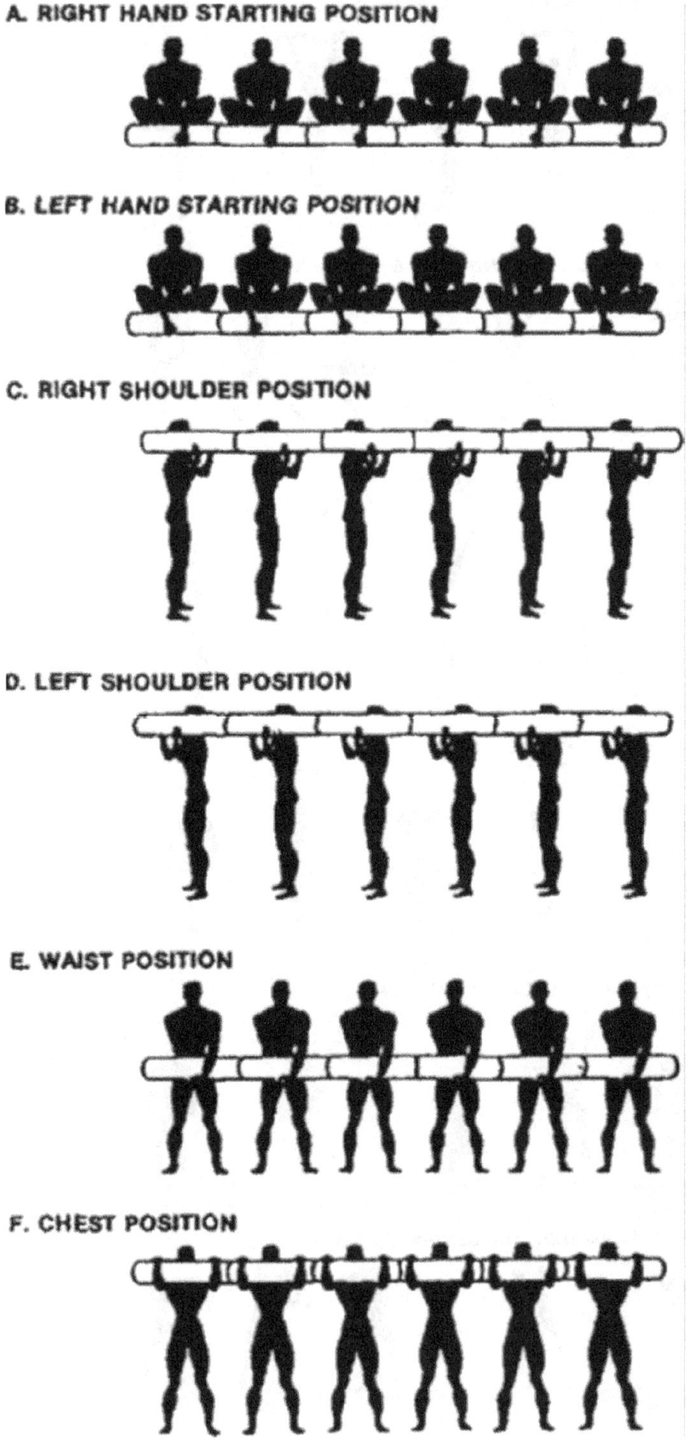

Figure 3-8. Starting Positions.

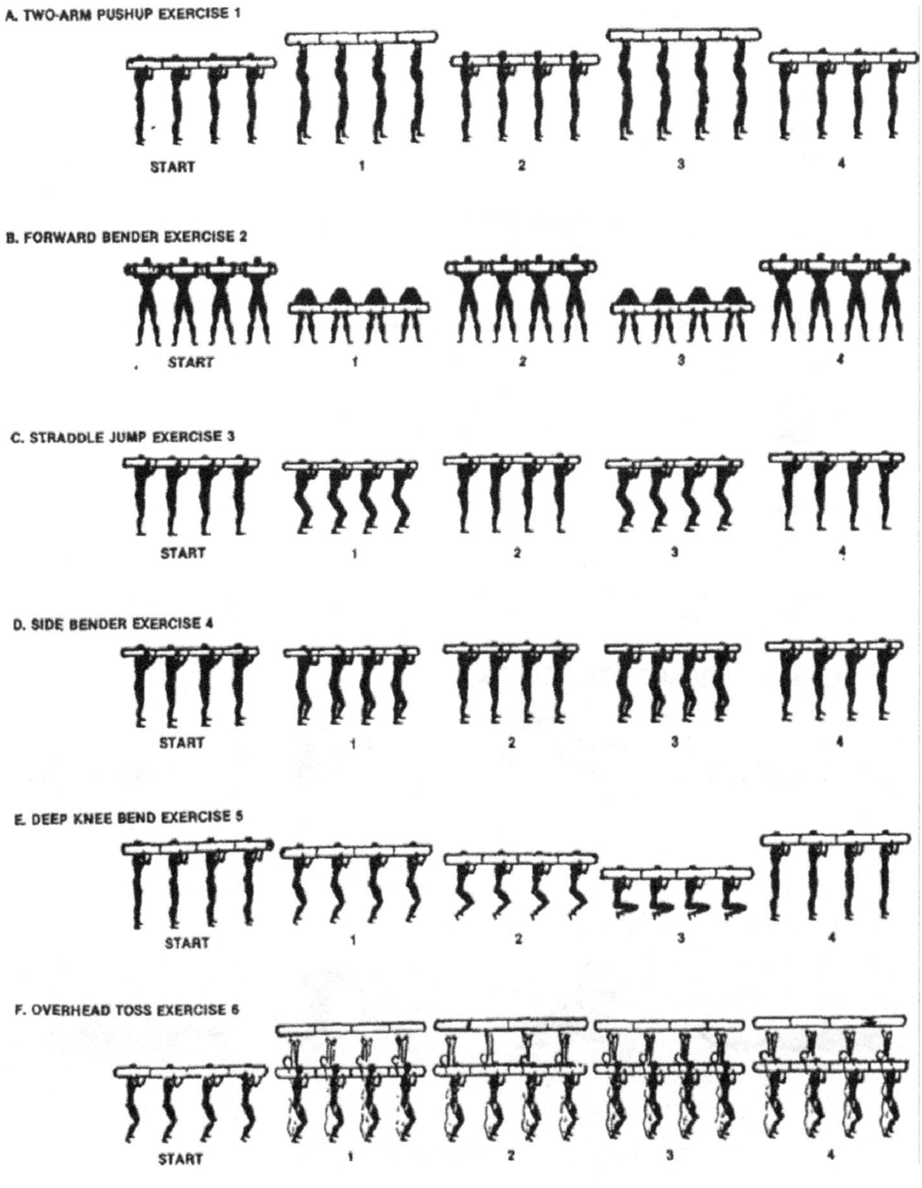

Figure 3-9. Log Drill

Figure 3-10.Basic Positions for Grass Drill.

Figure 3-11.Grass Drill One.

Figure 3-12.Grass Drill Two.

Figure 3-13.Guerrilla Table I.

Figure 3-14.Guerrilla Table II.

Figure 3-15.Proper Running Form.

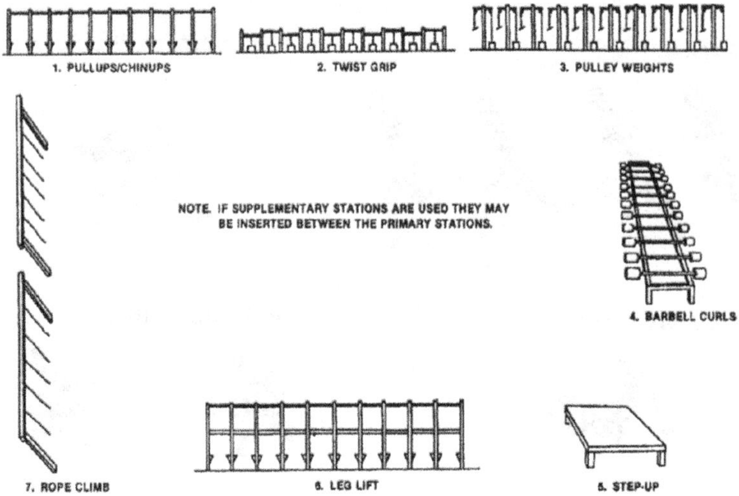

Figure 3-16. Fixed Strength Circuit.

STATION	ITEM	NO.	SPECIFICATION
1	BARBELL		1¼-INCH PIPE 5 FEET LONG WITH CONCRETE FILLED NO. 10 CANS
2	JUMP ROPE		¼- OR ⅜-INCH ROPE, 10 FEET LONG
3	TWIST GRIP		HANDLE 12 INCHES LONG, ROPE 4 FEET LONG, NO. 10 CAN CONCRETE FILLED.
4	INCLINE PLANE		¾-INCH PLYWOOD PLATFORM 2 FEET WIDE AND 6 FEET, 6 INCHES LONG, ELEVATED 10 INCHES AT ONE END. STRAP TO HOLD FEET DOWN.
5	WAR CLUB		HEAD IS 6 BY 12 INCHES, HANDLE IS 14 INCHES LONG BY 1¼ INCHES IN DIAMETER, ABOUT 20 POUNDS.
6	BICYCLE RIDE		PLYWOOD BOARD OR PLATFORM 2 BY 3 FEET WITH 2 BY 2 RUNNERS.
7	STEP—UP		A BOX OR STURDY PLATFORM 18 INCHES HIGH, 18 INCHES WIDE, 24 INCHES LONG.
8	ISOMETRIC PULL		TWO HANDLES 12 INCHES LONG WITH 4 FEET (BETWEEN HANDLES) OR LIGHT WIRE CABLE OR ¼-INCH ROPE

Figure 3-17.Movable Strength Circuit.

Figure. 3-18.Stations of Movable Strength Circuit.

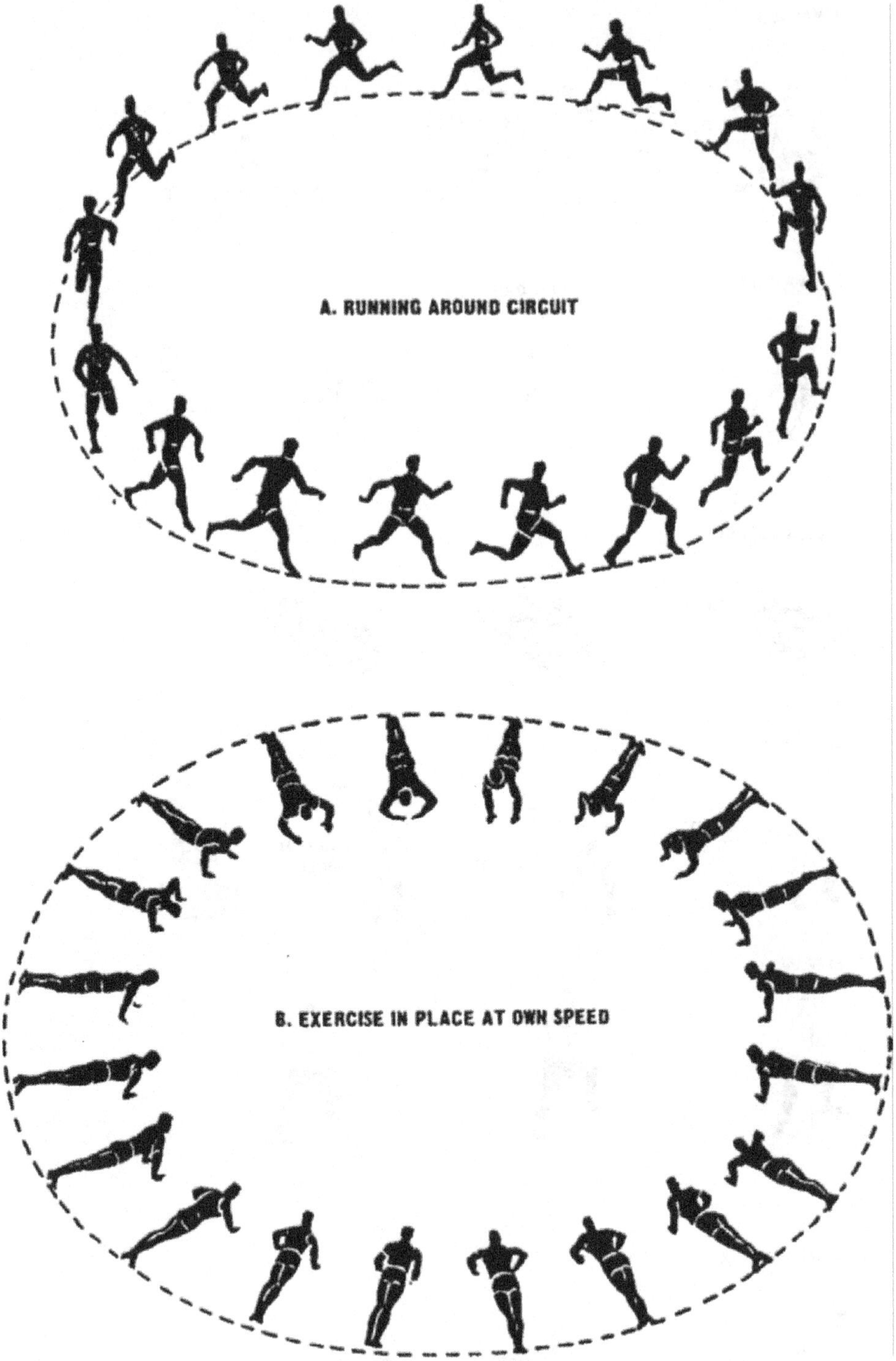

Figure 3-19. Circuit-IntervalTable.

Figure 3-20. Jumping.

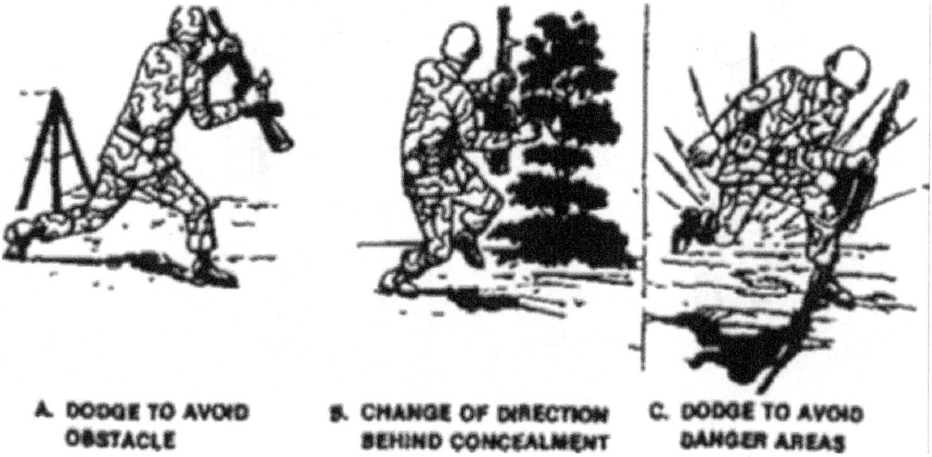

Figure 3-21.Dodging.

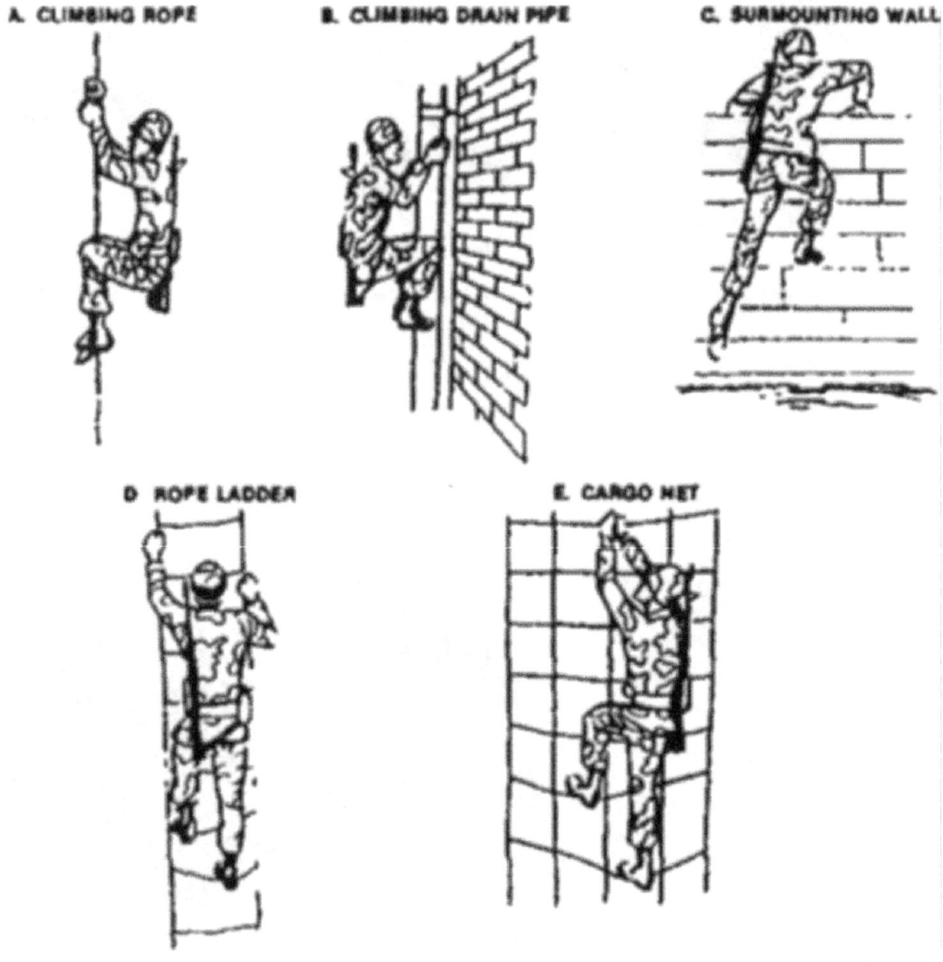

Figure 3-22.Climbing and Surmounting Vertical Objects.

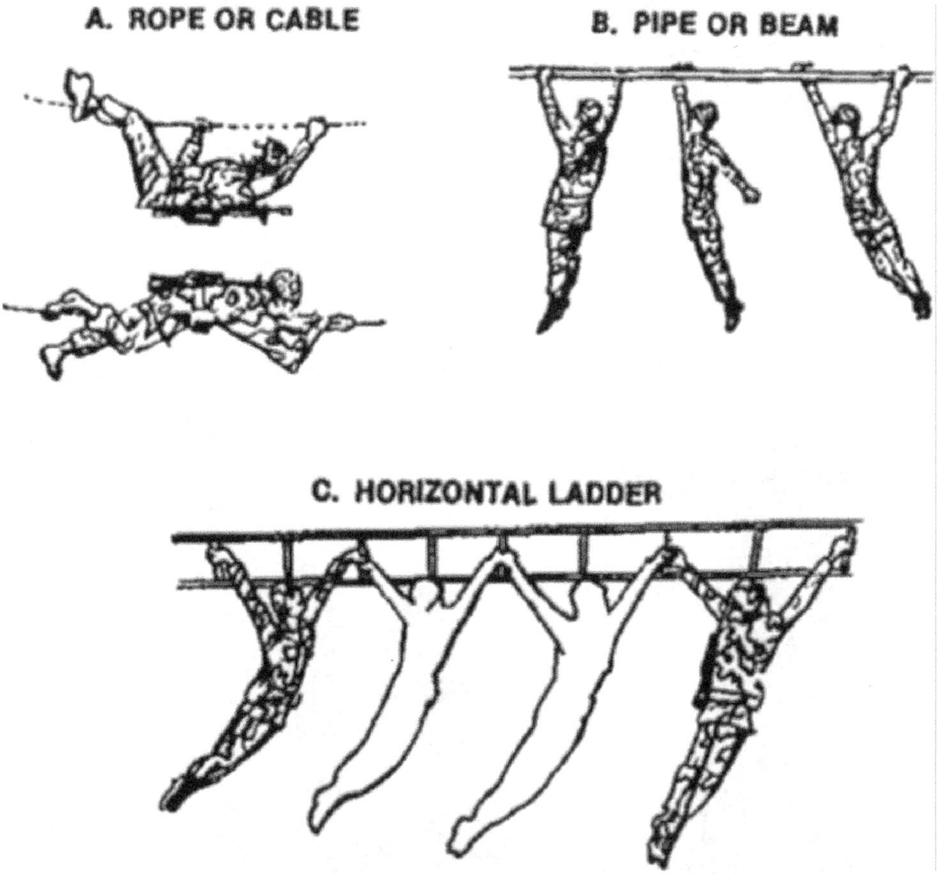

Figure 3-23. Traversing Horizontal Objects.

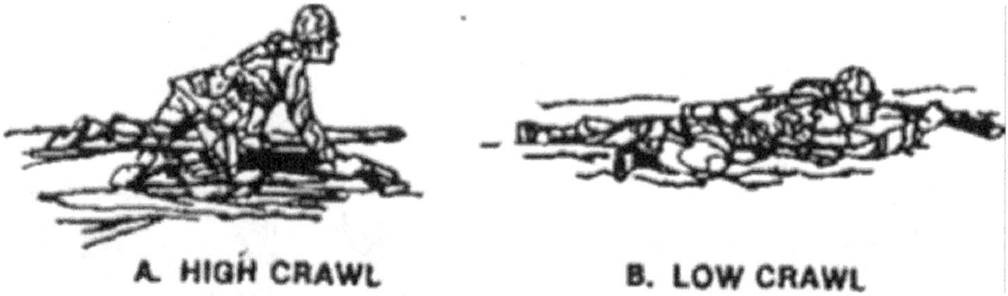

Figure 3-24.Crawling.

Figure 3-25.Throwing.

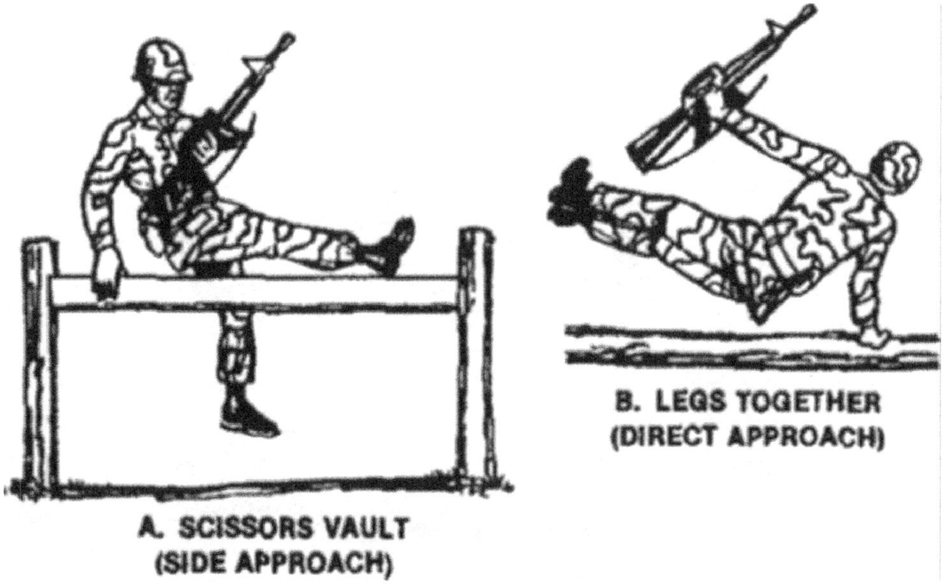

Figure 3-26. Vaulting

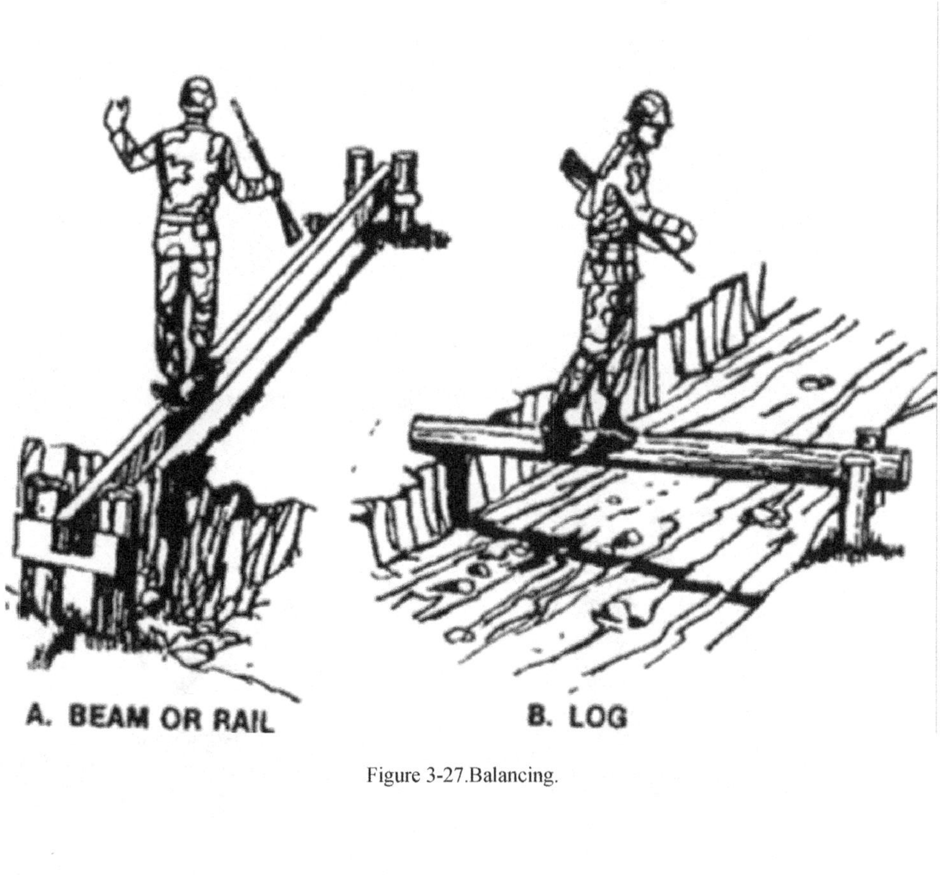

Figure 3-27.Balancing.

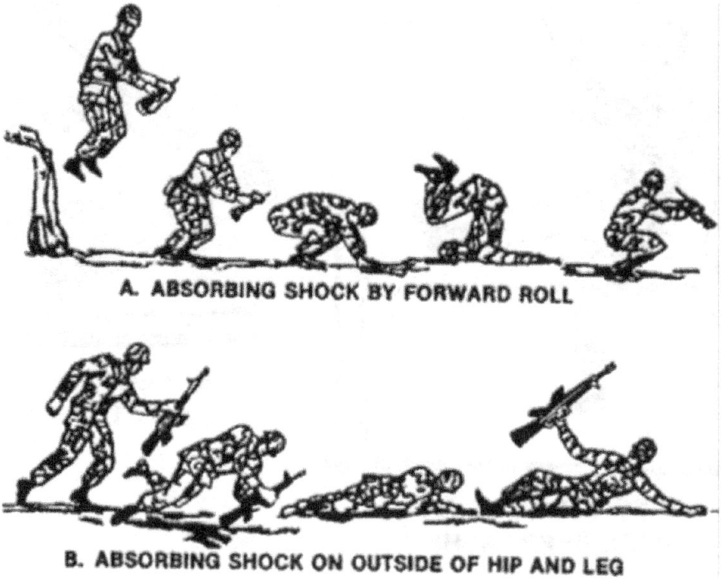

Figure 3-28.Falling.

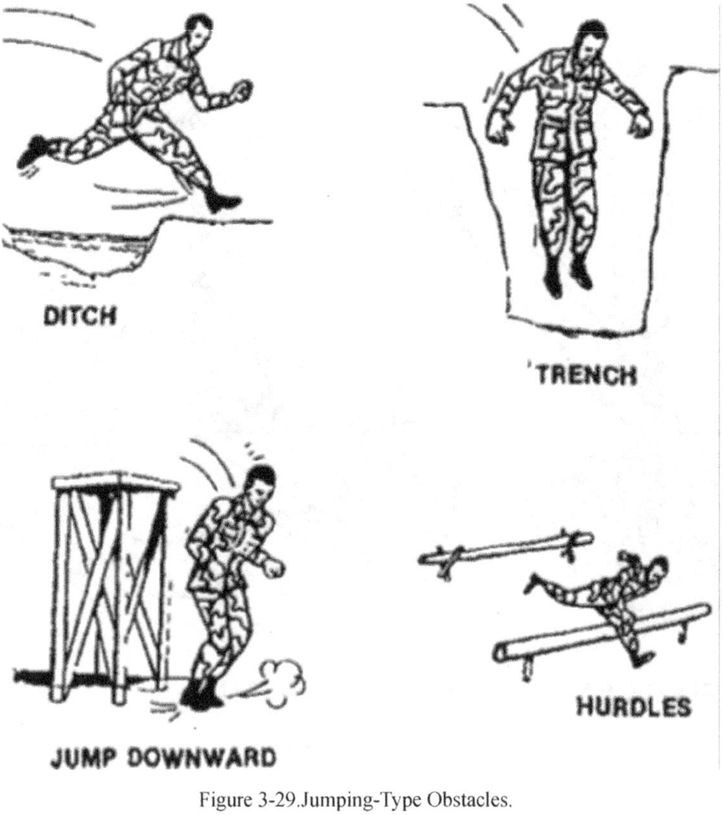

Figure 3-29.Jumping-Type Obstacles.

Figure 3-30.Dodging-Type Obstacles.

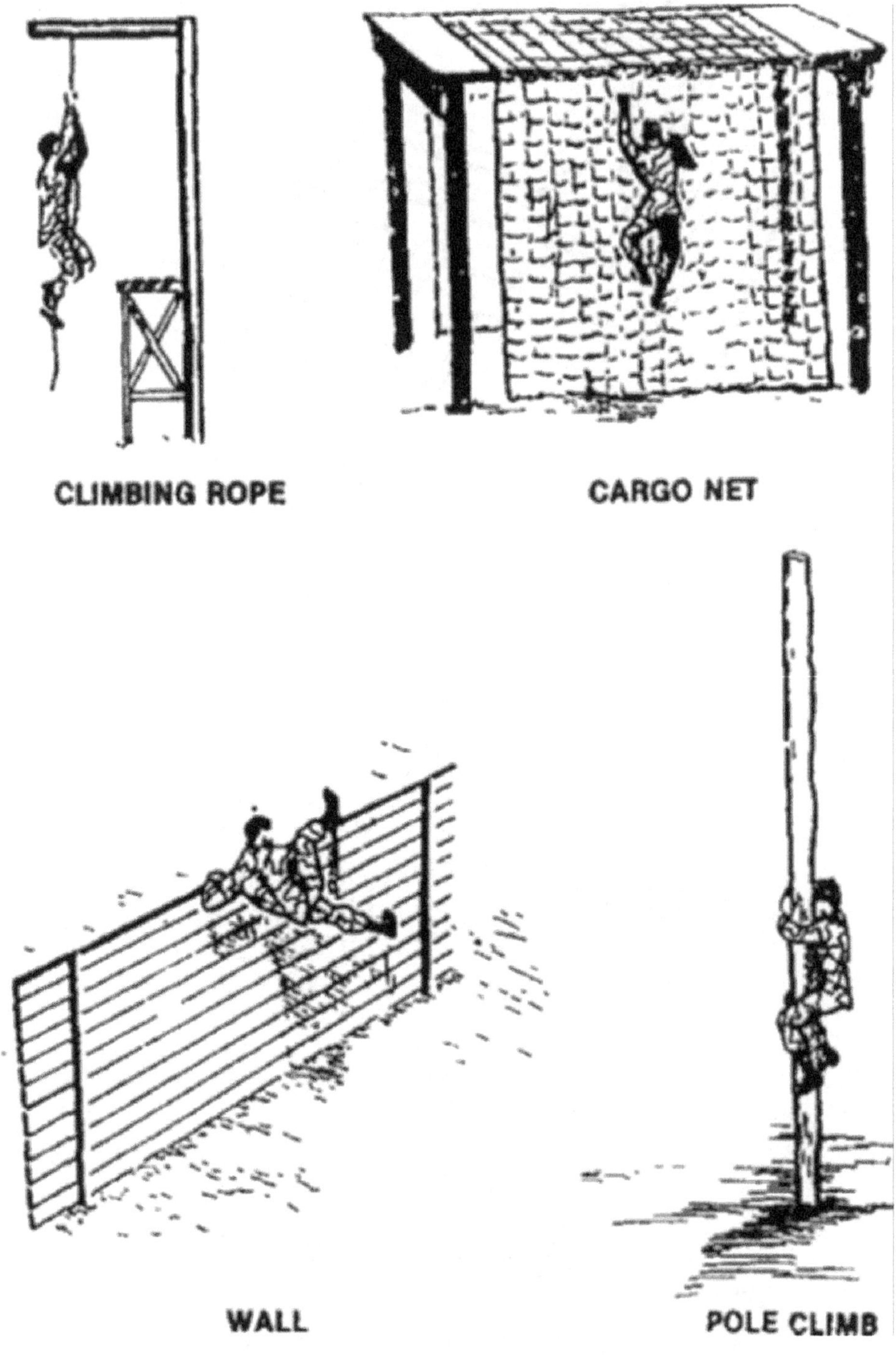

Figure 3-31. Vertical Climbing Obstacles.

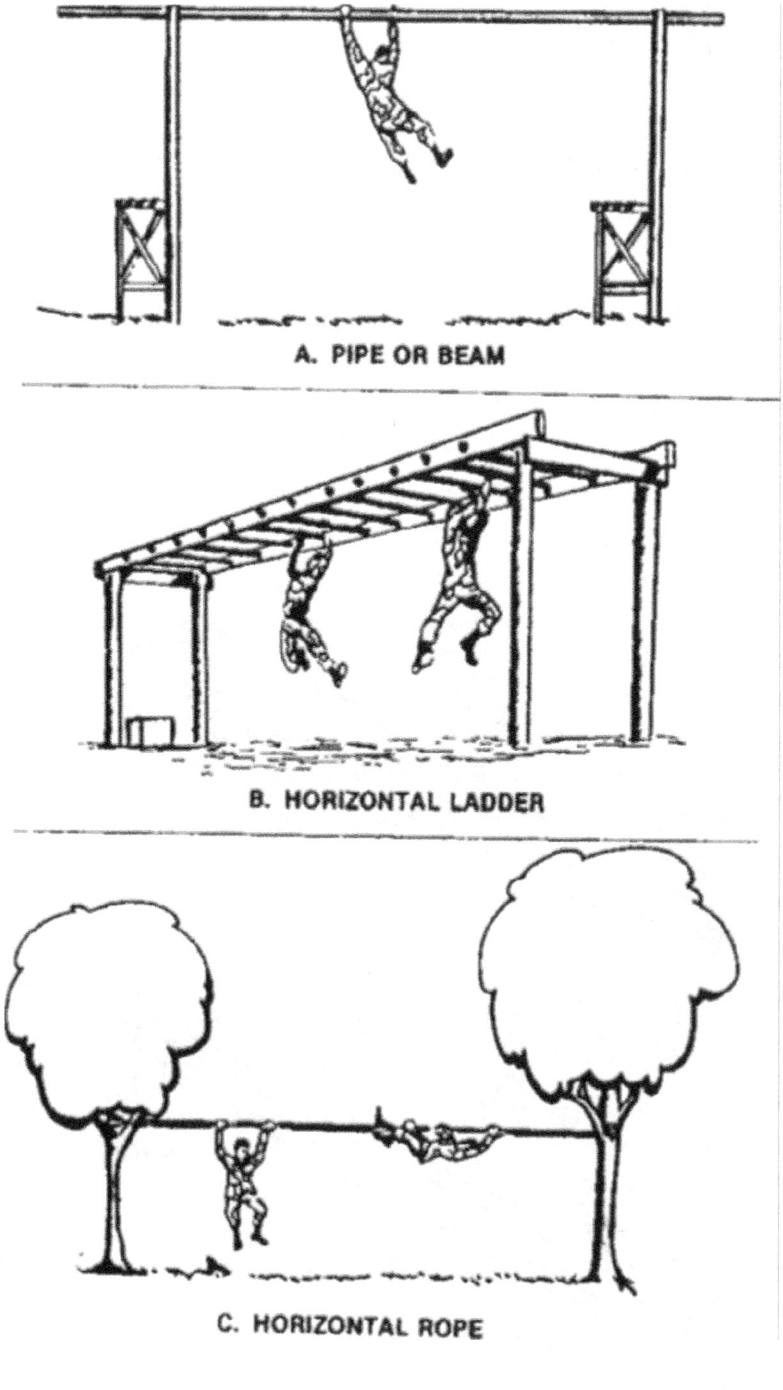

Figure 3-32.Horizontal Type Obstacles.

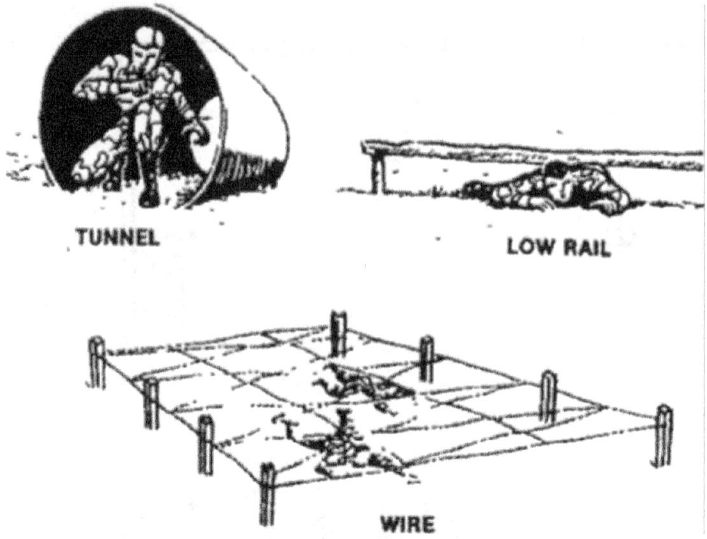

Figure 3-33.Crawling-Type Obstacles.

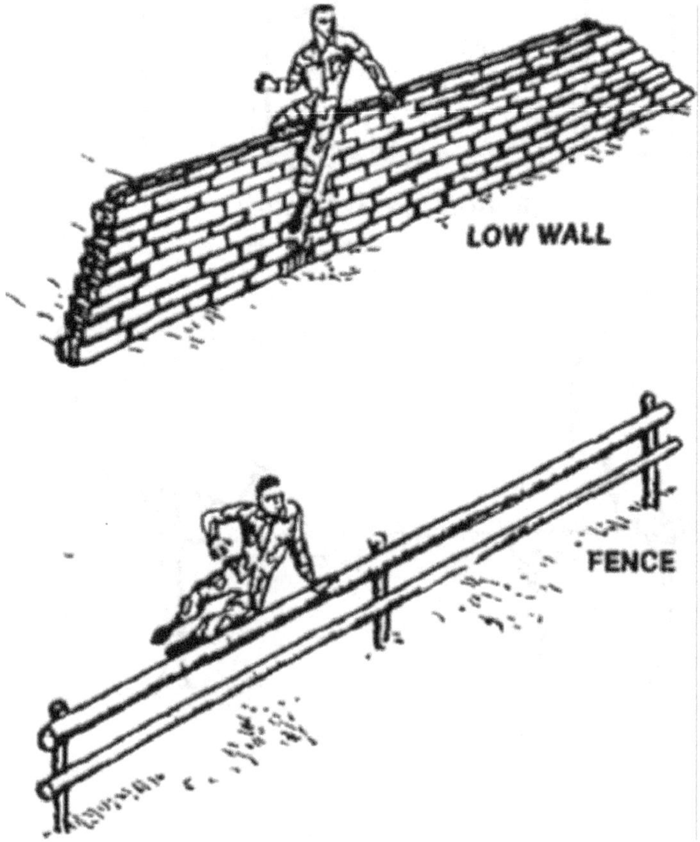

Figure 3-34. Vaulting Obstacles.

Figure 3-35.Balance-Type Obstacles.

A. THE BELLY BUSTER

B. REVERSE CLIMB

C. THE WEAVER

D. HIP—HIP

E. BALANCING LOGS

F. ISLAND HOPPER

Figure 3-36.Red Group.

Figure 3-37.White Group.

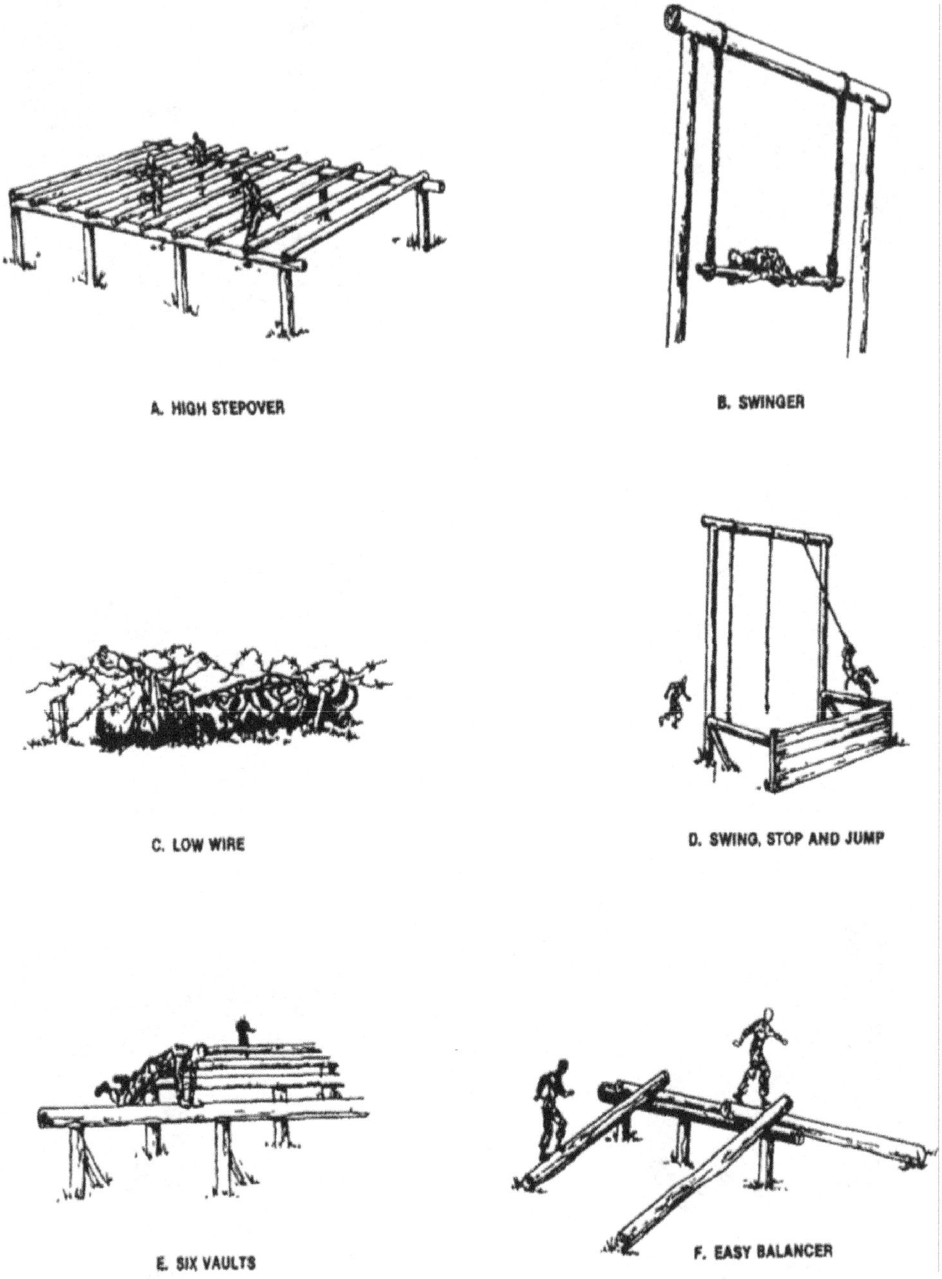

Figure 3-38.Blue Group.

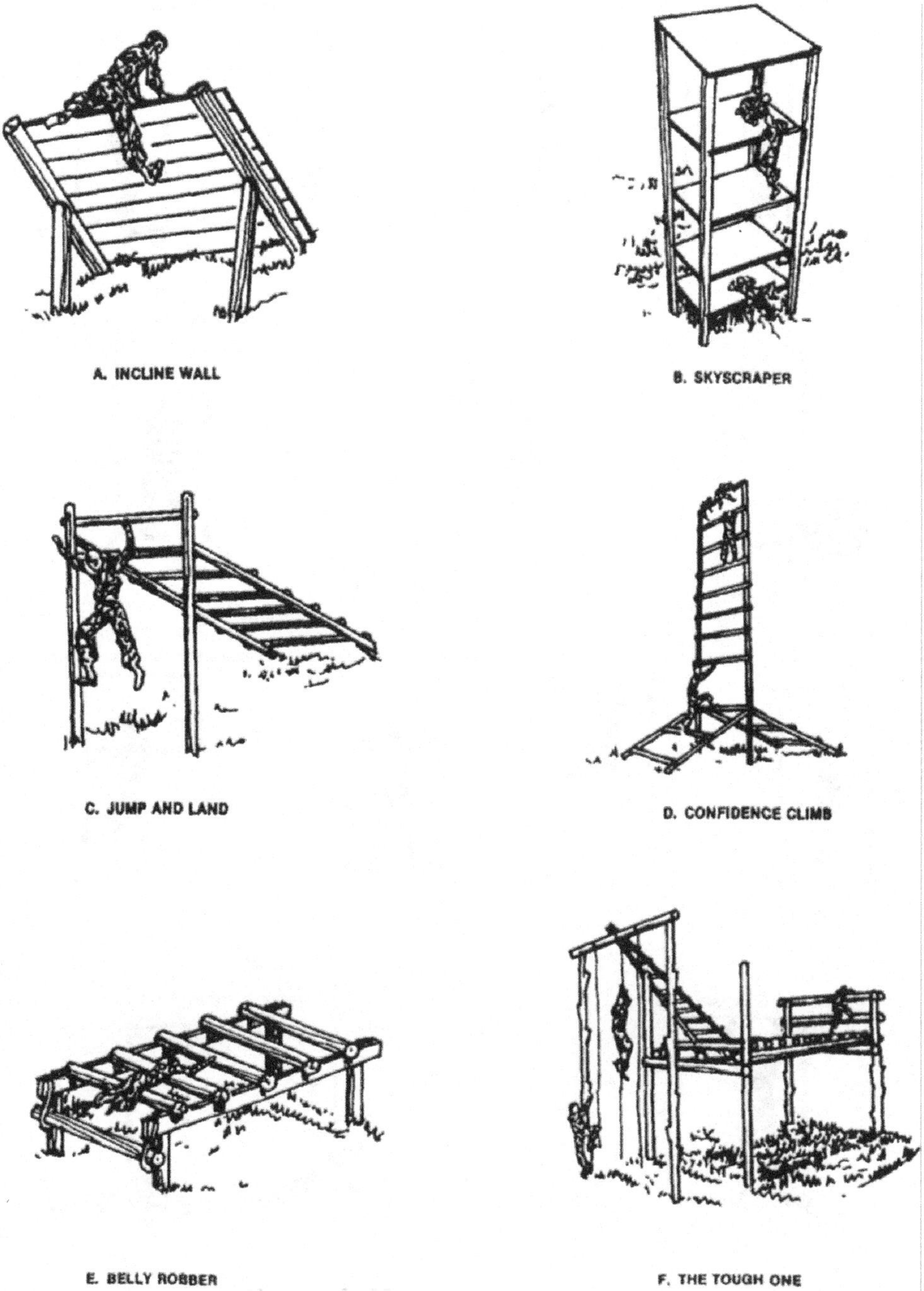

Figure 3-39.Black Group.

Figure 3-40. Bench Conditioning, Table I.

Figure 3-41. Bench Conditioning, Table II.

PROGRESSION GUIDE

AGE GROUP	LEVEL	EXERCISES 1	2	3	4	5	6
17	A	15	18	14	15	15	250
to	B	13	16	13	13	13	235
29	C	11	14	12	11	11	[illegible]
30	A	13	14	12	13	13	200
to	B	11	13	11	11	11	185
39	C	9	12	10	9	9	165
40	A	11	11	10	11	11	150
to	B	9	10	9	9	9	135
44	C	7	9	8	7	7	120
45	A	9	8	8	9	9	100
to	B	7	7	7	7	7	90
49	C	5	6	6	5	5	80
50	A	7	6	6	7	7	75
to	B	5	5	5	5	5	70
59	C	3	4	4	3	3	60
60	A	4	5	4	4	4	50
and	B	3	4	3	3	3	40
over	C	2	3	2	2	2	30
Minutes for each exercise		2	1	1	1	2	5

Table I. Progression Guide.

Table I. Exercises.

PROGRESSION GUIDE

AGE GROUP	LEVEL	EXERCISES 1	2	3	4	5	6
17	A	17	17	17	9	19	300
to	B	15	15	15	8	17	270
29	C	13	13	13	7	15	245
30	A	15	15	15	8	17	235
to	B	13	13	13	7	15	210
39	C	11	11	11	6	13	190
40	A	13	13	13	7	15	175
to	B	11	11	11	6	13	155
44	C	9	10	9	5	11	135
45	A	11	11	11	6	13	125
to	B	9	9	9	5	11	110
49	C	7	7	7	4	9	100
50	A	9	9	9	5	11	95
to	B	7	7	7	4	9	85
59	C	5	5	5	3	7	75
60	A	6	7	7	4	9	70
and	B	5	5	5	3	7	60
over	C	4	4	4	2	5	50
Minutes for each exercise		1	1	1	1 1/2	1 1/2	6

Table II. Progression Guide.

Table II. Exercises.

PROGRESSION GUIDE

AGE GROUP	LEVEL	EXERCISES 1	2	3	4	5	6
17	A	10	19	19	16	10	350
to	B	9	17	17	15	9	315
29	C	8	15	15	14	8	280
30	A	9	17	17	14	9	270
to	B	8	15	15	13	8	240
39	C	7	13	13	12	7	210
40	A	8	15	15	12	8	200
to	B	7	13	13	11	7	180
44	C	6	11	11	10	6	160
45	A	7	13	13	10	7	150
to	B	6	11	11	9	6	135
49	C	5	9	9	8	5	120
50	A	6	11	11	8	6	115
to	B	5	9	9	7	5	105
59	C	4	7	7	6	4	95
60	A	5	9	9	7	5	90
and	B	4	7	7	6	4	80
over	C	3	5	5	4	3	70
Minutes for each exercise		1 1/2	1	1	1 1/2	1	6

Table III. Progression Guide.

Table III. Exercises.

PROGRESSION GUIDE

AGE GROUP	LEVEL	EXERCISES 1	2	3	4	5	6
17	A	12	9	12	24	25	400
to	B	11	8	11	22	23	380
29	C	10	7	10	21	21	360
30	A	11	8	11	23	23	305
to	B	10	7	10	21	21	290
39	C	9	6	9	20	20	275
40	A	10	7	10	20	21	225
to	B	9	6	9	18	18	215
44	C	8	5	8	16	16	205
45	A	8	6	8	16	14	175
to	B	7	5	7	14	14	165
49	C	6	4	6	12	12	155
50	A	6	5	6	13	13	135
to	B	5	4	5	11	11	130
59	C	4	3	4	10	10	120
60	A	5	4	5	10	10	100
and	B	4	3	4	9	9	95
over	C	3	2	3	8	8	90
Minutes for each exercise		1	2	1	1	1	6

Table IV. Progression Guide.

Table IV. Exercises.

PROGRESSION GUIDE

AGE GROUP	LEVEL	EXERCISES 1	2	3	4	5	6
17	A	14	13	28	14	30	450
to	B	13	12	27	13	28	430
29	C	12	11	26	12	26	410
30	A	12	12	25	12	26	350
to	B	11	11	24	11	24	330
39	C	10	10	23	10	22	310
40	A	11	11	23	11	23	250
to	B	10	10	21	10	21	240
44	C	9	9	19	9	19	230
45	A	9	9	20	9	20	200
to	B	8	8	18	8	18	190
49	C	7	7	16	7	16	180
50	A	7	7	16	7	16	170
to	B	6	6	14	6	14	155
59	C	5	5	12	5	12	140
60	A	6	6	12	6	12	115
and	B	5	5	11	5	10	110
over	C	4	4	9	4	9	105
Minutes for each exercise		2	1	1	2	1	5

Table V. Progression Guide.

EXERCISE 1

EXERCISE 2

EXERCISE 3

EXERCISE 4

EXERCISE 5

EXERCISE 6

Table V. Exercises.

PROGRESSION GUIDE

AGE GROUP	LEVEL	EXERCISES 1	2	3	4	5	6
17	A	17	15	32	32	35	500
to	B	16	14	30	30	33	480
29	C	15	13	28	28	31	460
30	A	15	13	30	30	31	400
to	B	14	12	28	28	29	380
39	C	13	11	26	26	27	360
40	A	13	10	27	27	27	310
to	B	12	9	25	25	25	285
44	C	11	8	23	23	23	265
45	A	11	9	23	23	23	250
to	B	10	8	21	21	21	230
49	C	9	7	19	19	19	210
50	A	9	8	19	19	19	200
to	B	8	7	17	17	17	190
59	C	7	6	15	15	15	175
60	A	8	7	15	15	17	140
and	B	7	6	13	13	15	130
over	C	5	5	10	10	12	120
Minutes for each exercise		2	1	1	1	1	6

Table VI. Progression Guide.

Table VI. Exercises.

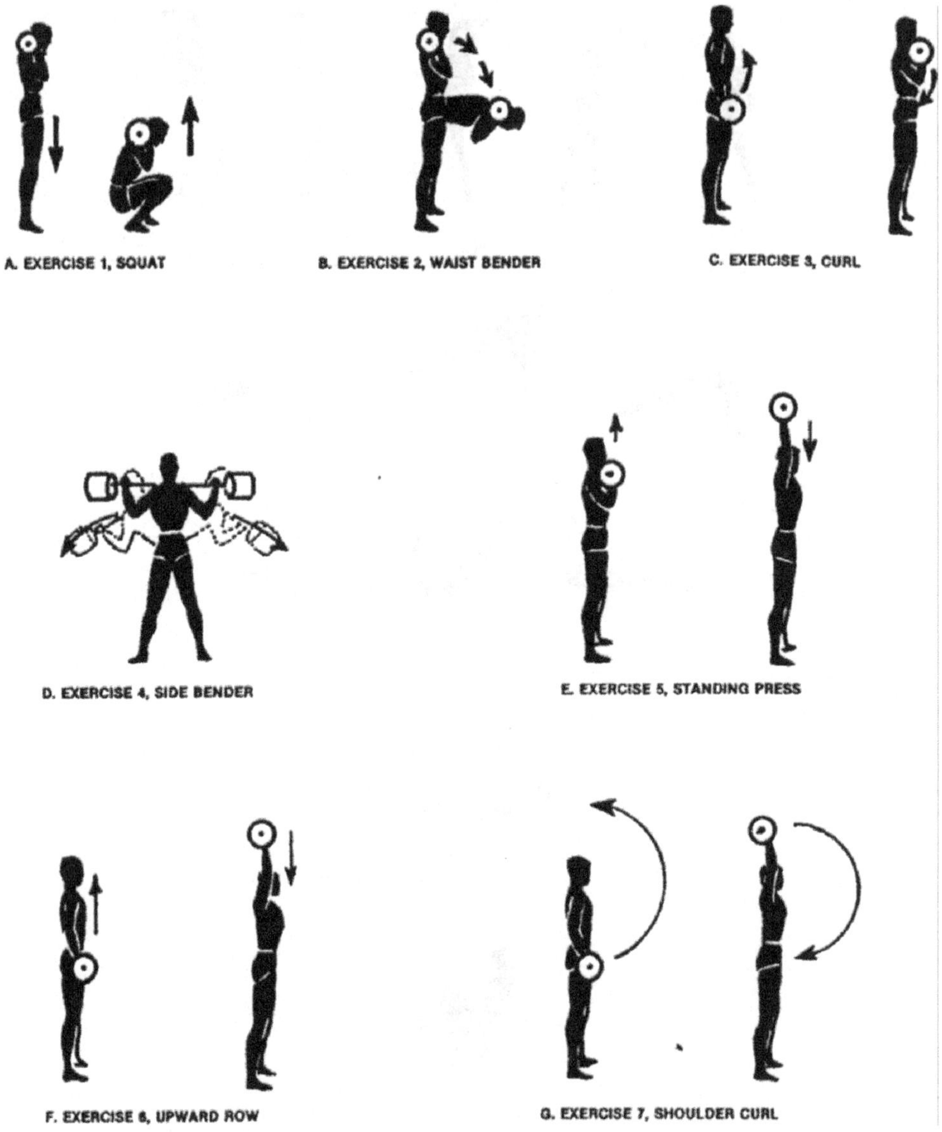

Figure 3-42. Weight Training.

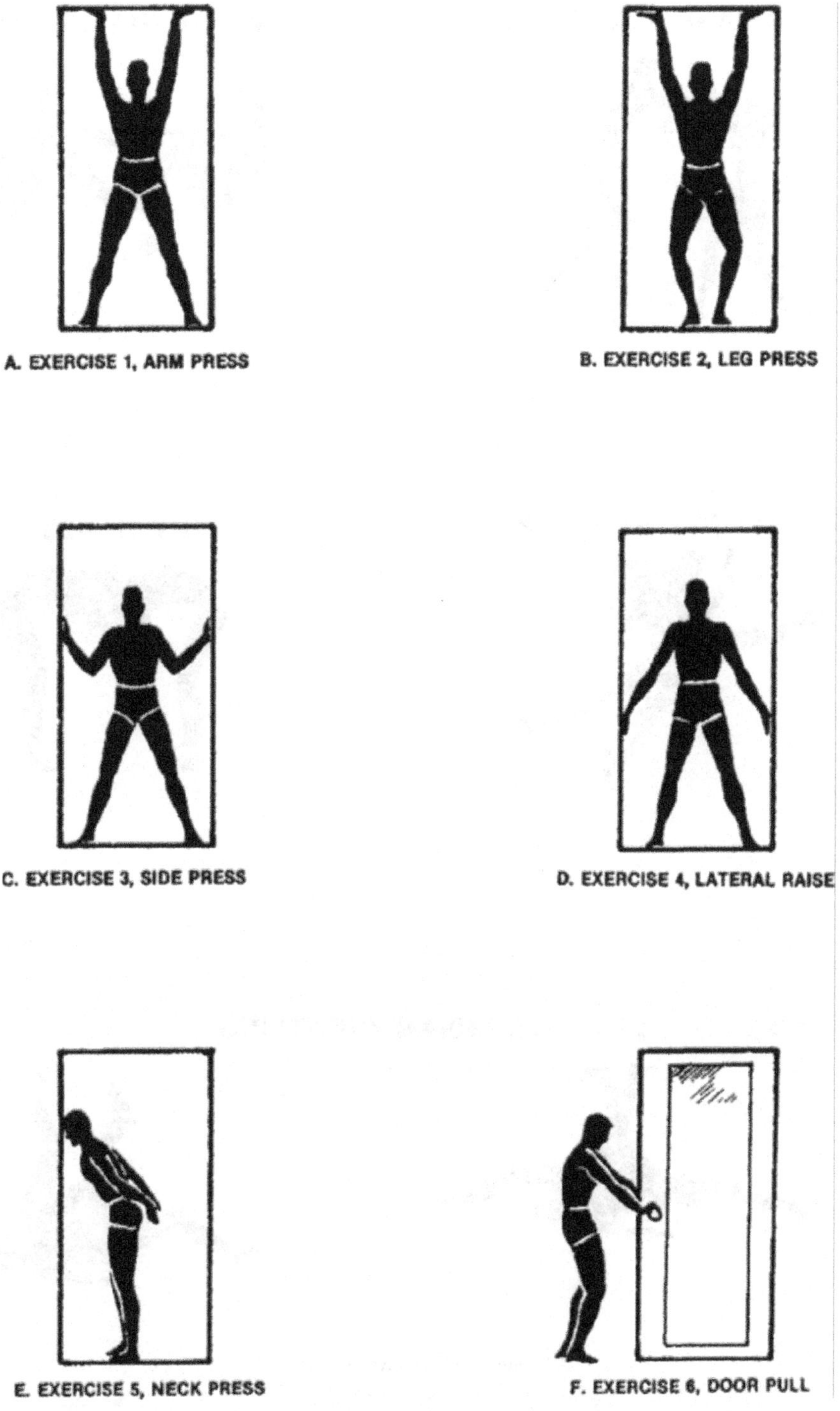

Figure 3-43. Isometric Exercises.

Figure 5-1.Combatives Table I.

Figure 5-2.Combatives Table II.

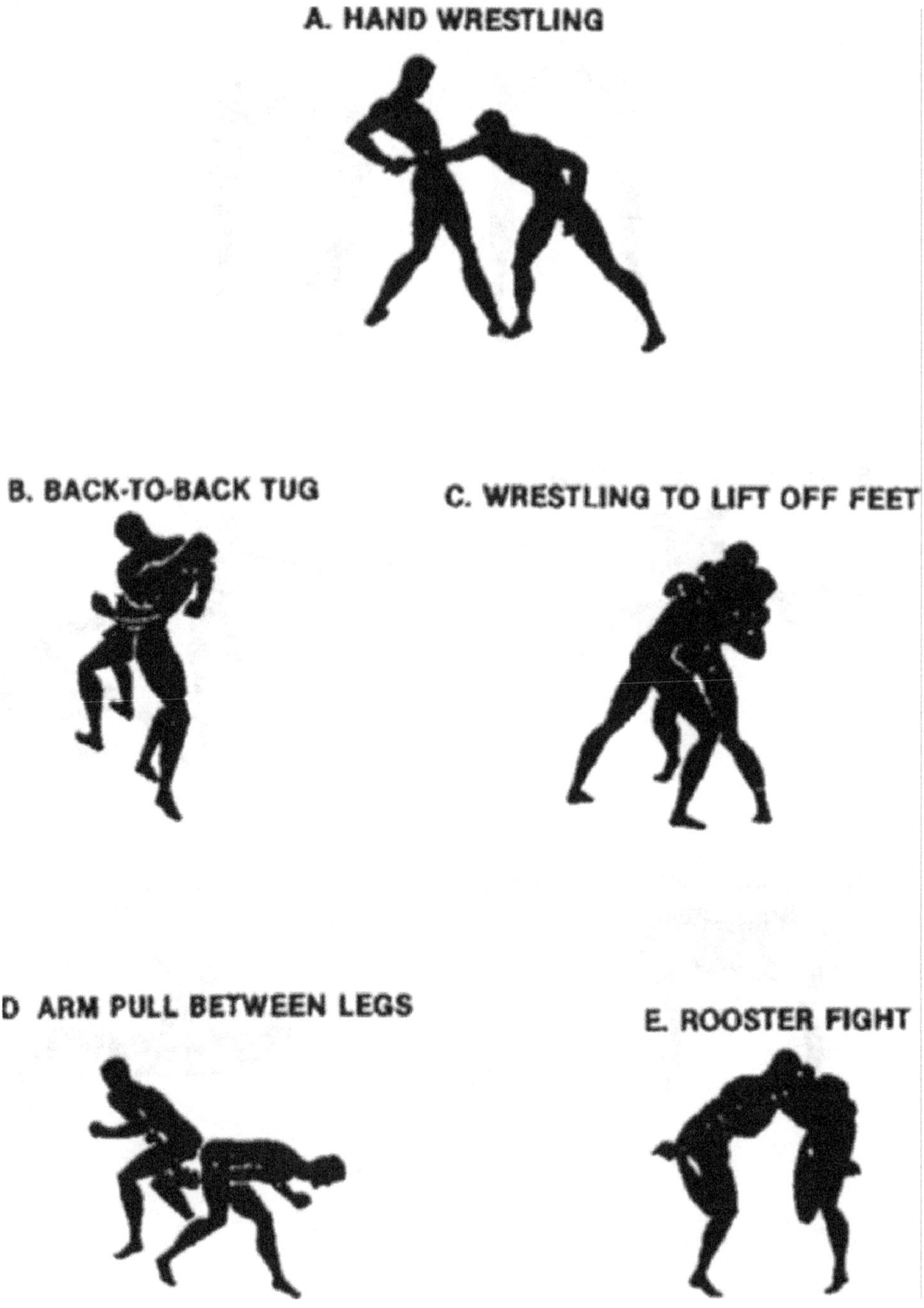

Figure. 5-3.Combatives Table III.

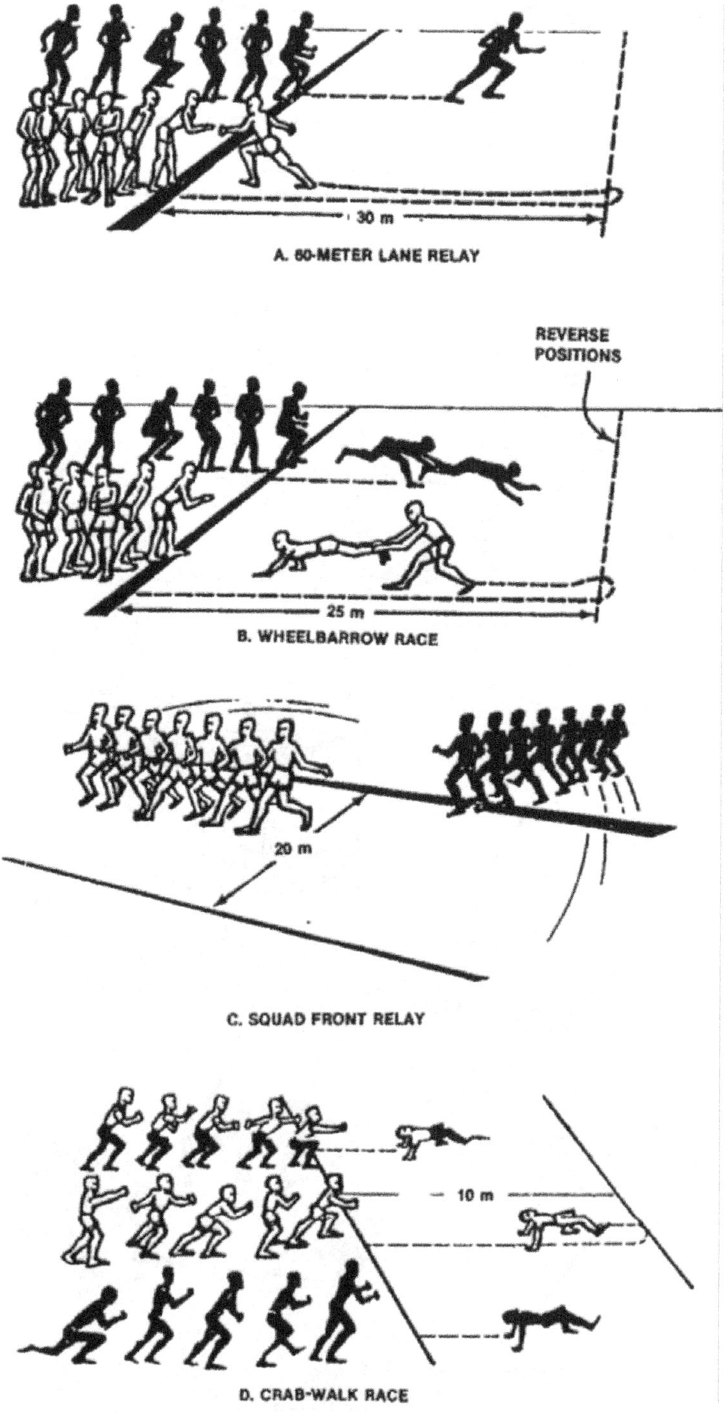

Figure 5-4.Relay Table I.

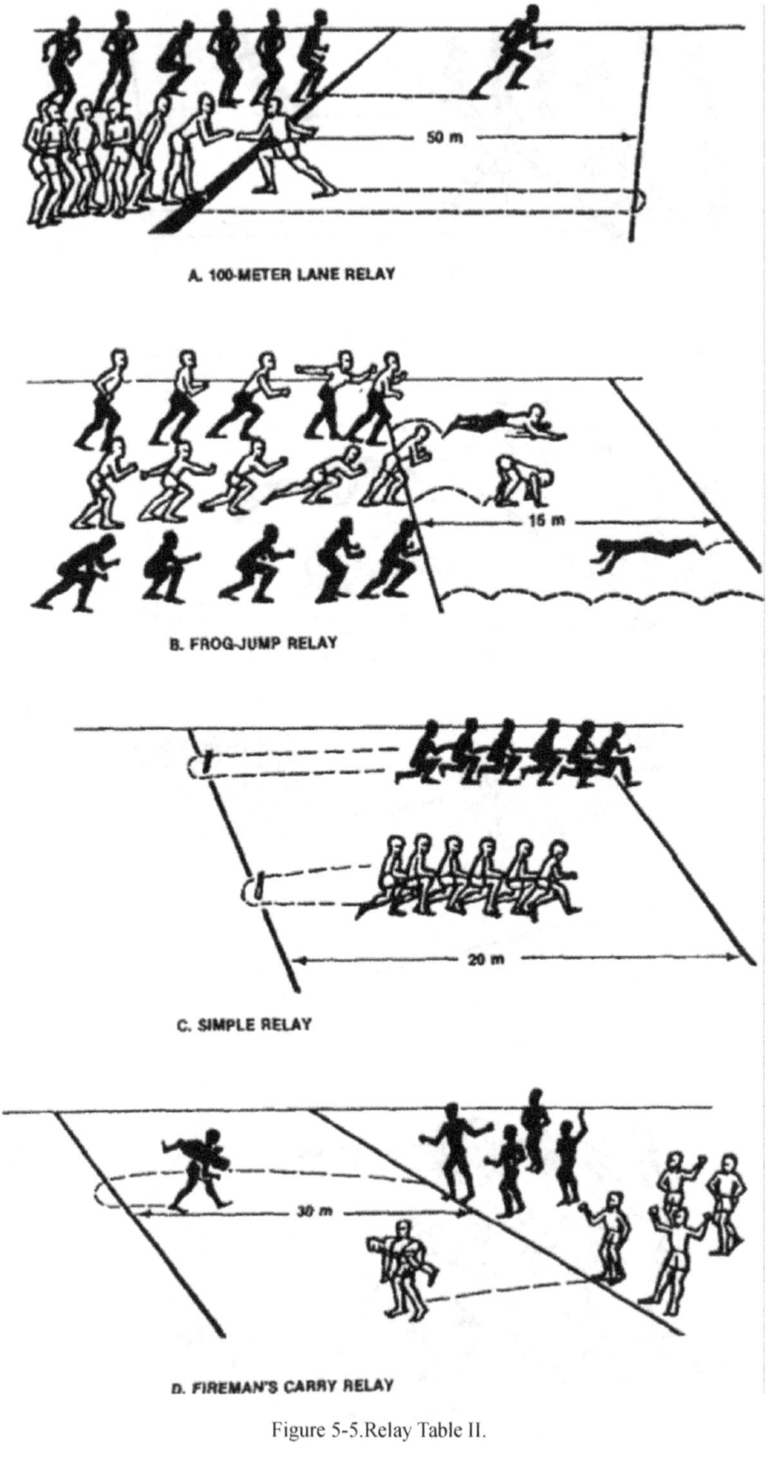

Figure 5-5.Relay Table II.

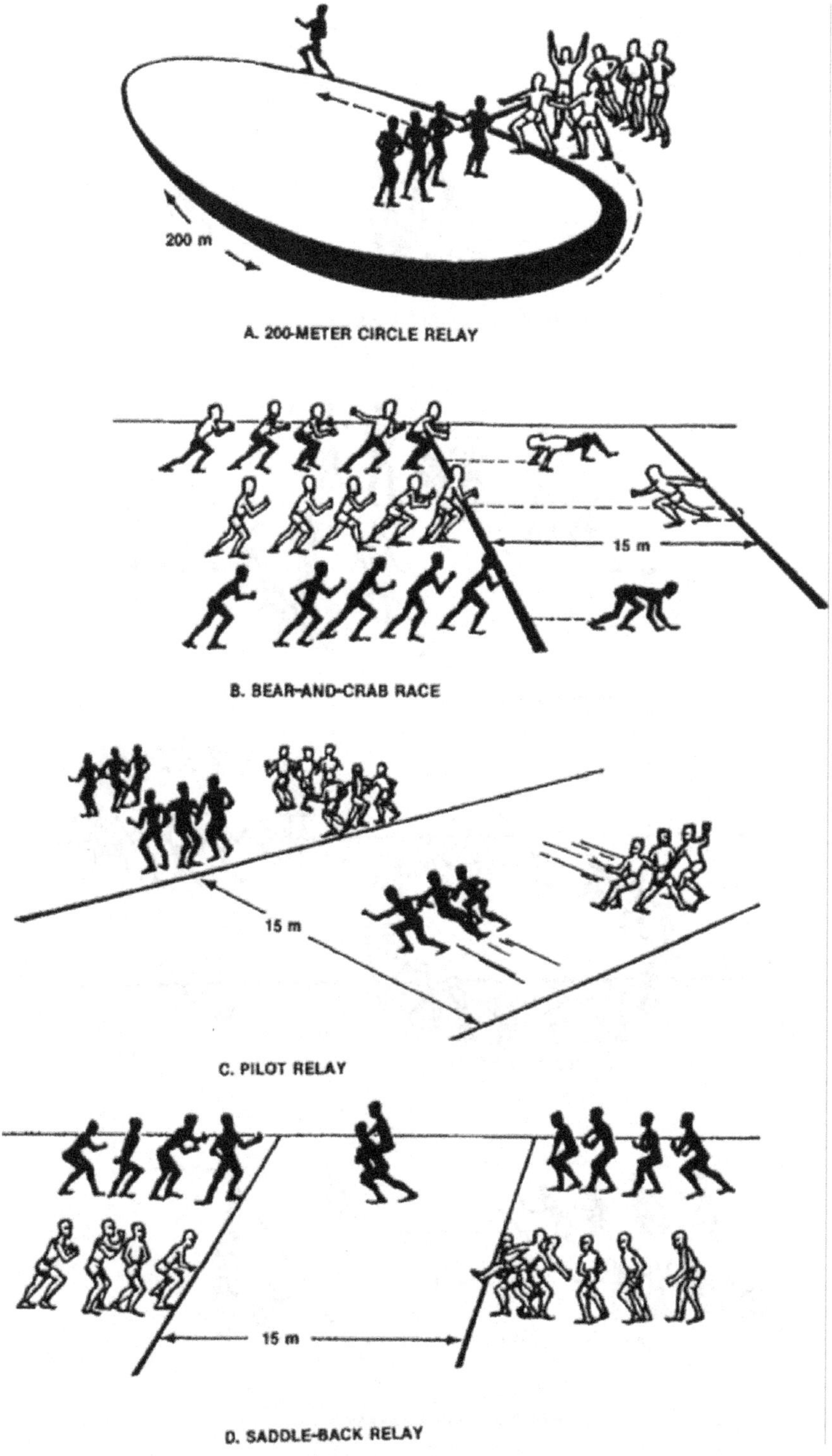

Figure 5-6.Relay Table III.

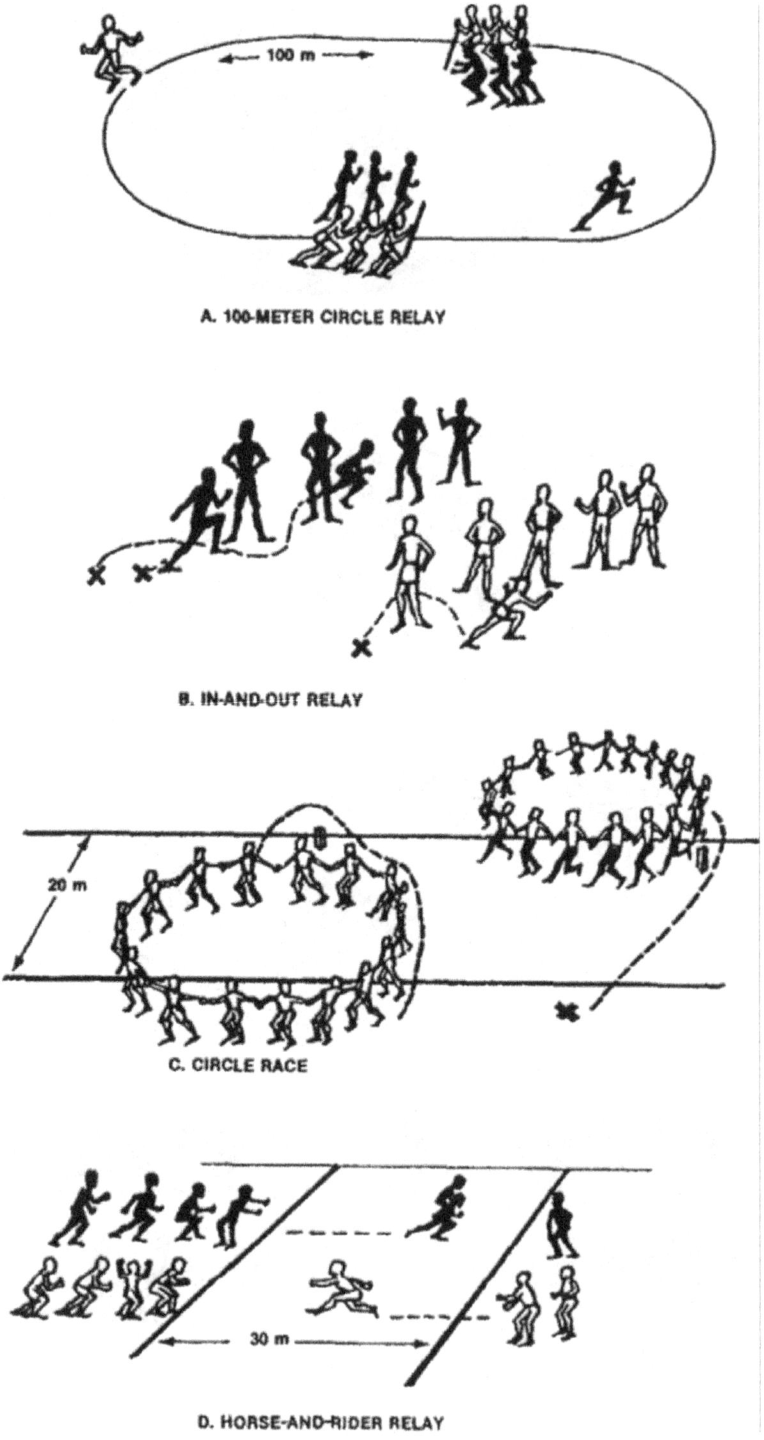

Figure 5-7.Relay Table IV.

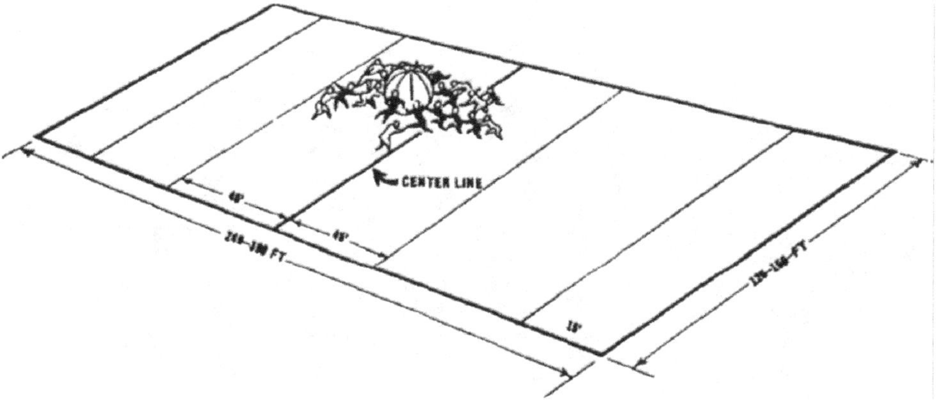

Figure 5-8.Pushball.

Figure 5-9.Line Rush.

Figure. 5-10. Human Tug-of-War.

Figure 5-11. Master of the Ring.

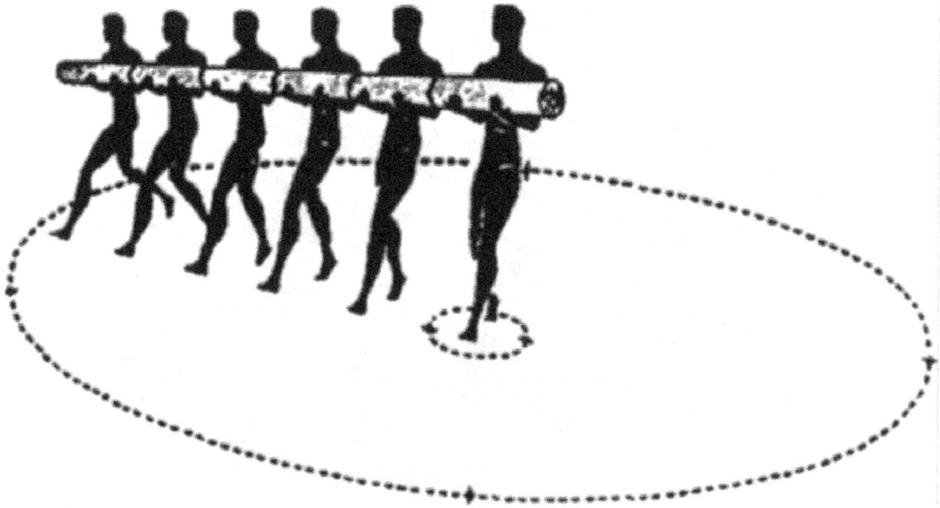

Figure 5-12.Log Pivot Circle.

Figure 5-13.Log-Rolling Race.

Figure 5-14. Prone Pushing Contest.

Figure 5-15. Shuttle Relay Race.

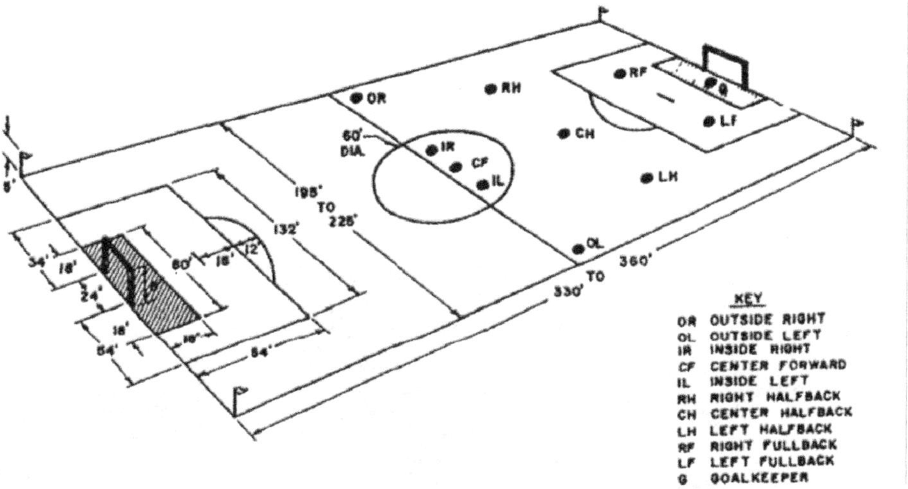

Figure 5-16. Soccer.

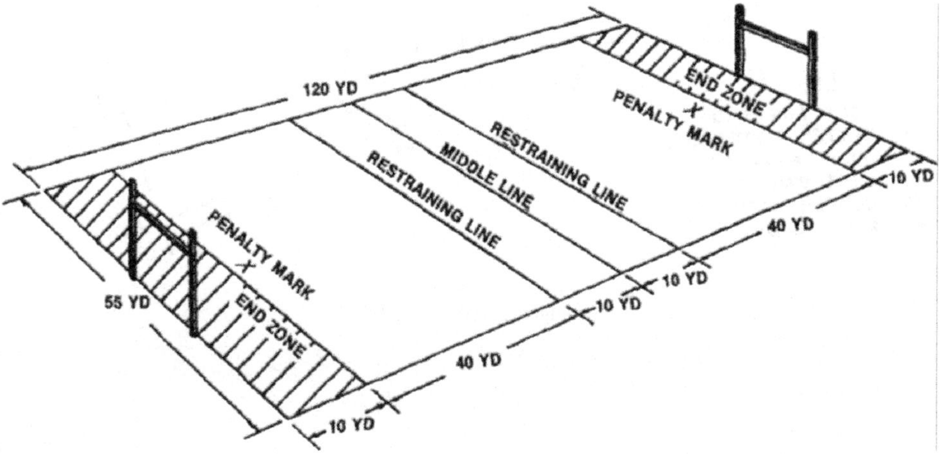

Figure 5-17.Speedball Field.

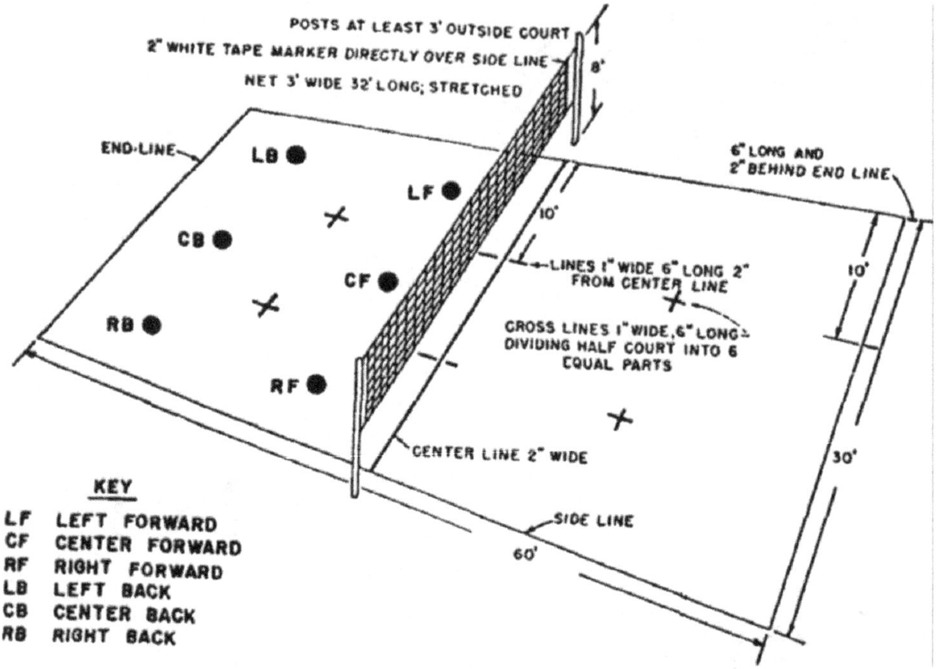

Figure 5-18. Volleyball.

PTS	UPHILL CLIMB	ROPE CLIMB	EVACUATION	FIRE AND MANEUVER	FORCED MARCH
100	1:20-BELOW	0:10-BELOW	0:25-BELOW	1:20-BELOW	26:00-BELOW
99	1:21-1:22	0:11		1 21-1:22	26:01-26:12
98	1:23-1:24	0:12	0:26	1:23-1:24	26:13-26:24
97	1:25-1:26	0:13		1:25-1:26	26:25-26:36
96	1:27-1:28	0:14	0:27	1:27-1:28	26:37-26:48
95	1:29-1:30	0:15		1:29-1:30	26:49-27:00
94	1:31-1:32	0:16	0:28	1:31-1:32	27:01-27:12
93	1:33-1:34	0:17		1 33-1.34	27:13-27:24
92	1:35-1:36	0:18	0:29	1:35-1.36	27:25-27:36
91	1:37-1:38	0:19		1:37-1:38	27:37-27:48
90	1:39-1:40	0:20	0:30	1:39-1:40	27:49-28:00
89	1:41-1:42	0:21	0:31	1:41-1:42	28:01-28:12
83	1:43-1:44	0:22	0:32	1:43-1:44	28:13-28:24
87	1:45-1:46	0:23	0:33	1:45-1:46	28:25-28:36
86	1:47-1:48	0:24	0:34	1.47-1:48	28:37-28:48
85	1:49-1:50	0:25	0:35	1.49-1:50	28:49-29:00
84	1:51-1:52	0:26	0:36	1:51-1:52	29:01-29:12
83	1:53-1:54	0:27	0.37	1:53-1:54	29:13-29:24
82	1:55-1:56	0:28	0:38	1:55-1:56	29:25-29:36
81	1:57-1:58	0:29	0:39	1:58-1:58	29:37-29:48
80	1:59-2.00	0:30	0:40	1:59-2:00	29:49-30:00
79	2.01-2:02	0:31	0:41	2:01-2:02	30:01-30:12
78	2:03-2:04	0:32	0:42	2:03-2.04	30:13-30:24
77	2:05-2:06	0.33	0:43	2:05-2:06	30:25-30:36
76	2:07-2:08	0.34	0.44	2:07-2:08	30:37-30:48
75	2:09-2:10	0:35	0:45	2:09-2:10	30:49-31:00
74	2:11-2:12	0:36	0:46	2:11-2 12	31.01-31:12
73	2:13-2:14	0:37	0.47	2.13-2:14	31:13-31:24
72	2:15-2:16	0:38	0:48	2.15-2:16	31:25-31:36
71	2:17-2:18	0:39	0:49	2:17-2:18	31:37-31:48
70	2:19-2:20	0:40	0:50	2:19-2.20	31:49-32.00
69	2.21-2:22	0:41	0:51	2:21-2:22	32:01-32:12
68	2:23-2:24	0:42	0:52	2:23-2:24	32:13-32:24
67	2:25-2.26	0:43	0:53	2:25-2:26	32:25-32:36
66	2:27-2:28	0:44	0:54	2.27-2:28	32.37-32:48
65	2:29-2:30	0.45	0.55	2.20-2 30	32.49-33:00

Figure 6-1.Physical Readiness Test Scoring Table for Male.

	EXAMPLE 1	EXAMPLE 2	EXAMPLE 3
Event #1	100 pts	100 pts	99 pts
Event #2	100 pts	99 pts	86 pts
Event #3	85 pts	75 pts (F)	80 pts
Event #4	90 pts	79 pts (F)	100 pts
Event #5	79 pts	85 pts	90 pts
Pts	454 pts	438 pts	455 pts
Final Score	79.0% (F)	69.0% (F)	91.0%
(F) = Failure			

Figure 6-2.Scoring Examples.

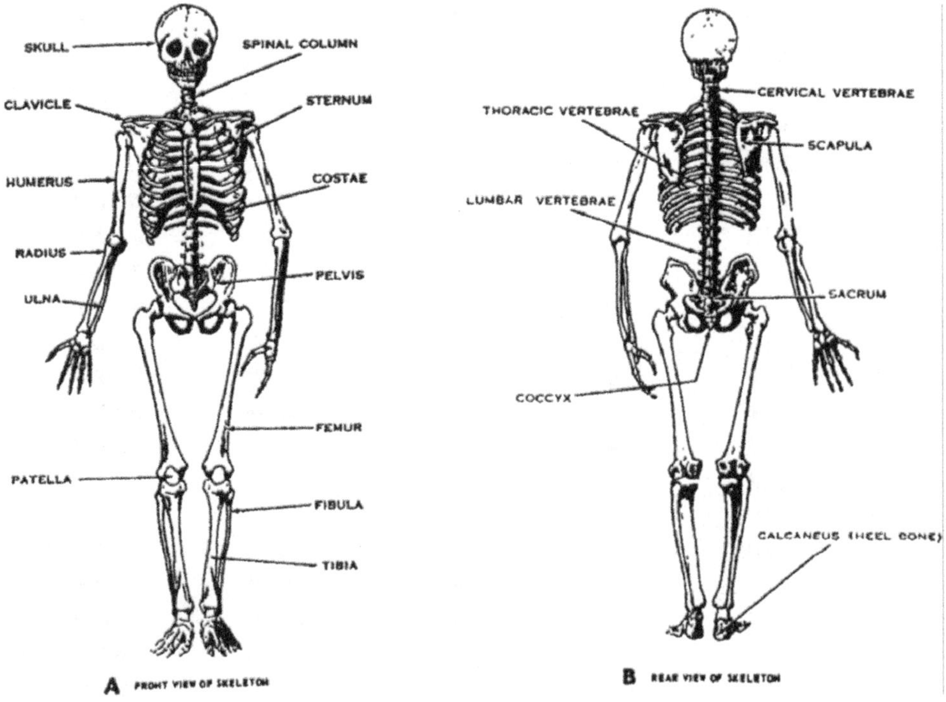

Figure 7-1.Front and Rear Views of Skeleton.

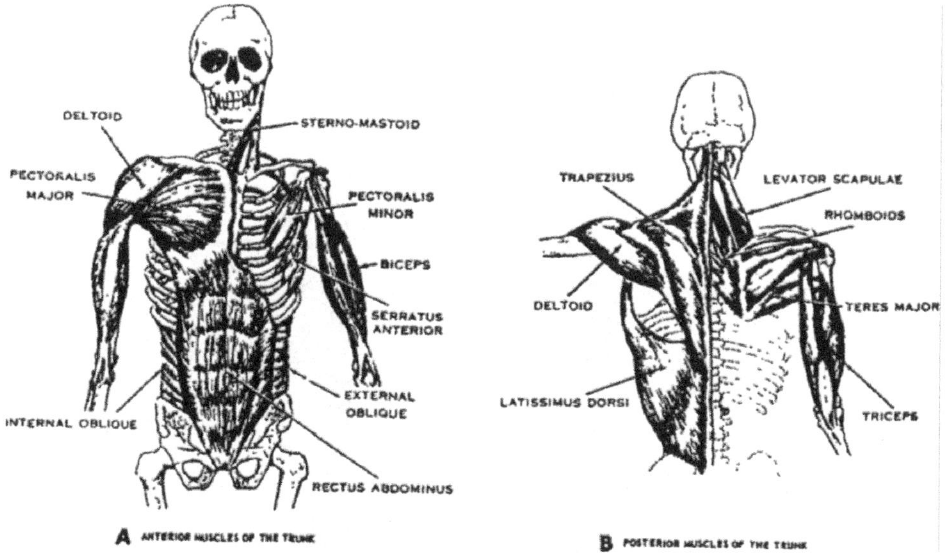

Figure 7-2.Muscles of the Trunk.

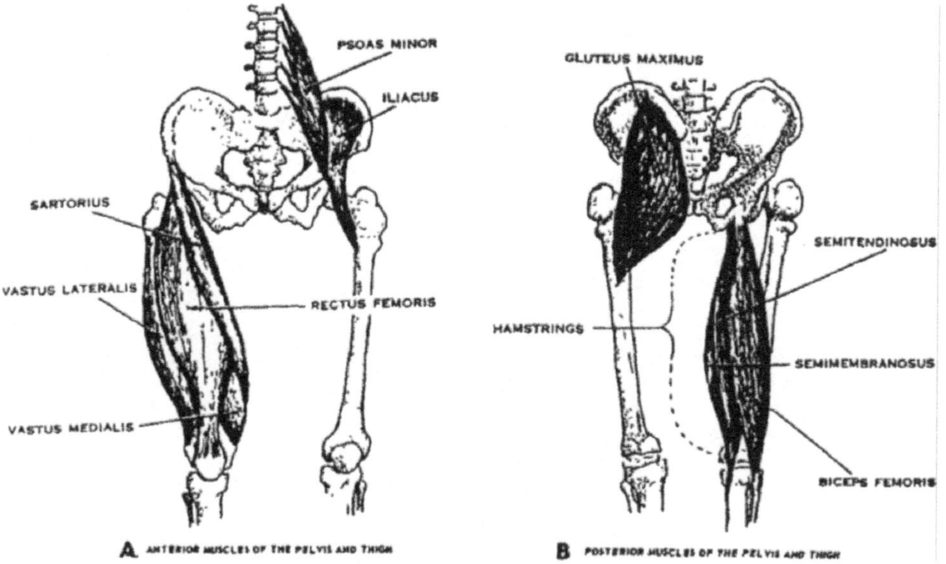

Figure 7-3.Anterior and Posterior Muscles of the Pelvisand Thigh.

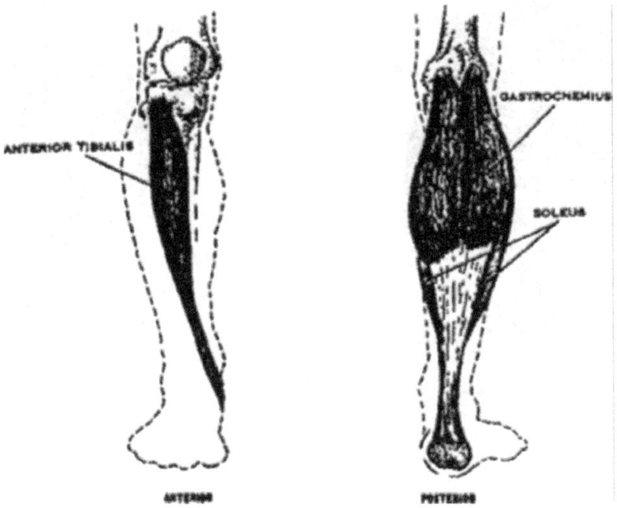

Figure 7-4. Muscles of Lower Leg.

Figure 7-5.Circulation of Blood from Heart to the Body.

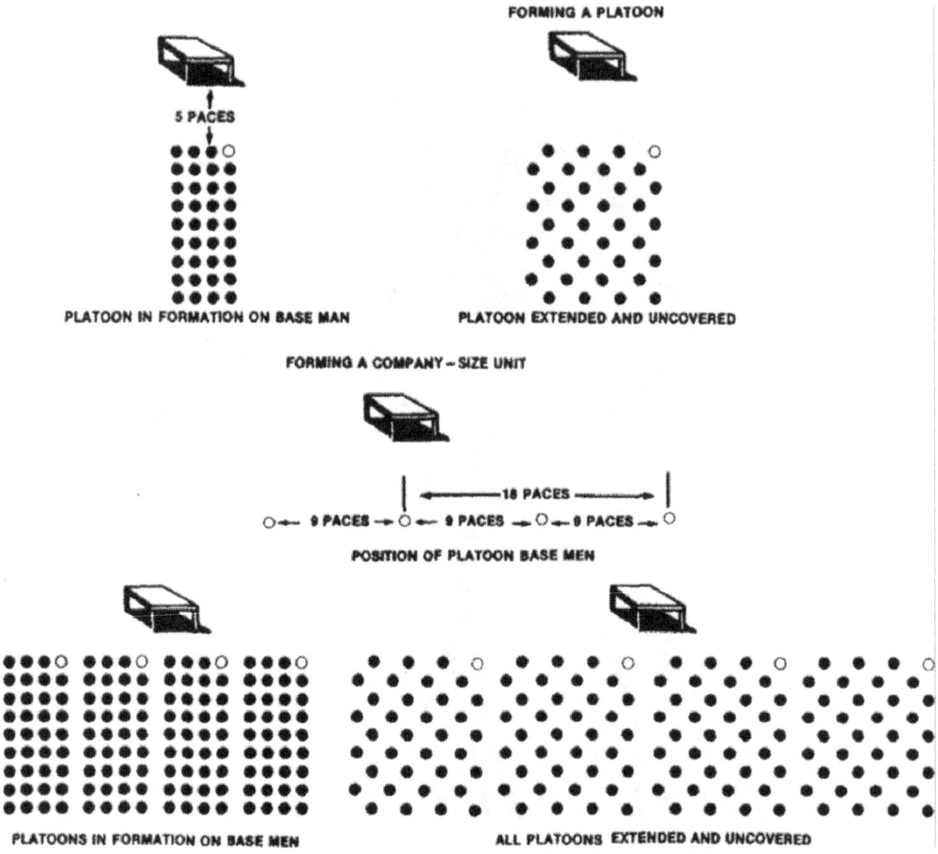

Figure A-l. Forming the Extended Rectangular Formation.

Figure A-2.Circle Formation.

*TC 90-1

Training Circular

No. 90-1

Headquarters

Department of the Army

Washington, DC, 19 May 2008

Training for Urban Operations

Contents

Distribution Restriction: Approved for public release; distribution is unlimited.

*This publication supersedes TC 90-1, 1 April 2002.

Figures

Tables

Chapter 1
Introduction

This chapter discusses definitions, training strategy, prerequisite training, individual task training, and collective task training for urban operations (UO). Military operations on urbanized terrain (MOUT) and UO differ in ways that reflect the changes in the way we engage in war today.

Section I. DEFINITIONS

This section defines and compares MOUT and UO.

> *Note:* The terms "urban operations" and "UO" replaces all instances of "military operations on urbanized terrain" and "MOUT."

MILITARY OPERATIONS ON URBANIZED TERRAIN

1-1. In the past, Army doctrine characterized urban areas as a type of specialized terrain, like jungles and mountains, and called operations in this type of terrain *military operations on urbanized terrain* (MOUT). Doctrine emphasized tactical warfighting tasks on this terrain, which was characterized by densely situated buildings and streets, and assumed the enemy consisted of conventional (Soviet-type) forces. Defeating or destroying those forces overrode all other considerations.

URBAN OPERATIONS

1-2. The concept of UO assumes that the urbanized terrain is populated, and that the populace must be a foremost consideration. UO recognizes that the populace can help either side, and that it requires a working infrastructure. Urban operations, whether against an insurgent or unconventional enemy, thus vastly differ from MOUT (Table 1-1).

Table 1-1. Comparison of MOUT and UO.

	MOUT	*UO*
Enemy force	Conventional, Soviet type	Insurgent or unconventional
Mission	Defeat or destroy enemy above all other considerations	Defeat the enemy Gain and retain the good will of the populace
Populace	Not considered	Important consideration

1-3. Although certain techniques and tactics are common to both types of operations, UO goes beyond MOUT, focusing not only on defeating the enemy, but also on gaining and retaining the good will of the population. UO consider the complex and dynamic interactions and relationships between the urban area's key components--the terrain (natural and manmade), the population, and the supporting infrastructure--as an overlapping and interdependent system of systems.

1-4. A single urban operation may include the full range of Army operations--offensive, defensive, stability and support (Appendix C)--that may be executed, either sequentially or simultaneously (FM 3-06, Chapter 1). Furthermore, to accomplish an objective, UO demand working with sister services, US government agencies, nongovernmental organizations, and regional and international organizations.

Section II. TRAINING STRATEGY

The UO training strategy is primarily organized by environment and phase. Training environments include live, virtual, and constructive (Figure 1-1), and are based on the unit's METL. Training phases are initial, intermediate, and culminating. Table 1-2 shows the relationship between the training environments and the first three training phases. (See also Appendix D, which discusses leader training.)

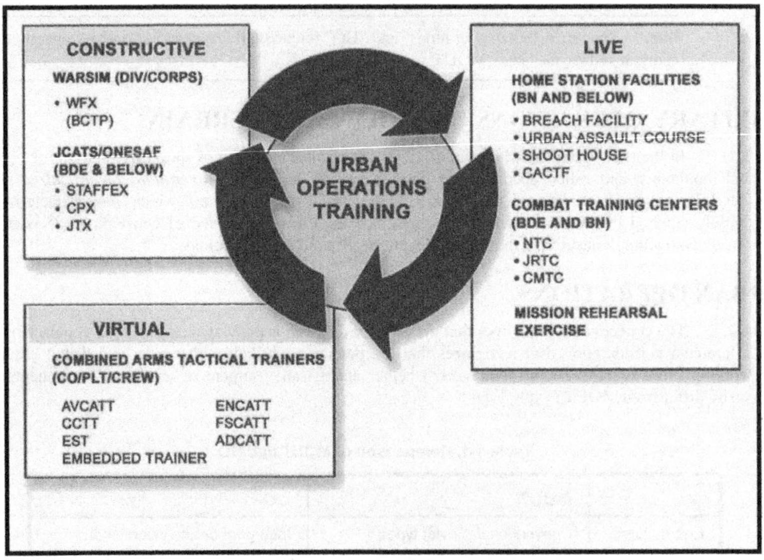

Figure 1-1. Urban Operations Training Support Strategy.

TRAINING PHASES

1-5. The three training phases are initial, intermediate, and culminating. Each of these includes some live, virtual, and constructive training.

INITIAL PHASE

1-6. This phase, which trains individual and collective tasks, is the first step in the building-block approach. It forms the basis of unit or collective training. For Soldiers and leaders, mastery of individual and crew skills is a prerequisite for UO training. The tasks train basic marksmanship skills and other Soldier tasks needed to perform any tactical operation. Included are training in a live environment (precision marksmanship) and virtual environment (Engagement Skills Trainer [EST], crew simulators, and CCTTs).

Table 1-2. Training environment/training phase matrix.

TRAINING PHASE / TRAINING ENVIRONMENT	INITIAL	INTERMED	CULMINATING
LIVE	Precision Marksmanship (Indiv / Sqd)	Breach Facility, Urban Assault Course, Shoot House (Team / Squad / Platoon)	CACTF, Platoon STX, Company STX, Battalion FTX Deployment CTC Rotation Mission Rehearsal Exercise
VIRTUAL	EST (Indiv / Sqd) Crew Simulators and CCTT (Crews Plt Ldrs, Co Cdrs)		
CONSTRUCTIVE			WARSIM (division/corps), which includes warfighting exercise (battle command training program). JCATS/OneSAF (brigade and below), staff exercises (STAFFEXs), command post exercises (CPXs), and joint training exercises (JTX)

Intermediate Phase

1-7. This phase of training is characterized by collective task training in varied conditions such as limited visibility; different entry techniques and points; incorporation of combatants, noncombatants, and OPFOR; use of MILES; and live-fire exercises. It is the vital link between individual tasks and mission accomplishment. It should be conducted at, but is not limited to, the *Breach Facility*, *Live-Fire Shoot House*, and *UAC*. (Section III discusses these facilities in detail.) Many types of organizations can use these facilities. Trainers *must* be certified before conducting demolition training IAW local standing operating procedure (SOP) or ACOM policy. AR 385-63 prescribes safety requirements. Except for Aviation, Armor, and other branches that have no practical use for these facilities, this phase must be completed to advance to the culminating phase.

Culminating Phase

1-8. The Combined Arms Collective Training Facility (CACTF, Figure 1-2) is where the culminating phase of urban operations training ends. This phase can include multiechelon, combined arms, and branch specific STX training may be conducted. Urban operations leader training occurs in leader professional development courses (Appendix D). Units have developed internal urban warfare training programs and TRADOC schools have assembled technical mobile training teams (MTTs) for marksmanship, combatives, and breaching techniques.

Education Phase

- Urban warfare reading/video list.
- Urban operations equipment.
- Review of Infantry doctrinal publications (FM 3-21.10, FM 3-06.11, for Infantry), and corresponding publications for other branches.
- Small-unit urban movement techniques and considerations.

Planning and Execution Phase

- Prepare company OPORD or WFF-related products.
- Brief platoon/company OPORD/WFF-related products to commander or staff representative.
- Participate as briefer in urban environment TEWT.

Sustainment Phase

- Participate in company/platoon TEWT or other urban operations event.
- Execute/facilitate a company or platoon TEWT.

TRAINING ENVIRONMENTS

1-9. Both live and virtual training environments are used to train UO.

Live Environment

1-10. The live training environment consists of home station training using the Urban Assault Course (UAC), Shoot House (SH), Breach Facility (BF), Combined Arms Collective Training Facility (CACTF), Combat Training Center (CTC) rotations, and any other live training that supports the UO training.

VIRTUAL ENVIRONMENT

1-11. The virtual training environment consists of simulators such as Close Combat Tactical Trainer (CCTT) or Engagement Skills Trainer (EST), which are used to train small-unit leaders and crews. The constructive environment consists of battle staff training for battalion and above staff and leaders such as Joint Conflict and Tactical Simulation (JCATS)/OneSAF.

CONSTRUCTIVE ENVIRONMENT

1-12. The constructive training environment consists of the following, which all occur in the culminating training phase:

- WARSIM (division/corps), which includes warfighting exercise (battle command training program).
- JCATS/OneSAF (brigade and below), staff exercises (STAFFEXs), command post exercises (CPXs), and joint training exercises (JTX)

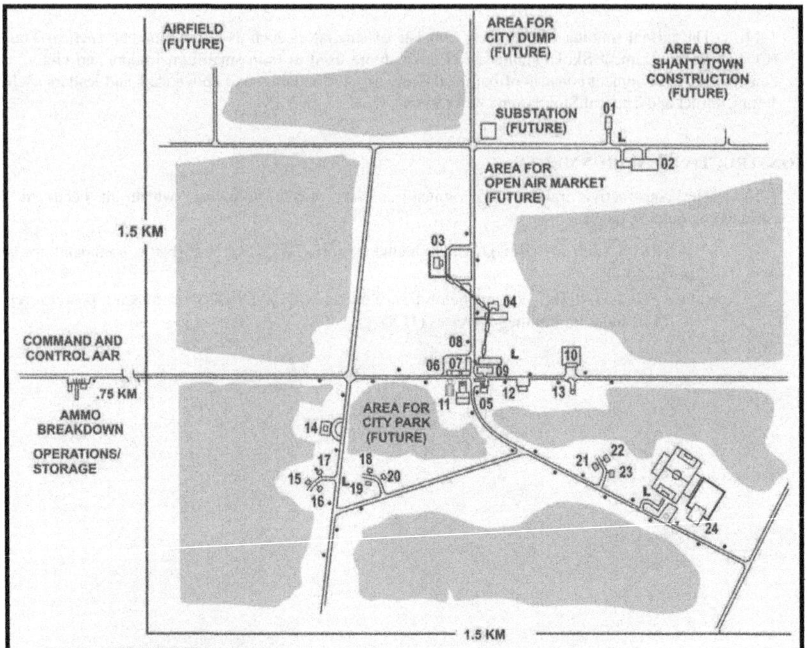

BLDG #	SIZE AND TYPE
01	1 STORY WAREHOUSE
02	1 STORY WAREHOUSE
03	3 STORY OFFICE (EMBASSY)
04	2 STORY OFFICE
05	1 STORY SERVICE STATION
06	1 STORY BUSINESS
07	1 STORY BUSINESS
08	1 STORY BUSINESS
09	3 STORY HOTEL
10	1 STORY POLICE STATION/JAIL
11	CHURCH AND CEMETARY
12	1 STORY BUSINESS
13	1 STORY BANK
14	2 STORY TOWNHOUSE
15	1 STORY RESIDENT
16	1 STORY RESIDENT
17	1 STORY RESIDENT
18	1 STORY RESIDENT
19	1 STORY RESIDENT
20	1 STORY RESIDENT
21	1 STORY RESIDENT
22	1 STORY RESIDENT
23	1 STORY RESIDENT
24	2 STORY SCHOOL

LEGEND

- FENCE WALL
- WALL
- TUNNEL SYSTEM
- TREE LINE
- • UTILITY POLE
- L LATRINE

SCALE: 1" = 500'

Figure 1-2. Combined Arms Collective Training Facility.

MISSION-ESSENTIAL TASK LIST

1-13. FMs 7-0 and 7-1 explain how to determine a unit's mission-essential task list (METL) and battle tasks, and how to plan, resource, and execute training events. Because training facilities are limited, leaders must use all available time and resources for training and maintaining combat readiness.

- System-specific FMs and TCs teach doctrine, conduct of operations, and employment of weapon systems.
- Tactics, techniques, and procedures (TTP) are lessons learned from training and combat.

1-14. The standards for evaluating training are accessible at the Digital Training Management System (DTMS), https://dtms.army.mil/DTMS.

BATTALION INTEGRATED TRAINING PLAN

1-15. The commander uses the battalion integrated training plan to set goals for training density. He focuses on combined arms operations, emphasizing dismounted Infantry maneuver. Tanks, BFVs, Strykers, attack aviation, and artillery normally provide supporting fires for the dismounted elements. UO training in the CACTF, Shoot House, and Breach Facility should be conducted semiannually and quarterly for the UAC. Army training calendars are divided into two cycles of training: *mission* and *support*. Figure 1-3 and Figure 1-4 show how the urban training strategy and facility usage might be programmed into unit cycles.

RISK ASSESSMENT

1-16. Risk assessment means making operations safe without compromising the level of realistic training (FM 7-1). Commanders continuously assess the risk of training conditions to prevent unnecessary loss of Soldiers and equipment. The degree of risk depends on conditions. Just as this principle applies to other training and warfighting, it also applies to UO. In a training situation, for example, commanders determine whether Soldiers and leaders have conducted this type of training before, whether conducting initial training at night or during limited visibility, and whether Soldiers are tired or exhausted. Risk assessment ensures UO training is not only tough and realistic but also safe. Training enhancers, such as live fire, blank ammunition, simulators, fires, smoke, and other pyrotechnics, must be fully integrated into the training to achieve the training objectives. Soldiers are required to wear earplugs, protective vests, and protective eyewear in order to protect themselves from noise and debris.

1-17. The commander considers risk and safety in urban operations training situations such as--.

- Soldiers engaging in close combat urban operations need to be safety conscious when using blanks, booby traps, and practice grenades. They may be unaware of the exact location of the OPFOR and may be susceptible to OPFOR flash suppressors, booby traps, and practice hand grenades.
- Soldiers should use ladders to enter buildings at upper stories of buildings to prevent unnecessary risks.

1-18. The commander is the safety officer and is responsible for ensuring that his unit, both leaders and Soldiers, comply with safety regulations and the units tactical SOP. All leaders must--.

- Use mission, enemy, terrain, troops, time available, and civilians (METT-TC) factors to identify risks.
- Assess possible losses and their costs.
- Select and develop risk-reduction measures.
- Implement controls by integrating them into plans and orders, SOPs, training performance standards, and rehearsals.
- Supervise and enforce risk reduction measures and safety standards at all times.

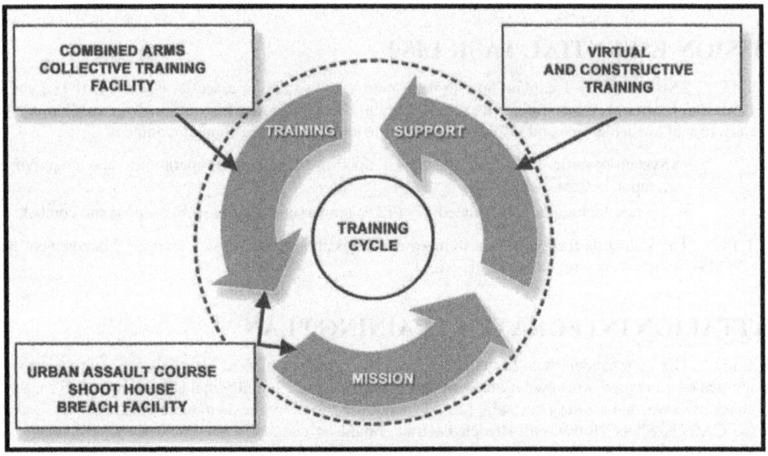

Figure 1-3. Active Army, Home Station, Urban Operations Training Strategy.

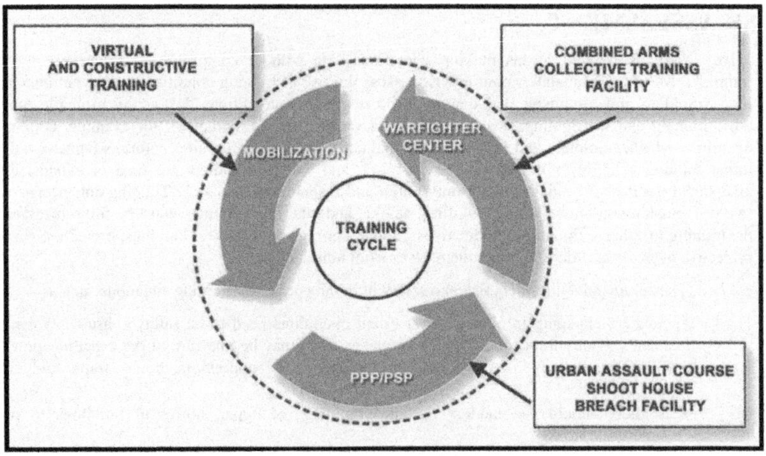

Figure 1-4. Army National Guard Urban Operations Training Strategy.

Section III. TRAINING FACILITIES

Training facilities include three that are discussed in this section: the Urban Assault Course (UAC, Chapter 2), Shoot House (SH, Chapter 3), Breach Facility (BF, Chapter 4). They also include the Combined Arms Collective Training Facility (CACTF, discussed in Chapter 5), Combat Training Centers (CTC), and any other live training that supports the UO training.

URBAN ASSAULT COURSE

1-19. The purpose of the UAC facility is to train a squad-sized force to *Enter a Building and Clear a Room*, improve grenadier gunnery, and validate subterranean TTP. However, the unit's training requirements determine the size of the unit as well as the tasks to be trained. The unit develops its organizational structure and steps for conducting the exercise based on its METL and on their training needs (Figure 1-5). (See Chapter 2 for a more detailed discussion.)

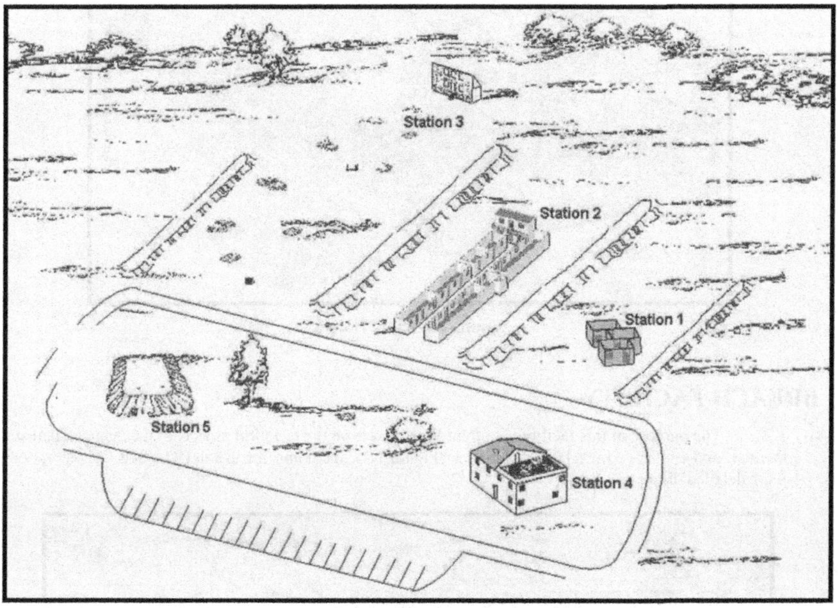

Figure 1-5. Urban Assault Course.

SHOOT HOUSE

1-20. The purpose of this facility is to provide the leader with a facility to train and evaluate fire teams, squads, and platoons during a live-fire exercise (Figure 1-6). Units are trained and evaluated on their ability to conduct tactical movement and engage targets, while practicing target discrimination in an urban environment. (See Chapter 3 for a more detailed discussion.)

> *Note:* The specific designs of the Shoot House and the other facilities may differ from one installation to another, due to variations in training requirements.

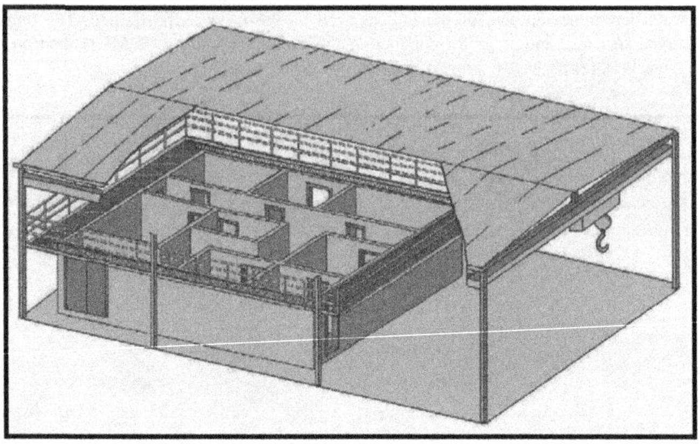

Figure 1-6. Shoot House.

BREACH FACILITY

1-21. The purpose of this facility is to train Warfighters on the technical aspects of mechanical, ballistic, thermal, and explosive breaching techniques (Figure 1-7), including for use in UO. (See Chapter 4 for a more detailed discussion.)

Figure 1-7. Breach Facility.

Section IV. CONDUCT OF TRAINING

Depending on unit proficiency, every training event should follow a *crawl-walk-run* sequence, based on clearly defined task, conditions, and standards. After each training event, leaders conduct an after-action review (AAR).

CRAWL

1-22. The trainer explains each training objective and performance standards, and then talks the Soldiers through the exercise, step by step, describing what each individual or unit must do. This training often includes a demonstration.

WALK

1-23. The Soldiers slowly practice each task to standard. Trainers coach Soldiers at this stage stopping as often as needed to correct mistakes and provide feedback. Soldiers practice each task repeatedly until they can perform it to standard.

RUN

1-24. Soldiers perform each task at full speed as if they are in combat. The Soldiers need feedback during this stage as well. The following can be used to enhance realism:

- Blank ammunition.
- Special Effects Small-Arms Marking System (SESAMS).
- Targetry.
- Pyrotechnics.
- Tactical engagement systems (MILES).
- Opposing forces (force on force).
- Nuclear, biological, and chemical simulation/weapons of mass destruction.
- Training in limited visibility conditions.

AFTER-ACTION REVIEW

1-25. Leaders conduct an after-action review (AAR) after each training event to help Soldiers and units improve their skills by providing immediate feedback. An AAR increases the benefits gained from each training exercise by allowing the leaders and the Soldiers to work together to analyze the performance of each task. Being involved in this analysis enhances the leaders and the Soldiers learning and retention ability. The AAR improves the performance of the leaders and Soldiers in later iterations by allowing interaction and feedback from these individuals during the debrief.

Section V. PREREQUISITE TRAINING

Prerequisite training teaches individuals, crews, units, leaders, and command and staff what they need to know before they are required to perform a task. Prerequisite training builds teamwork and individual confidence at all levels.

COMMAND AND STAFF TRAINING

1-26. The battalion battle staff and leadership must train before the battalion is required to conduct collective training. The battalion command and staff culminates their prerequisite training with a tactical exercise without troops (TEWTs) and command post exercises (CPXs). This sequence of training assists the commander and his staff develop better UO TTP and unit SOP.

LEADER TRAINING

1-27. A comprehensive leader's training program is the key to successful UO training (Appendix D). Time spent teaching and training the battalion leaders (trainers) during the planning and preparation phase will ensure success during the intermediate and culminating phases of training. Realistic training at the unit level will improve the leadership training officers and NCOs receive at the Basic Officer Leadership Course (BOLC), Captains Career Officer Course (CCOC), Basic Noncommissioned Officer Course [BNCOC], and Advanced Noncommissioned Officer Course [ANCOC]. Recommended training should include remedial training on those tasks that the unit had difficulty in during the previous training and evaluation.

INDIVIDUAL TRAINING

1-28. Many individual specific tasks and some collective tasks can be taught and practiced in garrison while most collective tasks are best conducted at the installation UO facilities and weapons ranges. The unit's preliminary individual training tasks should focus on three general areas: urban combat skills, urban marksmanship skills, and physical fitness. The following are example preliminary training tasks or TTP:

- Individual movement techniques.
- Battle drills.
- Weapon positioning in an urban area.
- Roadblock and vehicle search procedures.
- Dismounted and mounted urban navigation.
- Urban scanning, quick-fire, and assault fire techniques.
- Hasty urban firing positions.
- Preparation of urban firing positions.

1-29. Prior to urban collective training cycle, trainers should develop urban-specific firing conditions in conjunction with weapons firing ranges. Build facades on existing ranges, or use the unit's local UAC (if available). Urban combat skills include--

- Firing positions behind rubble, around corners, from windows or rooftops.
- Urban targets behind rubble, in windows and doorways.
- Fire commands.
- Lifting and shifting fires from lower to upper story windows.
- Fire control and fire discipline.
- Rapid firing engagements.
- ROE firing scenarios.

1-30. Urban physical fitness skills include--

- Endurance runs and upper body conditioning.
- Vaulting exercises.

- Urban-specific obstacle course negotiation.
- Leadership Reaction Course negotiation.

INDIVIDUAL TASK TRAINING

1-31. Urban operations require the Soldier to be proficient in those individual tasks that are unique to urban combat. Units should incorporate into their training regime those urban fighting TTP that have proven successful based on lessons learned from combat operations.

TASK FORCE SOLDIER--WARRIOR COMMON TASKS, SKILL LEVELS 1-3:

- 071-326-0541, *Perform Movement Techniques during an Urban Operation.*
- 071-326-0542, *Enter a Building during an Urban Operation.*
- 071-326-0550, *Prepare Positions for Individual and Crew-Served Weapons during an Urban Operation.*
- 071-326-0557, *Select Hasty Firing Positions during an Urban Operation.*
- 071-440-0027, *Execute Mechanical Breach in an Urban Operation.*
- 071-440-0028, *Engage Targets during an Urban Operation.*
- 071-440-0029, *Conduct a Breach during an Urban Operation.*
- 071-440-0003, *Conduct an Attack on a Building by a Squad during an Urban Operation.*
- 071-440-0006, *Conduct a Defense by Squad during an Urban Operation.*

FM 3-06.11, *COMBINED ARMS OPERATIONS IN URBAN TERRAIN*

Movement Techniques

- Crossing of a wall.
- Movement around corners.
- Movement past windows.
- Use of doorways.
- Movement parallel to buildings.
- Crossing open areas.
- Fire team employment.
- Movement between positions.
- Movement inside a building.

Entry Techniques

- Upper building levels.
- Use of ladders.
- Scaling of walls.
- Entry at lower levels.
- Hand grenades.

Firing Positions

- Hasty firing position.
- Prepared firing position.
- Target acquisition.
- Employment of snipers.

COLLECTIVE TASK TRAINING

1-32. Specific FMs teach doctrine, how to conduct operations, and how to employ weapon systems. TTP are lessons learned from the training and combat. The standards for conducting and evaluating collective training are accessible at the Digital Training Management System (DTMS), https://dtms.army.mil/DTMS. Those tasks that do not specifically address UO must be modified for use in an urban environment.

RESOURCES

1-33. Field manuals for the squad and platoon are--:

- FM 3-21.8, *Infantry Rifle Platoon and Squad.*
- FM 3-21.71, *Mechanized Infantry Platoon and Squad (BFV).*
- FM 3-21.9, *SBCT Infantry Rifle Platoon and Squad.*

1-34. Field manuals for the company are--.

- FM 3-21.10, *IBCT Rifle Company.*
- FM 3-21.11, *SBCT Rifle Company.*

1-35. Field manuals and mission training plans for the battalion are--.

- FM 3-21.20, *IBCT Infantry Battalion.*
- FM 3-21.21, *SBCT Infantry Battalion.*
- FM 3-06.11, *Combined Arms Operations in Urban Terrain.*
- TC 7-98-1, *Brigade and Battalion Stability and Support Operations.*

Chapter 2

Urban Assault Course

This chapter describes the Urban Assault Course (UAC), which incorporates doctrine and tactics from FM 3-06.11. The UAC contains five stations:

- Individual and Team Task/Technique Trainer Station 1
- Squad and Platoon Task/Technique Trainer Station 2
- Grenadier Gunnery Trainer Station 3
- Urban Offense/Defense Building Station 4
- Underground Trainer Station 5

These stations are designed for individual tasks and small-unit collective task training. Various types of organizations with a variety of missions can train at the UAC. Tactical operations in an urban environment may include combat to stability and support tasks. (Appendix C provides a SASO unit training matrix). The UAC is a dismounted training facility.

Note: The station discussions are arranged so that they may be removed from the book for ease of use.

PURPOSE

2-1. The purpose is to provide squad and platoon size units with a facility to train and evaluate urban operations tasks (Figure 2-1). Only Station 3, Grenadier Gunnery, is intended for live-fire training. It supports use of 40-mm TP and 5.56-mm service ammunition.

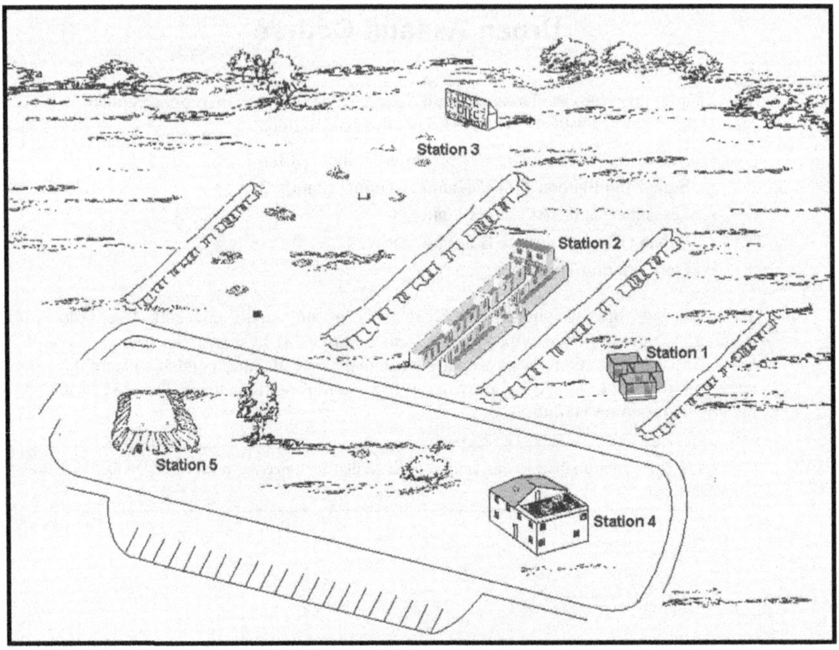

Figure 2-1. Urban Assault Course.

DESCRIPTION

2-2. The UAC has five training stations that facilitate the ***crawl-walk-run*** training concept:

Station 1, Individual and Team Trainer

2-3. This station has three rooms where team leaders and squad leaders train the basics of building and room clearing.

Station 2, Squad and Platoon Trainer

2-4. This station has four structures and multiple rooms. Squads build on tasks learned at Station 1, and begin to learn the concepts of clearing multiple buildings. This station can be used as individual buildings with a narrow street, or as rooms inside a building with a long, connecting hallway.

Station 3, Grenadier Gunnery Trainer

2-5. This is a live-fire station where M203 gunners master target engagements in an urban area, conduct tactical movement, and respond to the fire commands.

Station 4, Offense/Defense House

2-6. This is where a platoon trains to attack or defend a building. This station can also be divided into smaller training stations to reinforce training. It is also used to train tasks not yet trained at the other stations such as upper-level entry techniques.

Station 5, Underground Trainer

2-7. This station provides training for subterranean operations.

TARGETRY

2-8. Targets throughout the UAC are either 3D precision targets (for engagements less than 50 meters) or 2D nonprecision targets (for engagements greater than 50 meters).

Precision Targets

2-9. Precision targets support short distance engagement sensors. These sensors, located inside the precision targets, respond only to lethal shot placements.

- Torso lethal zone .... 18" wide x 8" high
- Neck-head lethal zone .... 4" wide x 8" high

Nonprecision Targets

2-10. Nonprecision targets support engagements at ranges greater than 50 meters. These targets are 2D pop-up type targets.

TRAINING AUDIENCE

2-11. The facility gives Infantry, Engineer, Military Police, Cavalry, and other units a facility for training individual Soldiers through platoon-level UO skills.

TRAINING INFORMATION

2-12. Combat units can use this facility as a mission rehearsal exercise site. Recommended training frequency is quarterly. For each UAC station, they should provide the purpose, a description, targetry, controller and safety requirements, individual UO tasks to be trained, training support requirements, and conduct of training information, if applicable:

Purpose

2-13. This UAC station [*state purpose of station*]. [*Units may use the station in the manner originally intended, or they may modify it to meet their unique training needs.*]

Description

2-14. This station is [*describe station use and design*].

Targetry

2-15. Recommended targetry for training at this station includes [*describe*]. Targets may be moved, modified, or added, depending on available target systems or training requirements.

Controller

2-16. Controller requirements for this station include [*describe*].

Safety

2-17. Safety requirements and considerations for this station include [*describe*]. [*The unit or installation range SOP may add more stringent safety requirements.*]

Individual Tasks or Skills Trained

2-18. These may be found in the Digital Training Management System located online at https://dtms.army.mil/DTMS.

Training Support

2-19. Minimum materiel requirements to support training at this station include [describe].

Conduct of Training

2-20. Example training scenarios and sequence of events for this station include [describe]. [The UAC can accommodate an assortment of training ammunition and pyrotechnics.]

STATION 1: INDIVIDUAL AND TEAM TASK/TECHNIQUE TRAINER

2-21. This station is used to train individual and collective tasks, tactics, techniques, and procedures, specifically *Enter a Building and Clear a Room* and *Engage Targets* (Figure 2-2). It has three rooms where team leaders and squad leaders train the basics of building and room clearing.

DESCRIPTION

2-22. Station 1 has a single wood structure. Its three rooms have one or more open doorways and windows, and a man-sized hole.

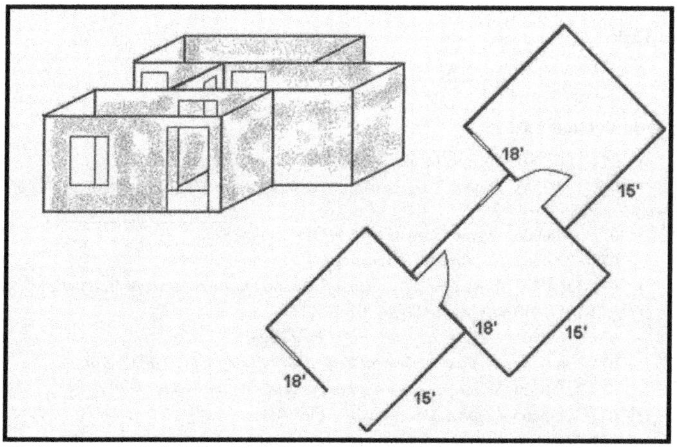

Figure 2-2. Individual and team task/technique trainer, Station 1.

TARGETRY

2-23. Targets present realistic close quarters combat situations (0 to 15 meters), which consist of a possible mixture of combatant and noncombatant targets. Target discrimination is trained by presenting up to six precision targets. Targetry sensors respond only to lethal shot placement. They also aid in training during limited visibility.

CONTROLLER REQUIREMENTS

2-24. These are based on unit requirements.

SAFETY REQUIREMENTS

2-25. Only smoke and practice grenades, or M84 stun grenades, may be used at this station.

> **WARNING**
>
> **To avoid injury, place the weapon on SAFE when climbing through obstacles such as windows, fences, and walls during training.**

INDIVIDUAL/COLLECTIVE TASKS TRAINED

2-26. This is noninclusive.

Collective Task

2-27. *Enter a Building/Clear a Room.*

Supporting Individual Tasks

- 071-311-2007, *Engage Targets with M16-Series Rifle/M4 Carbine Rifle.*
- 071-315-2308, *Engage Targets with M16-Series Rifle/M4 Carbine Using a Night Vision Sight AN/PVS-4.*
- 071-010-0006, *Engage Targets with M249 Machine Gun.*
- 071-325-4407, *Employ Hand Grenades.*
- 071-326-0503, *Move Over, Through, or Around Obstacles (except Minefields).*
- 052-193-1013, *Neutralize Booby Traps.*
- 071-326-0501, *Move as a Member of a Fire Team.*
- 071-326-0541, *Perform Movement Techniques during an Urban Operation.*
- 071-326-0557, *Select Hasty Firing Positions during an Urban Operation.*
- 071-326-5605, *Control Movement of a Fire Team.*
- 071-326-5611, *Conduct the Maneuver of a Squad.*
- 071-710-0008, *Operate Night Vision Goggles AN/PVS-7B.*

Related Tactics, Techniques, and Procedures

- FM 3-06.11, *Combined Arms Operations in Urban Terrain.*
- FM 3-22.9, *Rifle Marksmanship.*

TRAINING SUPPORT REQUIREMENTS

2-28. Use Table 2-1 for the minimum ammunition required to train at this station.

Table 2-1. Support requirements for Station 1.

Ammunition	*Rounds per Weapon*
5.56-mm Blank	10 rounds
5.56 mm Blank (Linked)	15 rounds
Practice Grenade Bodies	2 per team
Practice Grenade Fuses	2 per team

CONDUCT OF TRAINING

2-29. Table 2-2 lists the training sequence and times for Station 1.

Table 2-2. Sequence of events and times for Station 1.

Event	*Action*	*Estimated Time*
1	*Employ Hand Grenades*	30 minutes crawl, walk, run
	Perform Movement Techniques • *Engage Target with M16* • *Target Discrimination*	20 minutes crawl 20 minutes walk 10 minutes run
2	*Enter a Building*	20 minutes crawl 20 minutes walk 10 minutes run
3	*Enter and Clear a Room*	20 minutes crawl 20 minutes walk 10 minutes run
4	*Conduct an AAR*	40 minutes

Example Scenarios, Assault Team, Varied Entry Positions

2-30. Figures 2-3 and 2-4 show example scenarios. Rather than using the examples, units should tailor their training events based on their unit's METL and training needs.

Varied Entry Positions

2-31. The team leader first trains to execute *Varied Entry Positions* without targets, which is part of the *crawl* phase of training. The trainer has the assault teams position at various places to demonstrate how the start position can affect the final points of domination. (Figure 2-3).

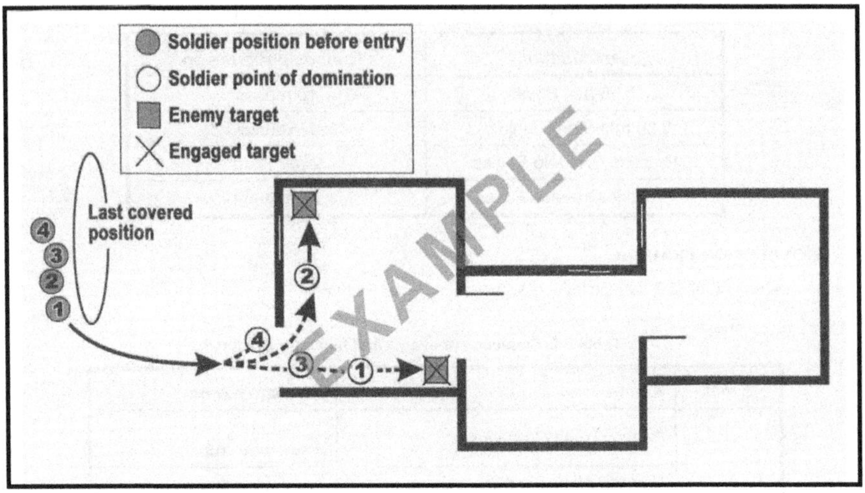

Figure 2-3. Example Station 1 scenario, *Varied Entry Positions*.

Multiple Room-Clearing Techniques

2-32. The team leader trains his Soldiers to execute *Varied Entry Positions* using targets in a multiple room scenario. This is part of the *walk* phase. The trainer has the assault teams enter the room and engage targets. He has them use using precision marksmanship techniques with the Tactical Engagement System (TES) or SESAMS (Figure 2-4). The assault team enters the first room and engages the targets. As the number two Soldier clears his area of responsibility, he sees and engages the target in the second room, but remains in place until the assault teams finishes clearing the first room. Once the room is clear, the assault team moves to the second room and continues clearing, or lets the next team move through.

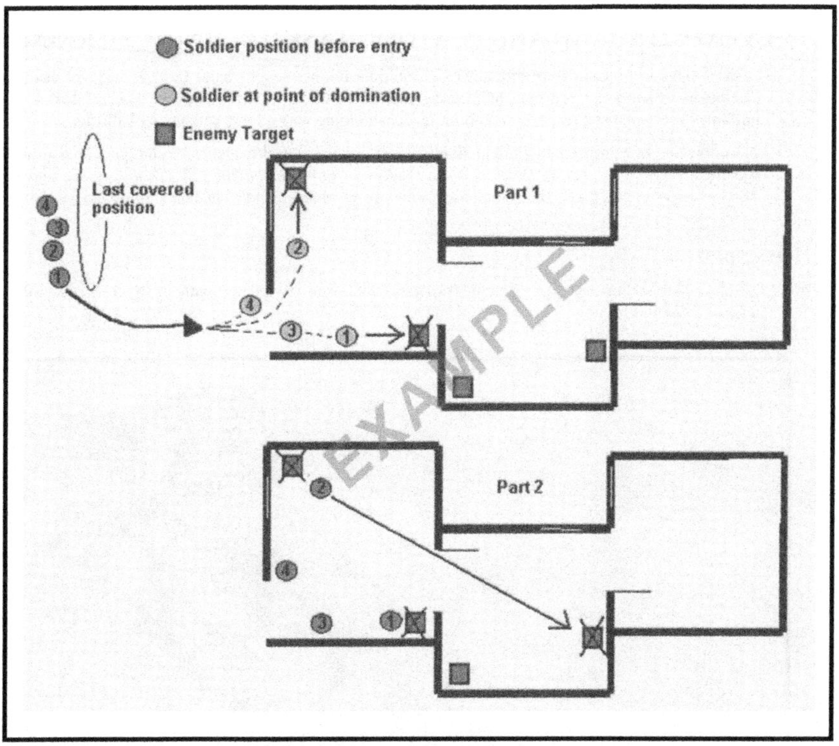

Figure 2-4. Example Station 1 scenario, *Multiple Room-Clearing Techniques*.

STATION 2: SQUAD AND PLATOON TASK/TECHNIQUE TRAINER

2-33. This station has four structures and multiple rooms. Squads build on tasks learned at Station 1, and begin to learn the concepts of clearing multiple buildings. This station can be used as individual buildings with a narrow street, or as rooms inside a building with a long, connecting hallway.

2-34. Station 2 is used to train individual and collective tasks, tactics, techniques, and procedures, specifically, *Move Tactically Down a Street (Hallway)*, *Enter a Building and Clear a Room*, and *Engage Targets*. Station 2 increases the complexity of command, control, and maneuver (Figure 2-5).

DESCRIPTION

2-35. This station has four separate structures that face inward to create a street. It also introduces a two-story building.

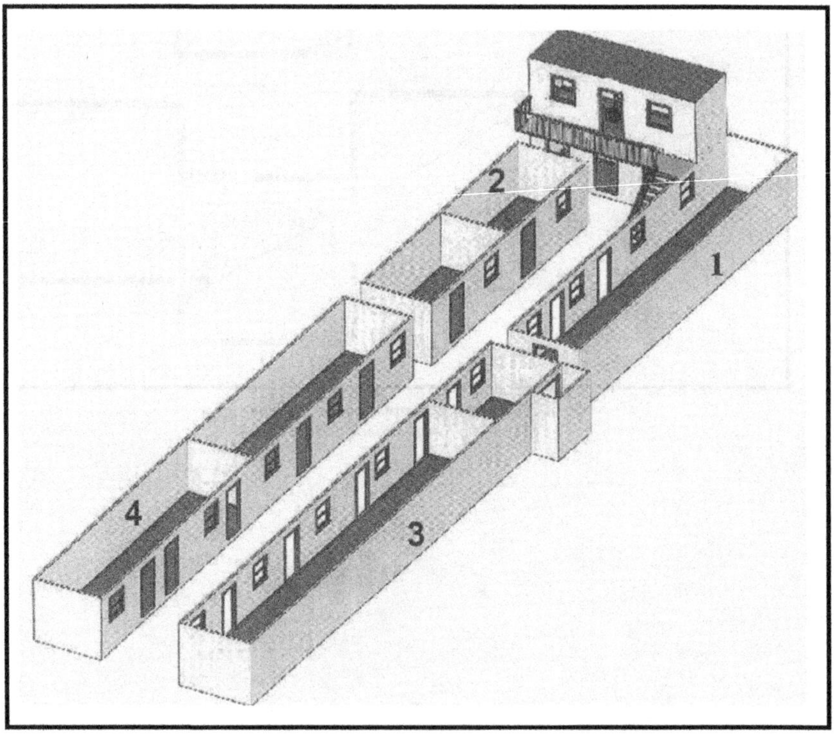

Figure 2-5. Squad and platoon task and technique trainer, Station 2.

> Building 1 is L-shaped.
>
> - The long leg of the building to the right looks like a single-story building.
> - The short leg of the building to the front has two floors. A landing along the length of the second floor has outside stairs. An inside stairway also joins the floors.
> - The two legs of the "L" are adjacent, but without interior access to each other.
>
> Building 2 has a single floor and two rooms, each with a door and window. An open area much like a street separates this structure from the first.
>
> Building 3 is a row of wooden buildings along a street. It has many windows and doorways. At the far end is an adjoining room.
>
> Building 4 is on the left. It mirrors Building 3, plus it has a wall on the side at the near end.

2-36. Units can place additional walls in any structure by using target cloth or other material. They can do this to add to the number of rooms or to increase the complexity of the training (Appendix A).

TARGETRY

2-37. Station 2 targets present realistic close quarters combat situations (0 to 15 meters), with the option to customize the mixture of combatant and noncombatant targets. Target discrimination is trained on up to ten precision targets. Targetry sensors respond only to lethal shot placement, and facilitate training during limited visibility.

CONTROLLER REQUIREMENTS

2-38. These are based on unit requirements.

SAFETY REQUIREMENTS

2-39. At this station, Soldiers may only use smoke grenades, practice grenades, and M84 stun grenades.

> **WARNING**
>
> **To avoid injury, place the weapon on SAFE when climbing through obstacles such as windows, fences, and walls during training.**

INDIVIDUAL AND COLLECTIVE TASKS TRAINED

2-40. This list is noninclusive.

Collective Task--*Enter and Clear a Building*

Related Collective Tasks

- *Conduct Consolidation and Reorganization (Company/Platoon).*
- *Treat Casualties.*
- *Evacuate Casualties.*
- *Defend MOUT Building.*

- *Conduct Tactical Movement.*
- *Enter and Clear a Room.*

Supporting Individual Tasks

- 052-193-1013, *Neutralize Booby Traps.*
- 071-010-0006, *Engage Targets with M249 Machine Gun.*
- 071-311-2007, *Engage Targets with M16-Series Rifle/M4 Carbine Rifle.*
- 071-315-2308, *Engage Targets with M16-Series Rifle/M4 Carbine Using a Night Vision Sight AN/PVS-4.*
- 071-325-4407, *Employ Hand Grenades.*
- 071-326-0501, *Move as a Member of a Fire Team.*
- 071-326-0503, *Move Over, Through, or Around Obstacles (except Minefields).*
- 071-326-0541, *Perform Movement Techniques during an Urban Operation.*
- 071-326-0557, *Select Hasty Firing Positions during an Urban Operation.*
- 071-326-0608, *Use Visual Signaling Techniques.*
- 071-326-5605, *Control Movement of a Fire Team.*
- 071-326-5611, *Conduct the Maneuver of a Squad.*
- 071-710-0008, *Operate Night Vision Goggles AN/PVS-7B.*
- 113-571-1022, *Perform Voice Communications.*
- 191-377-5250, *Handle Enemy Personnel and Equipment.*

Related Tactics, Techniques, and Procedures

- FM 3-06.11, *Combined Arms Operations in Urban Terrain.*
  -- Chapter 3, *Offensive Operation.*
  -- Chapter 5, *Fundamental Combat Skills.*
  -- Appendix M, *Field-Expedient Breaching of Common Urban Barriers.*
- FM 3-22.9, *Rifle Marksmanship.*

TRAINING SUPPORT REQUIREMENTS

2-41. Table 2-3 shows the minimum ammunition required to train at this station.

Table 2-3. Support requirements for Station 2.

Ammunition	Rounds per Weapon
5.56-mm blank	16
5.56-mm blank (linked)	24
Practice grenade fuses	8 per squad
Practice grenade bodies	8

CONDUCT OF TRAINING

2-42. Table 2-4 shows the sequence of events and times for Station 2. Figure 2-6 through Figure 2-28 show example scenarios. However, rather than use these examples, the unit should customize its training events based on its METL and training needs.

Table 2-4. Sequence of events and times for Station 2.

Event	Action	Estimated Time
1	*Conduct Tactical Movement in a Built-Up Area* *Move Across an Open Area* *Move Parallel to Buildings*	10 min Crawl 20 min Walk 10 min Run
2	*Enter and Clear a Room (Squad)* *Clear a Stairwell* *Clear a Hallway* *Conduct a Breach*	10 min Crawl 20 min Walk 10 min Run
3	*Enter and Clear a Building (Platoon)* *Clear Multiple Rooms* *Consolidate and Reorganize*	10 min Crawl 20 min Walk 10 min Run
4	*Conduct AAR (Squad)*	10 min
NOTE: Training should include techniques outlined FM 3-06.11.		

2-43. The squad is clearing Station 2 as if it were a single building, and the street a hallway. Fire Team A moves down the left side, throws the appropriate grenade IAW ROE, enters, engages the target, and clears the room on that side. The last Soldier remains outside and secures the rear Fire Team B prepares to enter the next room (Figure 2-6).

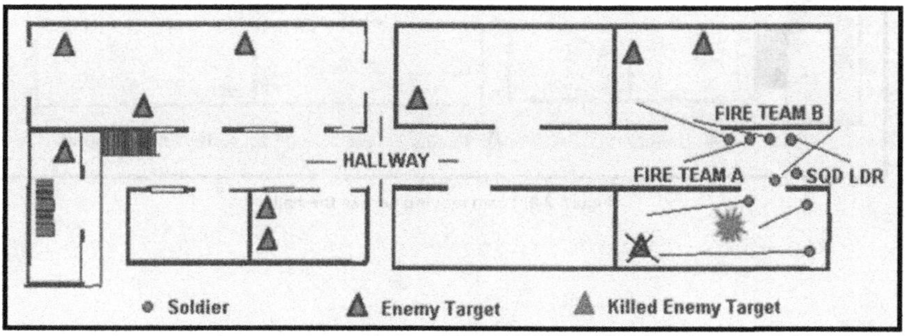

Figure 2-6. Conducting tactical movement.

2-44. Fire Team B throws the appropriate grenade, enters, engages the targets, and clears the room on the right while the last Soldier secures the rear Fire Team A prepares to continue movement to the next room (Figure 2-7).

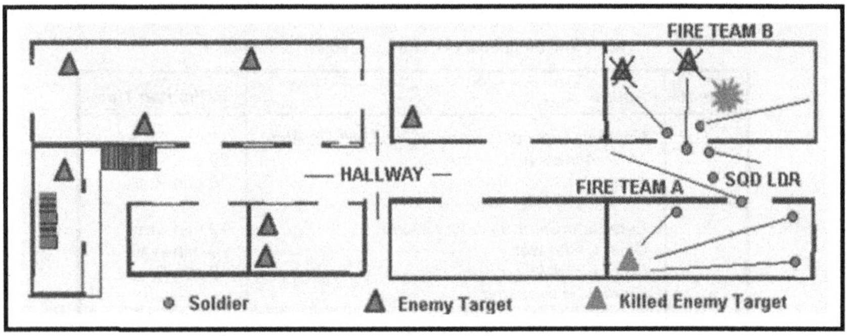

Figure 2-7. Team B--Clearing a building.

2-45. Fire Team A moves across the hall, throws the appropriate grenade, enters, engages the target, and clears the next room Three team members enter the room, the fourth remains in the doorway and secures the hallway (Figure 2-8).

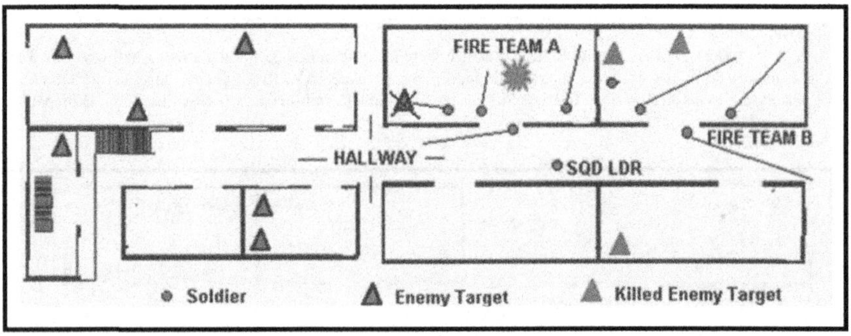

Figure 2-8. Team moving across the hall.

2-46. Fire Team B moves in a cross-cover formation down the hallway to the next room, throws the appropriate grenade, and clears the room. As Fire Team B passes the security man of Team A, he turns and becomes rear security. The last Soldier in Fire Team B remains by the door and secures the hallway (Figure 2-9).

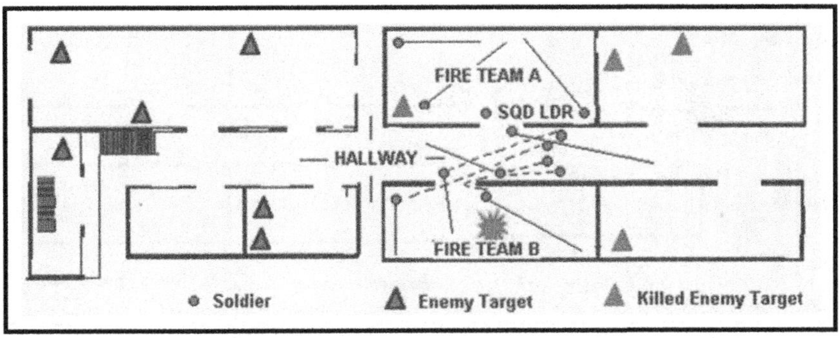

Figure 2-9. Team B--Moving in cross cover formation.

2-47. Fire Team A clears the hallway intersection, while Team B uses the Rolling-T technique to pass through Team A en route to the next room (Figure 2-10).

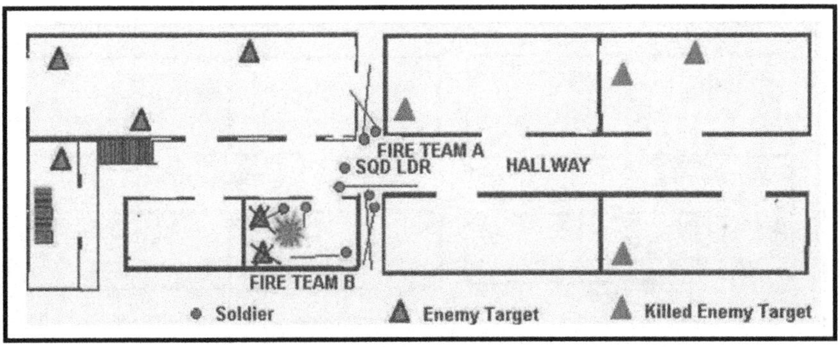

Figure 2-10. Team A--Clearing a hallway intersection.

2-48. Fire Team B throws the appropriate grenade, enters, engages the targets, and clears the room. Only three Soldiers enter, while the fourth Soldier provides rear security. Fire Team A prepares to move (Figure 2-11).

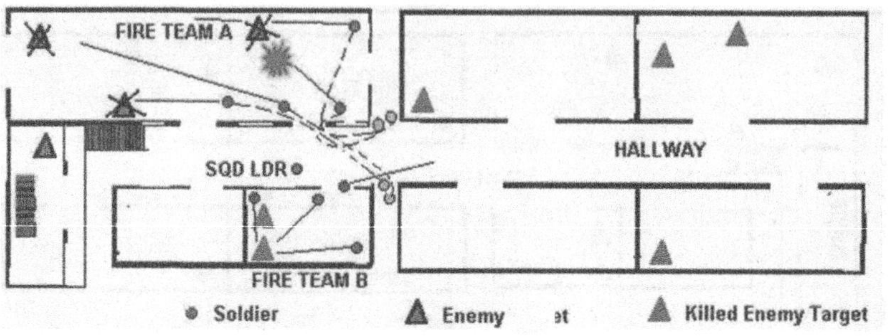

Figure 2-11. Team B--Clearing a room.

2-49. Fire Team A moves from the intersection and throws the appropriate grenade. The team enters, engages the targets, and then clears the next room. All team members enter, because of the size of the room. Fire Team B prepares to move (Figure 2-12).

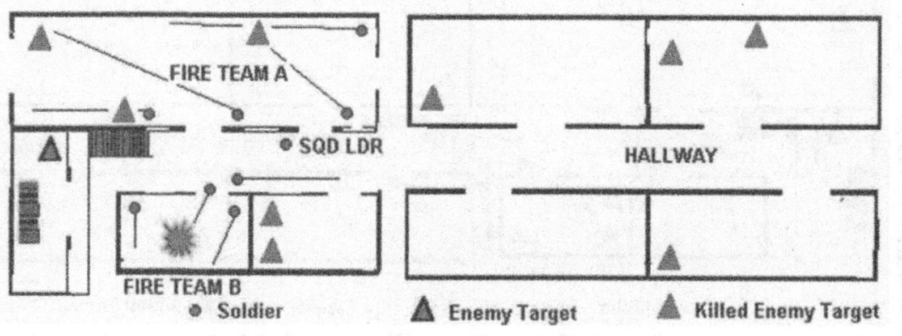

Figure 2-12. Team A--Clearing room by room.

2-50. Fire Team B employs the appropriate grenade, enters, and engages the target. Three Soldiers clear the next room. The fourth remains in the doorway and secures the rear. Fire Team A prepares to move up the stairs (Figure 2-13).

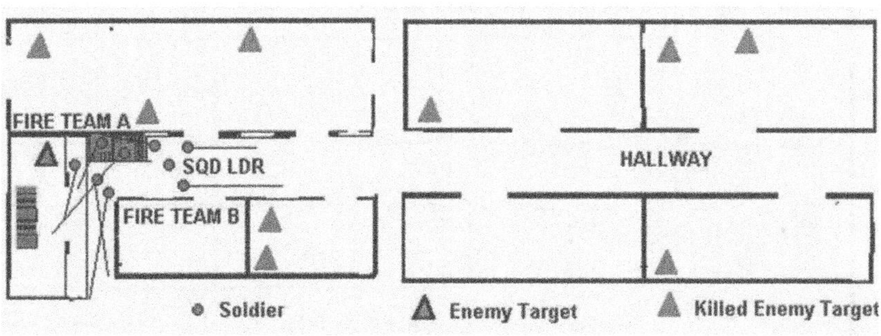

Figure 2-13. Team B--Clearing the next room.

2-51. Fire Team A moves up the stairs and onto the landing. The team prepares to enter the room off the landing. Fire Team B splits, and two members move around the corner and secure the area across from the stairs on the lower level. The other two members of Team B secure the hallway the squad came from (Figure 2-14).

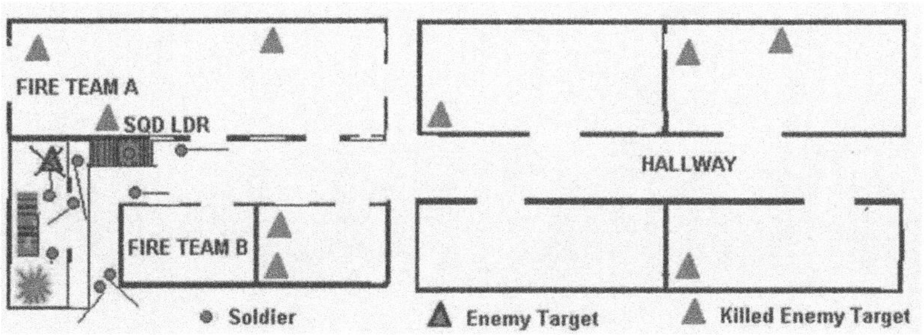

Figure 2-14. Securing the stairs and hallway.

2-52. Fire Team A members employ the appropriate grenade, enters the room, engages the target, and clears it. They find another set of stairs, and prepare to move up them. Fire Team B continues to secure the rear. The other team members move to the end of the short hallway and secure both directions (Figure 2-15).

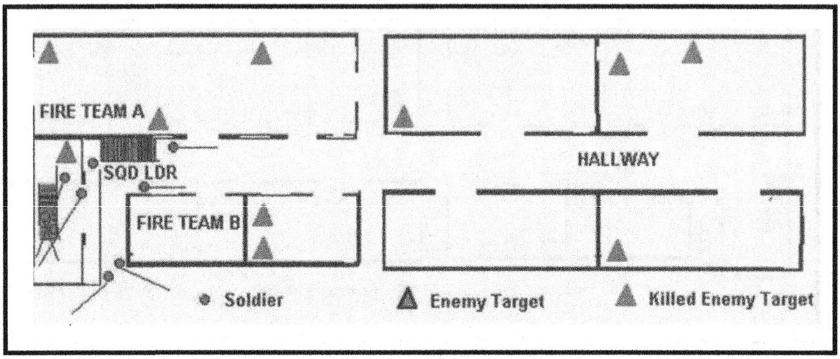

Figure 2-15. Preparing to clear more stairs.

2-53. Fire Team A clears the stairs and the landing. Fire team B continues to secure the ground level, ready to assist Team A, if necessary (Figure 2-16).

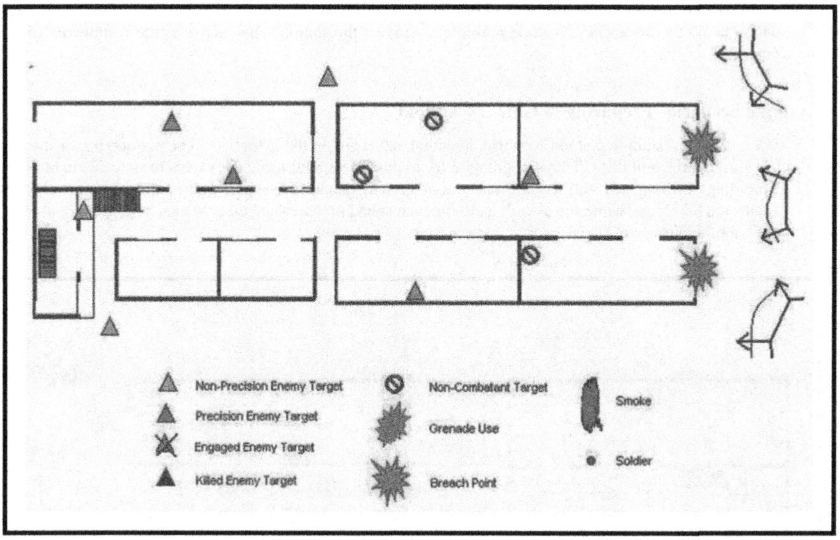

Figure 2-16. Team A--Clearing stairs and landing.

2-54. The previous example used no noncombatant targets, but used blanks/SESAMS and TES. This technique is the *crawl* phase for this station. Once the squad is ready to move on to more challenging exercises, the platoon leader/sergeant can add targets into the scenario. As proficiency increases, mixed combatant and noncombatant targets can be added. At any point, the platoon leader/sergeant can, until the platoon achieves proficiency, include multiple squads on the station. This increases the complexity of the scenarios.

Example Scenario--Preparing to Gain a Foothold

2-55. In this example platoon scenario, noncombatants are on the battlefield. The platoon has decided to clear through the buildings. They are going to try to do this without using the streets to move from building to building. Two squads will assault, and the other will remain in support/reserve. The support element isolates the buildings, while the assault team for each squad prepares and breaches each building in order to gain a foothold (Figure 2-17).

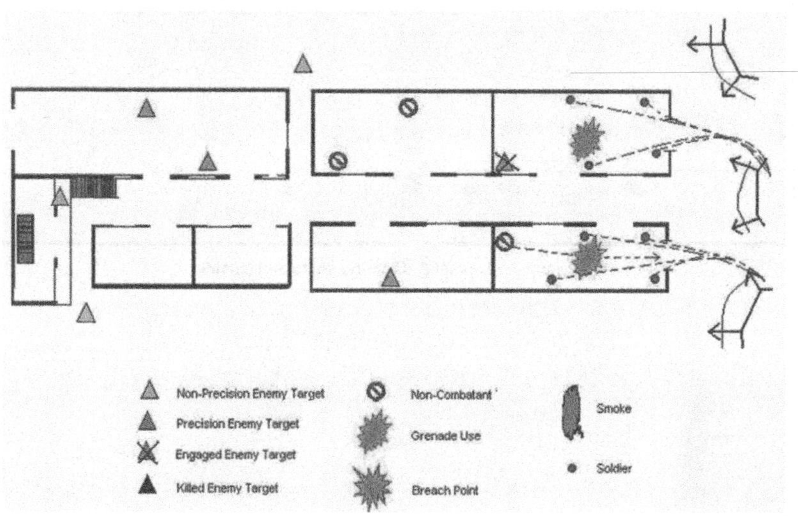

Figure 2-17. Preparing to gain a foothold.

2-56. The assault teams throw a grenade into the breach and their members follow, engaging enemy targets. They secure and move noncombatants to the collection point (CP) (Figure 2-18).

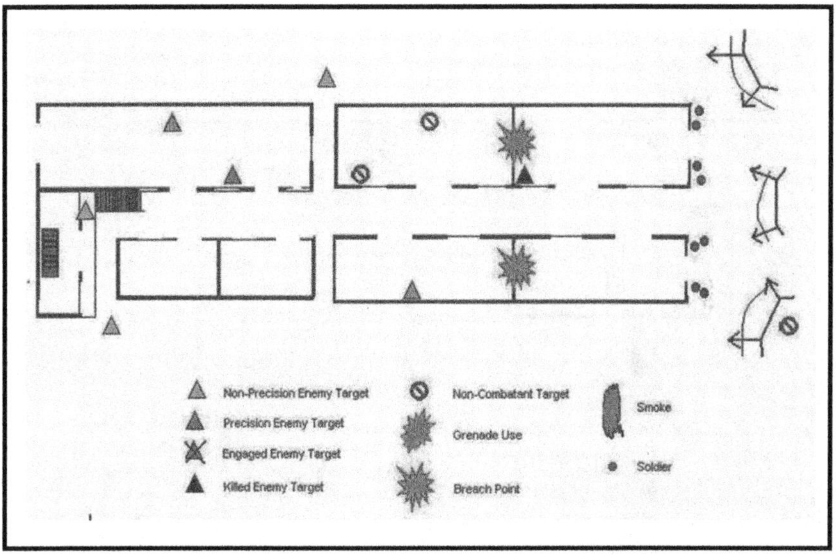

Figure 2-18. Securing and moving noncombatants to the command post.

2-57. The teams prepare and detonate wall breaches to continue movement through the building. Once the charges are ready for detonation, the assault teams move to a covered position and execute the breach (Figure 2-19).

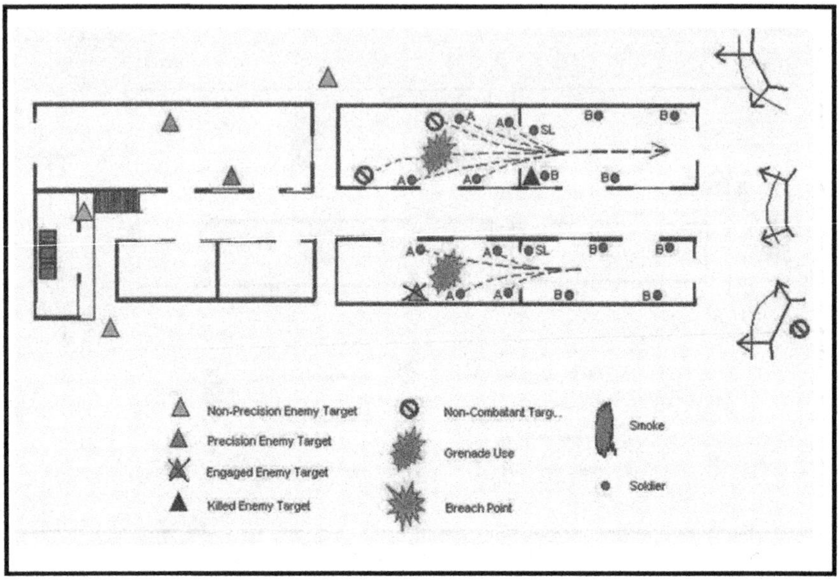

Figure 2-19. Executing breach.

2-58. The assault team employs grenades into the next room, enters the room, eliminates the enemy, and then secures and moves the noncombatants to a designated safe area. At the same time, the next assault team moves into the foothold (Figure 2-20).

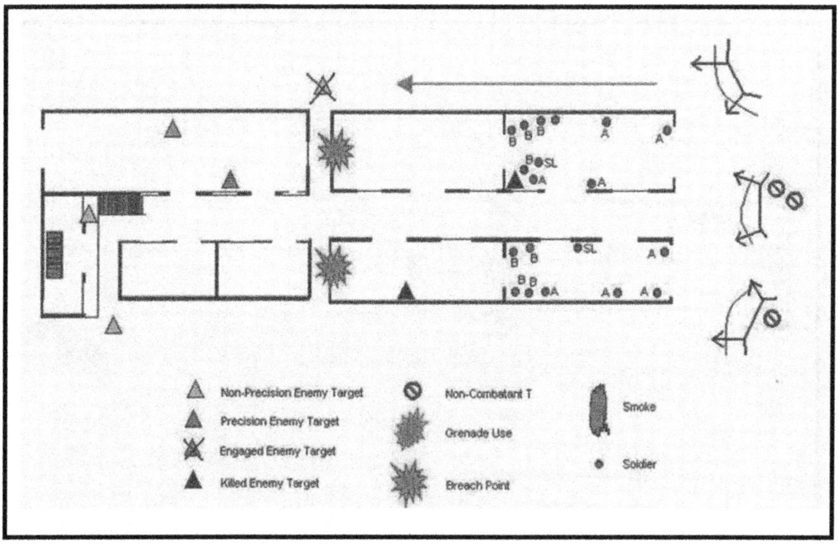

Figure 2-20. Moving to the foothold.

2-59. With the second room secured, another assault team moves forward and prepares to breach the wall Once the charges are placed, the teams move back to a covered position before detonating the charge. With the breach a success, the first nonprecision target is exposed and engaged by the support element (Figure 2-21).

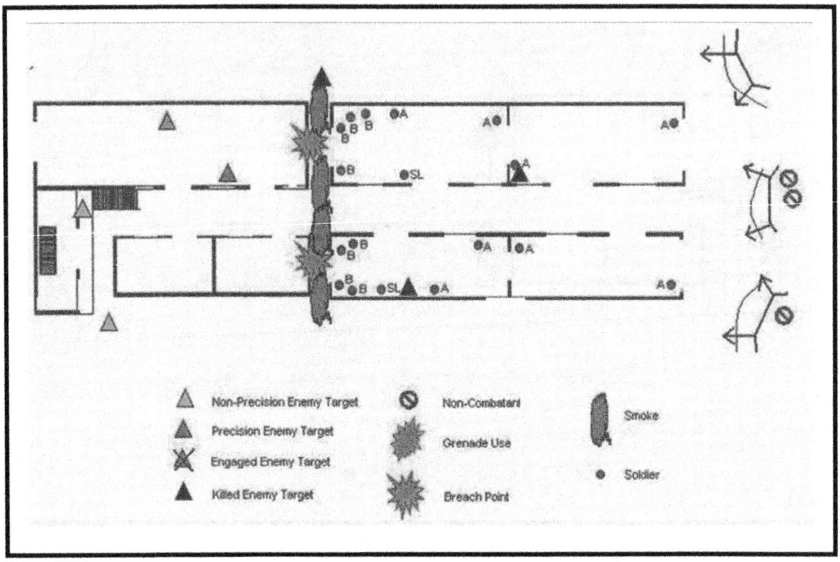

Figure 2-21. Breaching the wall.

2-60. Upon inspecting the breach, the assault teams find that they have an alley with another wall to breach. They employ smoke to obscure the teams placing the breaching charge. The first assault teams take up securing positions throughout the already cleared areas. Follow-on assault teams move to a covered position and execute the breach (Figure 2-22).

Note: Use smoke only in the alley, not inside the buildings.

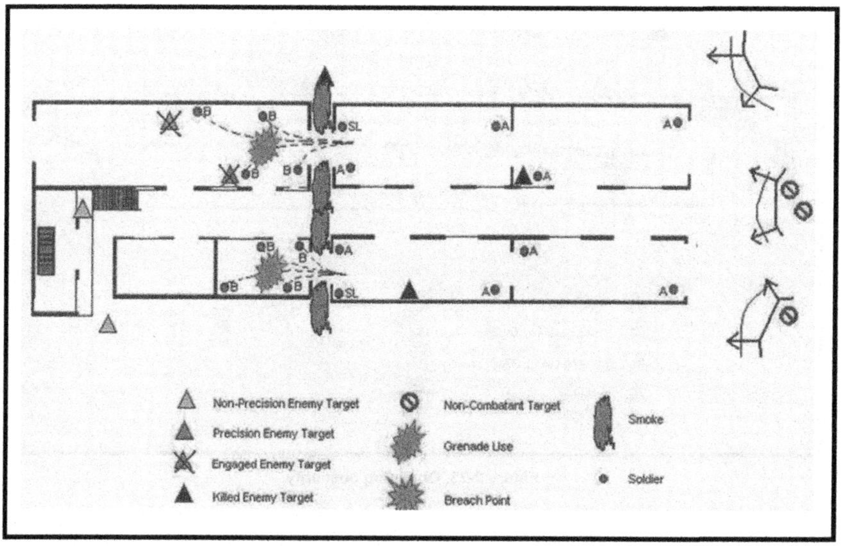

Figure 2-22. Securing a wall in an alley.

2-61. Once the assault teams execute the breach, they must determine if they need to use more smoke to cover their movement. Either way, once they have enough smoke to cover their movement, they throw in a grenade, enter, engage enemy targets, and clear the rooms (Figure 2-23).

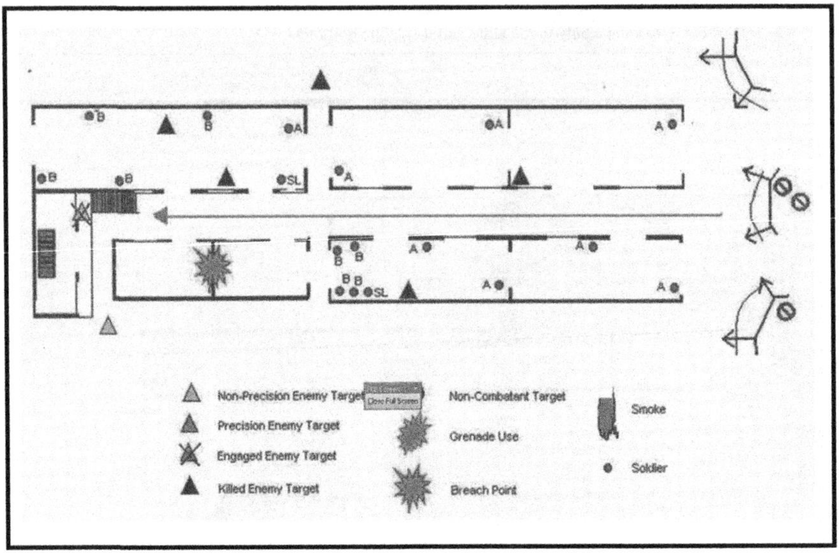

Figure 2-23. Obtaining obscurity.

2-62. While one team finds itself at the end of the building in their sector, the other team is faced with another wall and more clearing. One team prepares to breach the wall, and the other team takes up a hasty position and prepares for a counterattack. When the teams execute the breach, the next nonprecision target presents, and the support element engages it (Figure 2-24).

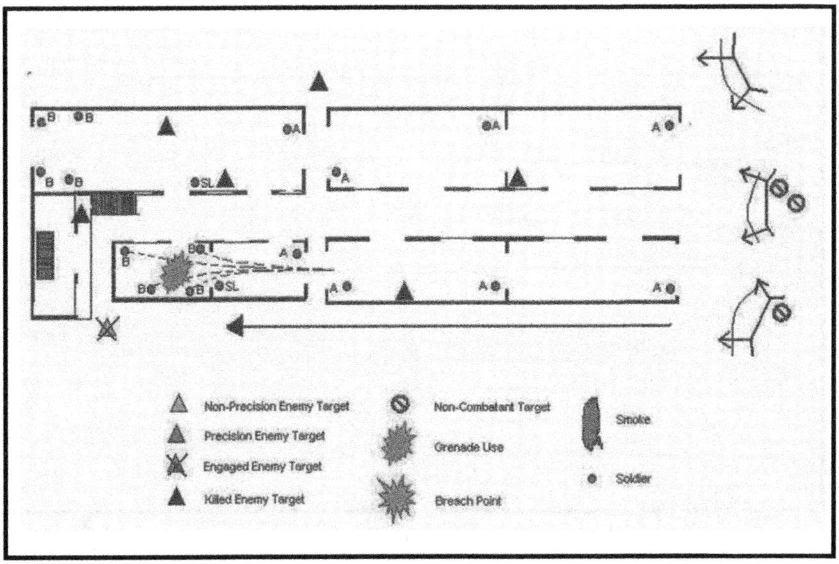

Figure 2-24. Clearing the wall.

2-63. Someone throws a grenade through the breach. The assault team enters and secures the room. When the grenade explodes inside the room, the final nonprecision target presents, and the support element engages it (Figure 2-25).

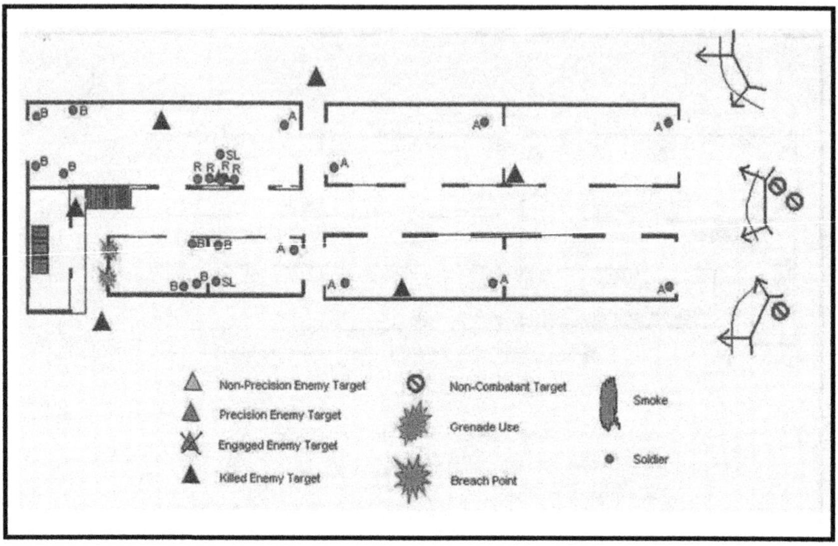

Figure 2-25. Securing a room.

2-64. With the first squad reaching its limit of advance, and with the second squad reaching the end of the building, the platoon leader calls another assault team forward from the reserve. The reserve assault team moves forward through the buildings, and positions itself to clear up the outside stairs of the two-story structure. The second squad places and detonates loophole charges on the far wall of the building. This supports the clearing of the last building (Figure 2-26).

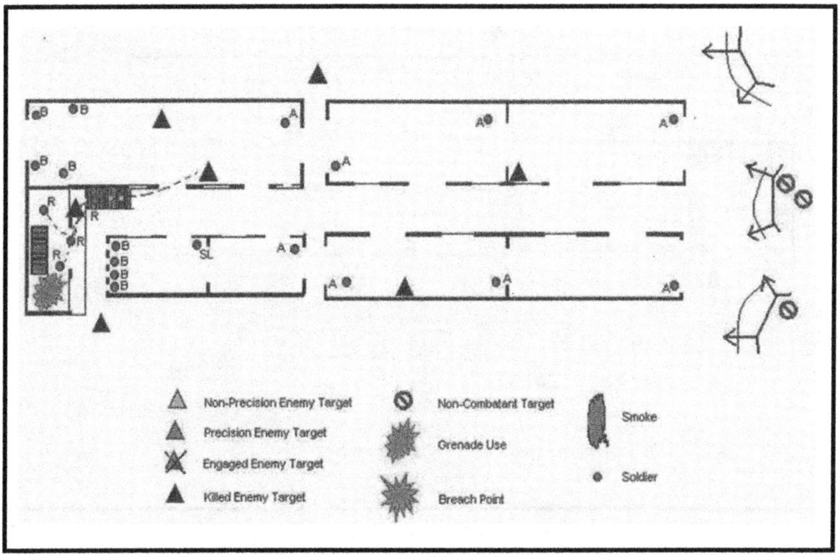

Figure 2-26. Reserve assault team moving forward.

2-65. Team B of the second squad moves into supporting fire positions. The new assault teams move up the outside stairs to the second floor of the final structure. They clear the balcony and the first room on the second floor. The support-by-fire position shift fires to the lower floor of the same structure (Figure 2-27).

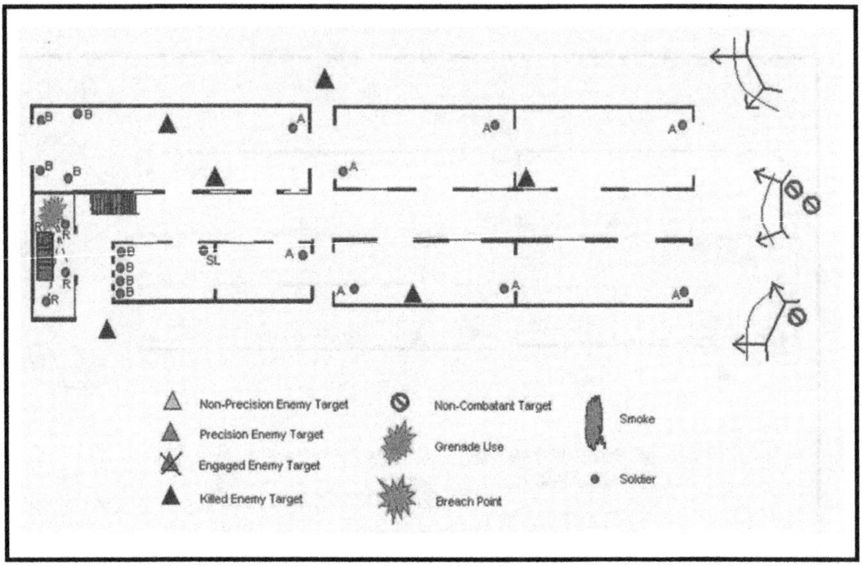

Figure 2-27. Team B, second squad--moving into the supporting fire position.

2-66. Once the floor is secure, the assault team moves down the stairs and clears the remaining floor (Figure 2-28). Meanwhile, the platoon repositions, and then consolidates and reorganizes.

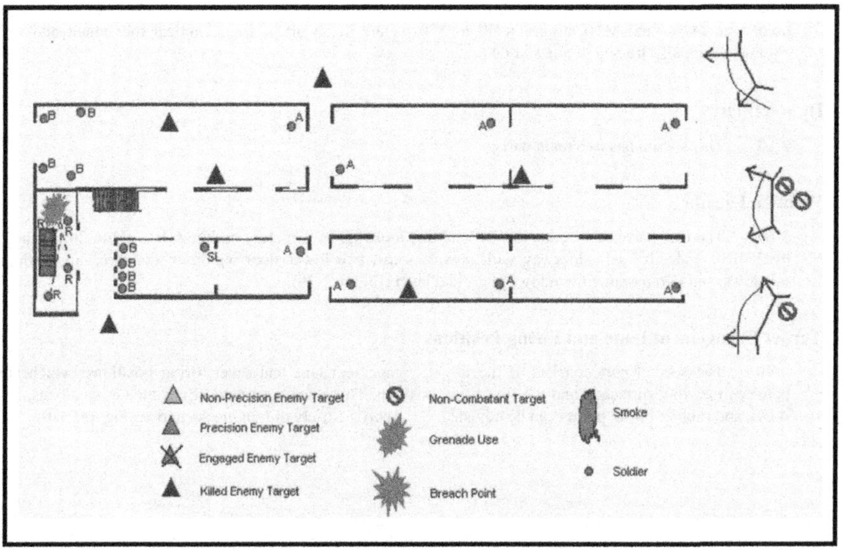

Figure 2-28. Consolidation and reorganization.

STATION 3: GRENADIER GUNNERY TRAINER

2-67. This is a live-fire station where M203 dual-purpose weapon gunners master the engagement of point-type M203 and M16 targets at 50 to 150 meters in an urban area. Tactical movement and fire commands are also trained at this station.

DESCRIPTION

2-68. This station has two main parts:

Wooden Facade

2-69. The first part is a wooden facade wall that looks like a two-story building. It is about 20 feet high by 30 feet wide. It has a doorway with outside steps, two lower floor windows and three upper floor windows, and two basement windows at ground level (Figure 2-29).

Target Engagement Lane and Firing Positions

2-70. The second part consists of the target engagement lane and seven firing positions, which are between the wooden façade and the first firing position. These positions have cover such as sandbags, log walls, and rubble. Firing points can be adjusted to support a variety of training scenarios (Figure 2-30).

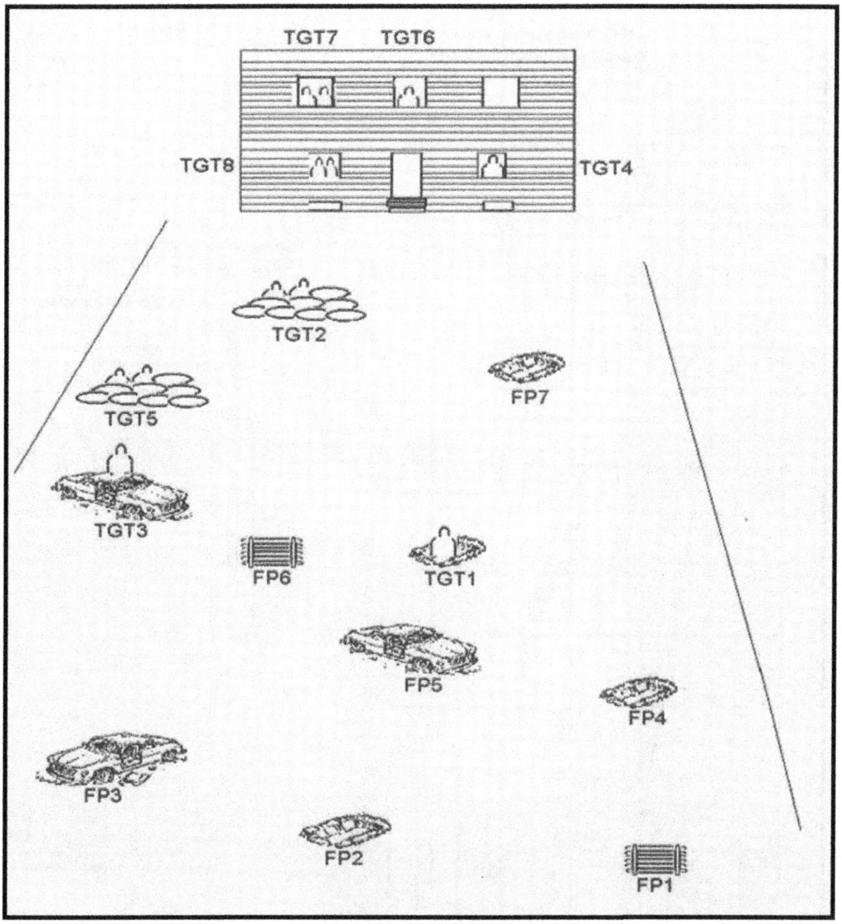

Figure 2-29. Part 1, Grenadier Gunnery Trainer, Station 3, wooden facade.

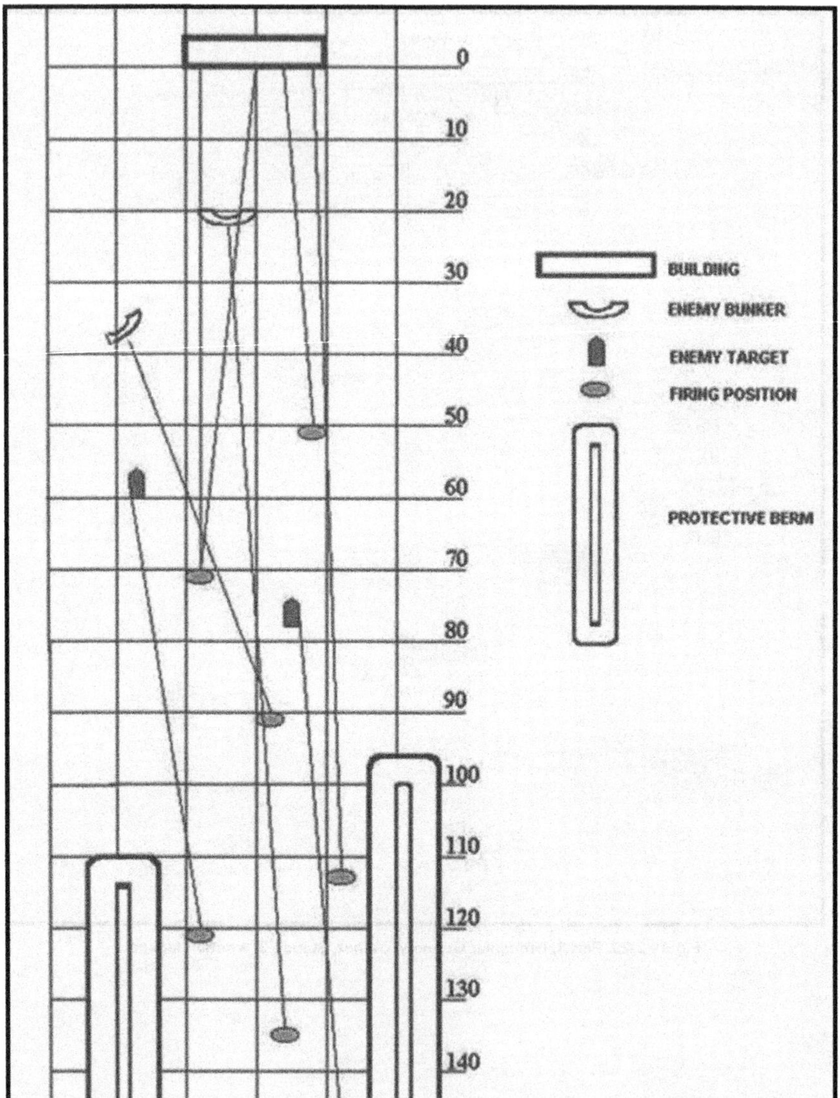

Figure 2-30. Part 2, Grenadier Gunnery Trainer, Station 3, target engagement lane and firing positions.

> **WARNING**
>
> **Never engage targets with the 40-mm at ranges less than 31 meters, which is the minimum safe range. Trainers ensure the station accommodates this requirement as well as the 14-meter arming range.**

TARGETRY

2-71. This station has twelve target mechanisms. These mechanisms are arranged so as to influence the gunner's decision as to which munition to use.

CONTROLLER REQUIREMENTS

2-72. As required by unit.

SAFETY REQUIREMENTS

2-73. Check the unit SOP and local safety regulations regarding live-fire ranges. Ensure all weapon systems are on SAFE before moving to the next firing position. Ensure that no 40-mm target is engaged at less than minimum arming range.

INDIVIDUAL AND COLLECTIVE TASKS TRAINED

2-74. These lists are noninclusive:

Supporting Individual Tasks

- 071-311-2007, *Engage Targets with M16-Series Rifle/M4 Carbine Rifle.*
- 071-311-2130, *Engage Targets with M203 Grenade Launcher.*
- 071-326-0503, *Move Over, Through, or Around Obstacles (except Minefields).*
- 071-326-0541, *Perform Movement Techniques during an Urban Operation.*
- 071-326-0557, *Select Hasty Firing Positions during an Urban Operation.*
- 071-326-0608, *Use Visual Signaling Techniques.*

Related Tactics, Techniques and Procedures

- FM 3-06.11, *Combined Arms Operations in Urban Terrain.*
  -- Chapter 3, *Offensive Operations.*
  -- Chapter 5, Fundamental Combat Skills.
- Appendix M, *Field Expedient Breaching of Common Urban Barriers.*

TRAINING SUPPORT REQUIREMENTS

2-75. Table 2-5 shows the minimum materials required to train at this station.

Table 2-5. Support requirements for Station 3.

Ammunition	*Rounds for each Weapon*	*Targets Required*
5.56-mm Ball	8	Four nonprecision human target mechanisms.
40-mm TP	8	Eight nonprecision human target mechanisms.

CONDUCT OF TRAINING

2-76. Whereas Table 2-5 showed the minimum amount of ammunition required for this station, Table 2-6 shows an example sequence of events and times, and Table 2-7 shows an example target scenario.

Table 2-6. Sequence of events and times for Station 3.

Event	*Action*	*Estimated Time*
1	*Perform Movement Techniques* *Select Hasty Firing Position* *Engage Targets With 5.56-mm (dry run)* *Engage targets with 40mm*	20 min
2	*Perform Movement Techniques* *Select Hasty Firing Position* *Engage Targets with 5.56mm (live-fire)* *Engage targets with 40mm*	10 min
NOTE: Training should include techniques outlined in FM 3-06.11		

Table 2-7. Example target scenario (presentation order) for Station 3.

Firing Position	*Weapon*	*Presentation Time*	*Range to Target*	*Target Presented*
1	5.56mm	6 sec	67 meters	Single Human Target (TGT 1)
2	40mm	2 min	105 meters	Multiple Human Targets (TGT 2)
3	5.56mm	6 sec	63 meters	Single Human Target (TGT 3)
4	5.56mm	6 sec	105 meters	Single Human Target (TGT 4)
5	40mm	2 min	57 meters	Multiple Human Targets (TGT 5)
6	40mm	2 min	70 meters	Multiple Human Targets (TGT 6)
6	5.56mm	6 sec	70 meters	Single Human Target (TGT 8)
7	40mm	2 min	50 meters	Multiple Human Targets (TGT 7)

STATION 4: URBAN OFFENSE/DEFENSE BUILDING, TASK/TECHNIQUE TRAINER

2-77. This station trains to attack or defend a building. Collective and individual TTP are trained. The platoon practices the task steps and performance measures for offensive and defensive tasks. This station increases the complexity of command, control, and maneuver (Figure 2-31).

Figure 2-31. Urban offense/defense building, Station 4.

DESCRIPTION

2-78. This station is a two-story building with a basement. Each floor has several rooms, windows, doorways, loopholes, and mouseholes. The gabled roof has windows and a hatch to the attic below and half of the rooftop is flat. The building has exterior and interior stairways.

TARGETRY

2-79. This station has ten target mechanisms.

CONTROLLER REQUIREMENTS

2-80. These are based on unit requirements.

Safety Requirements

2-81. Only smoke, practice, and M84 stun grenades may be used at this station. It is not a live-fire station. Training is conducted using TES with blank 5.56-mm, or TES or SESAMS. Units should consult with their supporting combat engineers for the proper demolition effects simulators (DES) to use against the blow panels throughout the station.

DANGER

TO AVOID ASPHYXIATION, NEVER DETONATE SMOKE GRENADES INSIDE BUILDINGS. THEY ARE *ALL* DANGEROUS IN AN ENCLOSED SPACE.

WARNINGS

1. **To avoid injury, place your weapon on SAFE when climbing through obstacles such as windows, fences, and walls.**

2. **To avoid inflicting carbon burns or more serious injury, never fire blank ammunition at other Soldiers within 1 meter.**

Demolition Effects Simulators

2-82. Demolition effects simulators (DES) replicate breaching to gain entry to buildings and rooms. The walls of the UAC, Shoot House, and CACTF facilities may be breached using wall breach charge DES. Generally, the interior and exterior walls in these facilities are made with blow panels, plasterboard, or similar material, and measure about 4 feet square.

2-83. The DES charges for use at the UAC, shoot house, and the CACTF are constructed to breach a predetermined entry point without damaging the facility.

Field-Expedient Charges

- Flex-linear shaped charge.
- Doorknob charge.
- E-silhouette charge.

Operational Charges

2-84. Unit trainers must consult with installation safety and combat engineers to determine the appropriate reduced charge. See also FM 3-06.11.

Individual or Collective Tasks Trained

Collective Task

- FM 3-21.8 (7-8).
  - -- 07-3-9018, *Clear a Building (Platoon/Squad).*
  - -- 07-6-1216, *Defend an Urban Area (Battalion/BCT).*

Related Collective Tasks

- 07-2-5027, *Conduct Consolidation and Reorganization (Company/Platoon).*
- 08-2-0003. *Treat Casualties.*
- 08-2-0004. *Evacuate Casualties.*
- 19-3-3100, *Process Detainees at a Collecting Point (CP).*
- 07-2-1342, *Conduct Tactical Movement (Company/Platoon).*
- 07-2-1189, *Conduct a Dismounted Tactical Road March (Company/Platoon).*
- 07-2-1198, *Conduct a Mounted Tactical Road March (Company/Platoon).*
- Warrior Battle Drill 12, *Enter and Clear a Room.*

Supporting Individual Tasks

- 052-193-1013, *Neutralize Booby Traps.*
- 071-010-0006, *Engage Targets with M249 Machine Gun.*
- 071-311-2007, *Engage Targets with M16-Series Rifle/M4 Carbine Rifle.*
- 071-315-2308, *Engage Targets with M16-Series Rifle/M4 Carbine Using a Night Vision Sight AN/PVS-4.*
- 071-325-4407, *Employ Hand Grenades.*
- 071-326-0501, *Move as a Member of a Fire Team.*
- 071-326-0503, *Move Over, Through, or Around Obstacles (except Minefields).*
- 071-326-0541, *Perform Movement Techniques during an Urban Operation.*
- 071-326-0557, *Select Hasty Firing Positions during an Urban Operation.*
- 071-326-0608, *Use Visual Signaling Techniques.*
- 071-326-5605, *Control Movement of a Fire Team.*
- 071-326-5611, *Conduct the Maneuver of a Squad.*
- 071-710-0008, *Operate Night Vision Goggles AN/PVS-7B.*
- 113-571-1022, *Perform Voice Communications.*
- 191-377-5250, *Handle Enemy Personnel and Equipment.*

Related Tactics, Techniques, and Procedures

- FM 3-06.11, *Combined Arms Operations in Urban Terrain.*
  -- Chapter 3, *Offensive Operations.*
  -- Chapter 5, *Fundamental Combat Skills.*
  -- Appendix M, *Field Expedient Breaching of Common Urban Barriers*
- FM 3-22.9, *Rifle Marksmanship.*

Training Support Requirements

2-85. Table 2-8 shows the minimum ammunition required to train at this station.

Table 2-8. Ammunition requirements for Station 4.

Ammunition	*Rounds for each Weapon*
5.56-mm Blank	16
5.55-mm Blank (Linked)	24
7.62-mm Blank	100
Practice Grenade Bodies	5
Practice Grenade Fuzes	10 per platoon
M84 Stun Grenade Bodies	5
M84 Stun Grenade Fuzes	10 per platoon
Smoke Grenades	4 per platoon

Conduct of Training

2-86. Table 2-9 shows the sequence of training for Station 4. Figures 2-32 and 2-33 show example scenarios of training events that can be conducted at this station. These are only examples. The unit should develop training events based on its METL and training needs.

Table 2-9. Sequence of events and times for Station 4.

Event	*Action*	*Estimated Time*
1	*Clear a Building*	2 Hours Crawl
		1 Hour Walk
		30 Minutes Run
2	*Defend Built-up Area/Building*	1 Hours Crawl
		1 Hour Walk
		2. Hour Run
3	*Conduct AAR*	1 Hour

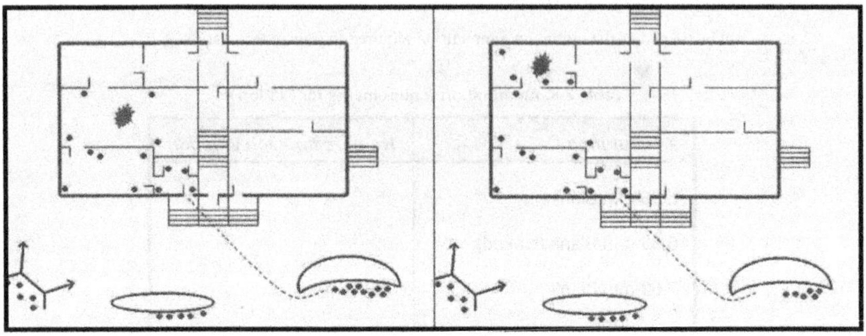

Figure 2-32. Event 1, *Clear a Building.*

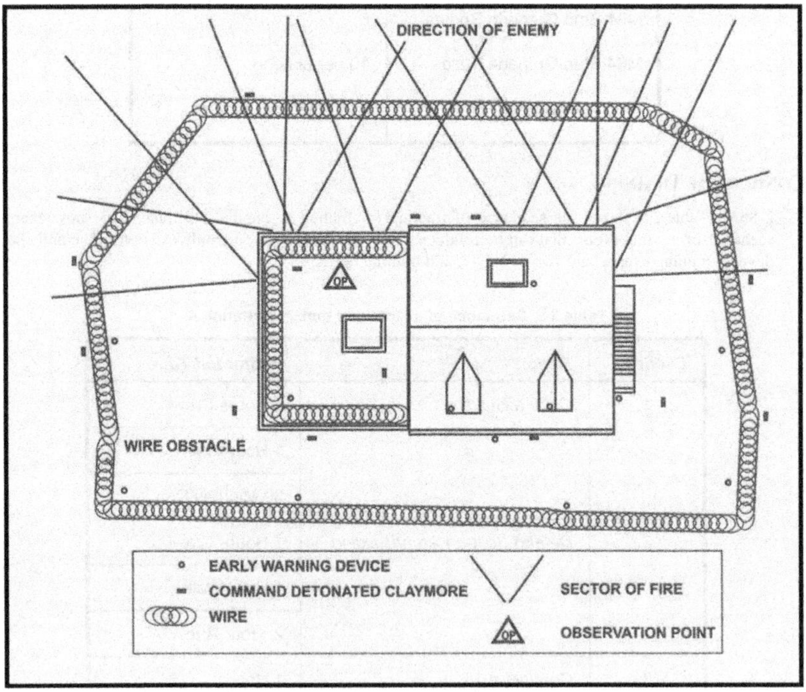

Figure 2-33. Event 2, *Defend a Building.*

STATION 5: UNDERGROUND TRAINER

2-87. This station provides training up to squad level for subterranean operations, specifically, the task *Clear and Move Tactically in a Subterranean Environment.*

DESCRIPTION

2-88. This station is a sewer system built in an irregular pattern, with four manhole covers. Two access tunnels lead from the outside tunnel to the underground loop, and the station is completely covered by dirt, only the manholes and ground level entrances are exposed (Figure 2-34).

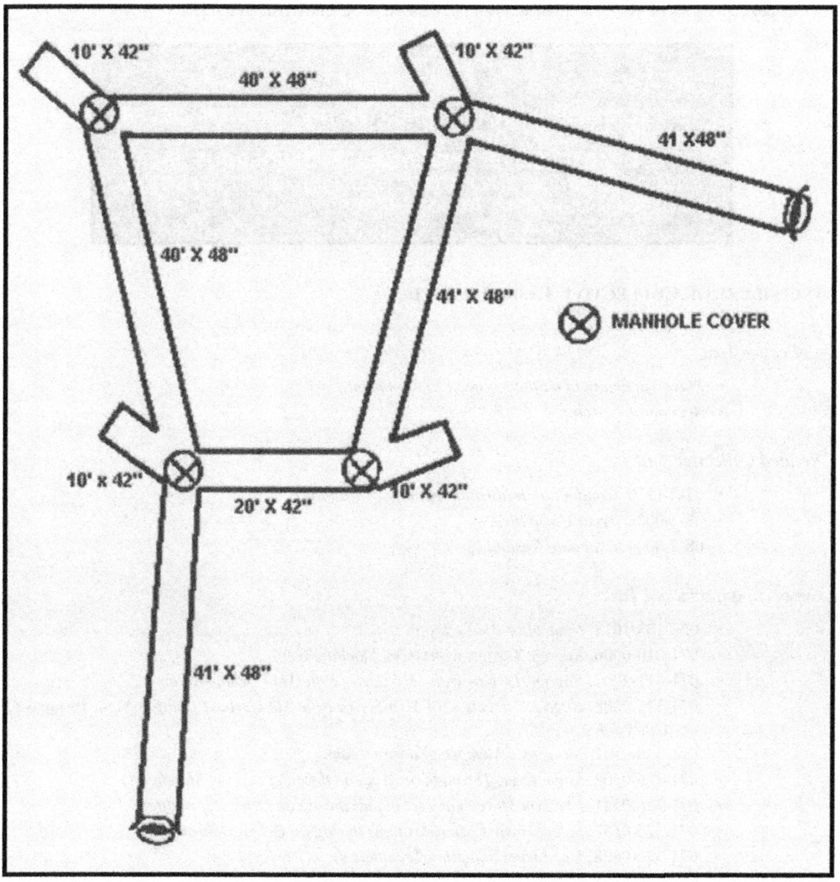

Figure 2-34. Underground Trainer, Station 5.

TARGETRY

2-89. Station 5 is a force-on-force training station and has no targetry.

CONTROLLER REQUIREMENTS

2-90. As required by unit.

SAFETY REQUIREMENTS

2-91. Soldiers should wear their helmets at all times when in the sewer system to avoid head injuries. Smoke grenades or M84 stun grenades *will not* be used in the underground trainer.

DANGER

TO AVOID ASPHYXIATION WHEN IN THE TRAINER, NEVER DETONATE SMOKE GRENADES IN THE SEWER. USING ANY TYPE OF EXPLOSIVE SIMULATOR IS DANGEROUS IN AN ENCLOSED SPACE. THE PROTECTIVE MASK IS NOT EFFECTIVE IN OXYGEN-DEFICIENT ATMOSPHERES.

INDIVIDUAL OR COLLECTIVE TASKS TRAINED

Collective Task

- *Move Tactically in a Subterranean Environment.*
- *Reconnoiter Area.*

Related Collective Tasks

- 07-2-1270, *Conduct an Infiltration (Company/Platoon).*
- 08-2-0003. *Treat Casualties.*
- 08-2-0004. *Evacuate Casualties.*

Supporting Individual Tasks

- 052-193-1013, *Neutralize Booby Traps.*
- 071-010-0006, *Engage Targets with M249 Machine Gun.*
- 071-311-2007, *Engage Targets with M16-Series Rifle/M4 Carbine Rifle.*
- 071-315-2308, *Engage Targets with M16-Series Rifle/M4 Carbine Using a Night Vision Sight AN/PVS-4.*
- 071-326-0501, *Move as a Member of a Fire Team.*
- 071-326-0503, *Move Over, Through, or Around Obstacles (except Minefields).*
- 071-326-0541, *Perform Movement Techniques during an Urban Operation.*
- 071-326-0557, *Select Hasty Firing Positions during an Urban Operation.*
- 071-326-0608, *Use Visual Signaling Techniques.*
- 071-326-5605, *Control Movement of a Fire Team.*
- 071-326-5611, *Conduct the Maneuver of a Squad.*
- 071-710-0008, *Operate Night Vision Goggles AN/PVS-7B.*

- 113-571-1022, *Perform Voice Communications.*
- 191-377-5250, *Handle Enemy Personnel and Equipment.*

Related Tactics, Techniques and Procedure

- FM 3-06.11, *Combined Arms Operations in Urban Terrain.*
  -- Chapter 3, *Offensive Operations.*
  -- Chapter 5, *Fundamental Combat Skills.*
  -- Appendix M, Field Expedient Breaching of Common Urban Barriers.
- FM 3-22.9, *Rifle Marksmanship.*

TRAINING SUPPORT REQUIREMENTS

2-92. Use Table 2-10 for the minimum ammunition required to train at this station.

Table 2-10. Ammunition requirements for Station 5.

Ammunition	*Rounds for each Weapon*
5.56-mm Blank	20
5.55-mm Blank (Linked)	40

CONDUCT OF TRAINING

2-93. Table 2-11 shows the sequence of training for this station. These are only examples. The unit should develop training events based on its METL and training needs.

Table 2-11. Sequence of events and times for Station 5.

Event	*Action*	*Estimated Time*
1	*Reconnoiter*	30 minutes (Crawl) 30 minutes (Walk) 30 minutes (Run)
2	*Conduct an Infiltration (Co/Plt)*	30minutes (Crawl) 30 minutes (Walk) 30 minutes (Run)
3	*Conduct AAR*	30 minutes

Chapter 3
Shoot House

This chapter describes the live-fire Shoot House and provides layouts and training scenarios. Fighting in built-up areas is fragmented; gains are often small. Therefore, units depend on the initiative, skill, and discipline of small-unit leaders and individual Soldiers. Leaders must be competent and confident in their urban operations skills and must have the courage to accomplish their missions while isolated from their parent units. Soldiers and leaders require mission-type orders that allow for decentralized execution.

PURPOSE

3-1. The Shoot House provides the leader with a facility to train and evaluate the unit during a live-fire exercise. Units are trained and evaluated on their ability to conduct tactical movement, engage targets, conduct breaches, and practice target discrimination in an urban environment.

DESCRIPTION

3-2. This facility has two parts, the live-fire facility and the AAR room:

LIVE-FIRE FACILITY

3-1. The live-fire facility is a single-level structure with a barn roof, catwalk, and eight adjoining rooms. Each room has video capability for AARs. The walls are bulletproof and prevent ricochets. Each side has an entrance. Replaceable blow panels throughout the facility can withstand reduced DES blasts. These also offer alternative entry points. A crane system in the roof allows trainers to selectively replace doorways and blow panels with solid wall sections. Before training explosive and mechanical breaching operations, trainers should place mobile exterior baffles or containment walls placed by each entry to prevent rounds and demolition effects from escaping. All of these capabilities let the unit vary its training scenarios. For safety, the Shoot House is windowless (Figure 3-1).

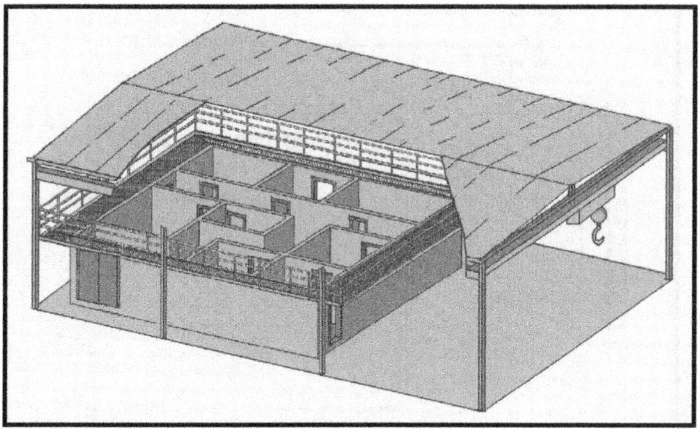

Figure 3-1. Shoot House.

AAR BUILDING

3-2. The AAR building has a conference/AAR room that can accommodate up to a platoon size element, an instrumentation and targetry control room, and an AAR edit and projection room that can handle a formal AAR (Figure 3-2).

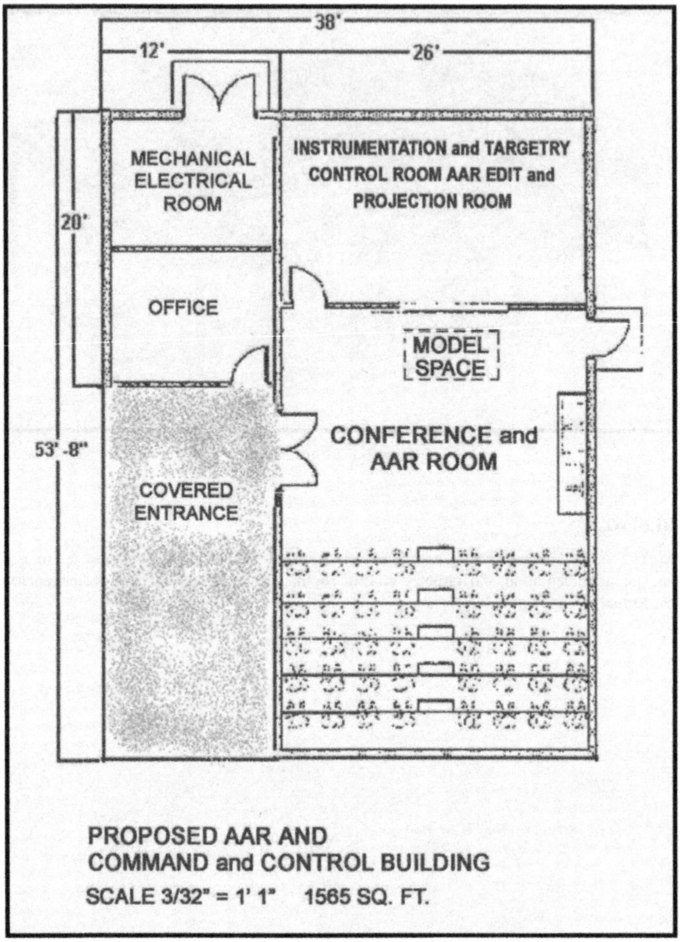

Figure 3-2. Command and control building and AAR room.

TRAINING AUDIENCE

3-3. The facility provides leaders the ability to train and evaluate the unit during a live-fire exercise. However, the unit's training requirements determine the size of the unit and the tasks to be trained at this facility. The unit develops its own organization and the steps for conducting the exercise based on its METL and training needs.

TRAINING INFORMATION

3-4. Tactical operations in an urban environment may range from combat missions to stability and support missions. Commanders should note that the Shoot House *is not* suited for the employment of live-fragmentation/concussion grenades, 40-mm HE/HEDP, antiarmor weapons, mortars, artillery, or air delivered munitions. Crew-served weapon systems, M2 and M1 tanks, and machine guns can be employed against supplemental targetry that is exterior to the Shoot House. The Shoot House has specially constructed areas that are used to conduct either mechanical or explosive breaching techniques. When conducting explosive breaching, units should consult with local Engineers for appropriate DES charges and consult Appendix A before conducting the training.

CONDUCT OF TRAINING

3-5. The unit develops realistic time frames for each operation or task. The times are based on the *crawl-walk-run* concept. The training event or STX can be arranged in any order as long as they follow a tactical scenario. *Assault a Building*, *Enter a Building and Clear a Room*, and *Clear a Building* are the primary collective tasks trained at this facility.

EXECUTION

3-6. Tables 3-1 through 3-5, pages 3-4 through 3-7, shows a sequence of events and times, and Figure 3-3 through Figure 3-6, pages 3-6 through 3-9, show example scenarios. These are only examples. The unit should develop training events based on its METL and training needs. However, the leader should do the following when conducting training at the live-fire Shoot House:

1. Identify the unit's critical missions (see commander's training guidance), and select those tasks that support specific tasks related to urban operations.
2. Arrange the related tasks in the order of their importance to mission accomplishment.
3. Assess the unit's current level of proficiency (strengths and weaknesses), and determine the degree of difficulty of each task. The training events and scenarios should start from the least difficult to the most difficult tasks. This gives the Soldiers confidence.
4. Identify the individual and leader tasks to be trained.
5. Conduct individual training first, because it is the first step in the building-block approach. It is also the basis for unit or collective task training. Mastery of individual skills is a precondition for effective training.
6. Set up training conditions based on unit proficiency. At first, keep the conditions basic, until Soldiers attain proficiency. Increase difficulty and realism as training progresses. Then, add limited visibility, CBRN factors, and live fires with service or short-range training ammunition (SRTA).
7. Understand that the Shoot House is a live-fire capable training facility. However, it can be used to train any level of proficiency, including blank, TES, SESAM, SRTA, and ball (no tracers). The leader assesses the unit's proficiency.

Table 3-1. Sequence of events and times for the platoon.

Element	*Event action*	*Estimated Time*
All 1st Squad 2d Squad 3rd Squad	1. Conduct precombat checks (for example; install, align, and troubleshoot MILES equipment; conduct inspection; draw equipment). 2. Support Operations. 3. Conduct Rehearsals. 4. Conduct Event 1 (Crawl).	2 Hours
3rd Squad 1st Squad 2d Squad	5. Support Operations. 6. Conduct Rehearsals. 7. Conduct Event 1 (Crawl).	2 Hours
2d Squad 3rd Squad 1st Squad	8. Support Operations. 9. Conduct Rehearsals. 10. Conduct Event 1 (Crawl).	2 Hours
1st Squad 2d Squad 3rd Squad	11. Support Operations. 12. Conduct Rehearsals. 13. Conduct Event 1 (Walk).	2 Hours
3rd Squad 1st Squad 2d Squad	14. Support Operations. 15. Conduct Rehearsals. 16. Conduct Event 1 (walk).	2 Hours
2d Squad 3rd Squad 1st Squad	17. Support Operations. 18. Conduct Rehearsals. 19. Conduct Event 1 (walk).	2 Hours
1st Squad 2d Squad 3rd Squad	20. Support Operations. 21. Conduct Rehearsals. 22. Conduct Event 1 (Run).	1 Hours
1st Squad 2d Squad 3rd Squad	23. Support Operations. 24. Conduct Rehearsals. 25. Conduct Event 1 (Run).	1 Hours
1st Squad 2d Squad	26. Support Operations. 27. Conduct Event 1 (Run).	1 Hours

Table 3-2. Event 1, Squad Sequence (*Crawl*).

Element	*Event Action*	*Estimated Time*
1 (CRAWL)	**Terminal Learning Objective** Familiarize the squad with how to enter and clear a room during the crawl phase of training **Tasks** *Enter and Clear a Room (Squad)* • *Perform Movement Technique* • *Select Hasty Firing Position* *Conduct a Breach (Mechanical)* • *Employ Hand Grenades* • *Engage Targets with M16A2/M4* • *Engage Targets with M249 Machine Gun*	**2 Hours**
NOTES: 1. Perform controlled pairs and target discrimination techniques (FM 3-22.9). 2. Include techniques (FM 3-06.11) and performance measures (FM 3-21.8 and DTMS).		

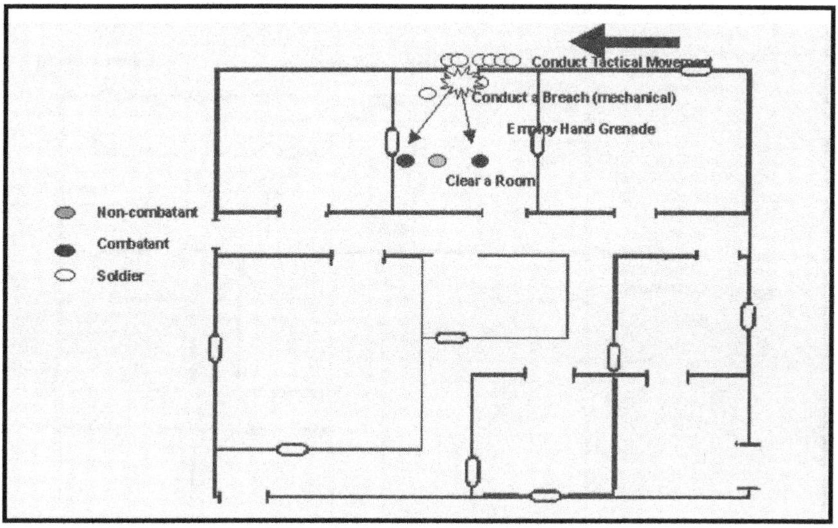

Figure 3-3. Event 1, *Squad Scenario (Crawl)*, force-on-targetry.

Table 3-3. Event 2, Squad Sequence (*Walk*).

Element	*Event Action*	*Estimated Time*
2 (WALK)	**Terminal Learning Objective** Familiarize the squad with how to enter and clear multiple rooms during the walk phase of training **Tasks** *Enter and Clear a Building (Squad)* • *Perform Movement Technique* • *Select Hasty Firing Position* *Conduct a Breach (Explosive DES)* • *Engage Targets with M16A2/M4* • *Engage Targets with M249 Machine Gun* *Conduct AAR*	2 Hours
NOTES: 1. Perform controlled pairs and target discrimination techniques (FM 3-22.9) 2. Include techniques (FM 3-06.11) and performance measures (FM 3-21.8 and DTMS).		

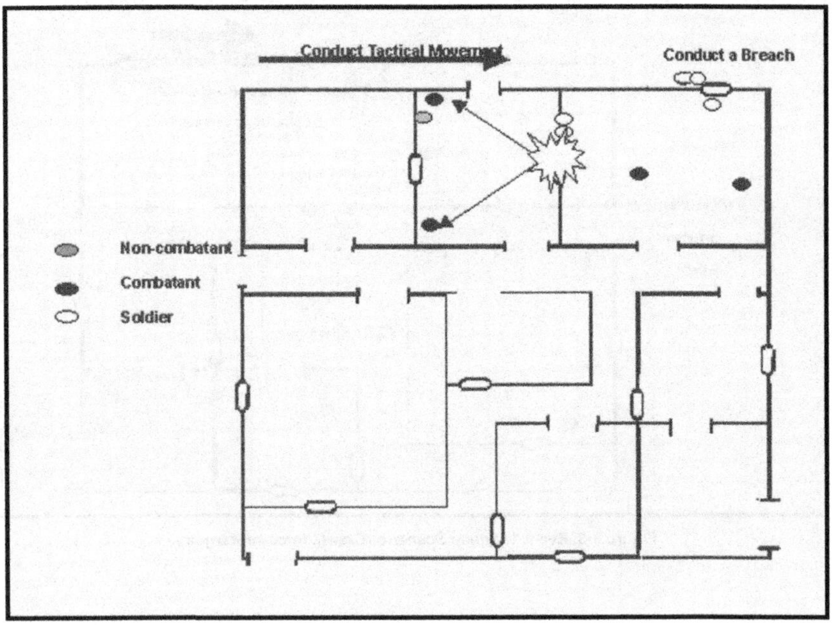

Figure 3-4. Event 2, *Squad Scenario (Walk)*, force-on-force.

Table 3-4. Event 3, Squad Sequence (*Run*).

Element	*Event Action*	*Estimated Time*
3 (RUN)	**Terminal Learning Objective** Refamiliarize squad with how to enter and clear multiple rooms during the run phase of training. **Tasks** *Enter and Clear a Building (Squad)* • *Perform Movement Technique* • *Select Hasty Firing Position* *Conduct a Breach (Mechanical)* • *Engage Targets with M16A2/M4* • *Engage Targets with M249 Machine Gun* *Clear a Hallway* • *Engage Targets with M16A2/M4* • *Engage Targets with M249 Machine Gun* *Conduct a Breach (explosive DES)* • *Engage Targets with M16A2/M4* • *Engage Targets with M249 Machine Gun* *Conduct AAR*	**1 Hour**

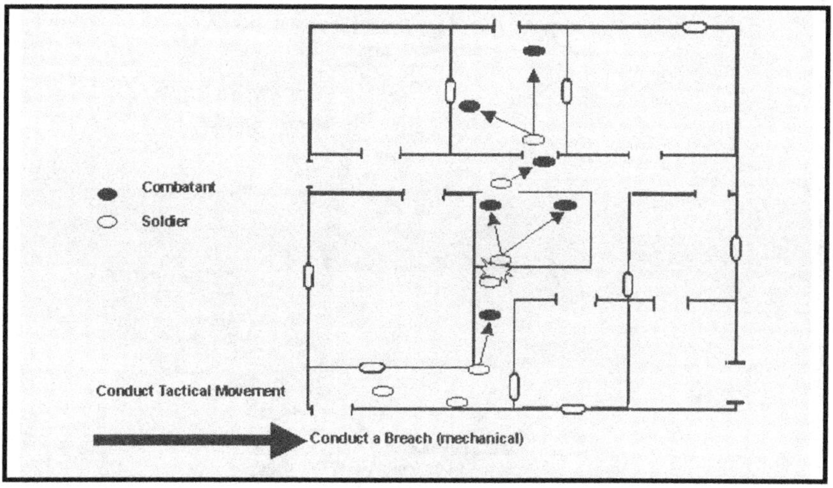

Figure 3-5. Event 3, *Squad Scenario (Run)*, force-on-targetry.

Table 3-5. Event 4, Platoon Evaluation Sequence (*Run*).

Element	*Event Action*	*Estimated Time*
4 (RUN)	*Occupy Assembly Area*	1 Hour
	Develop and Communicate a Plan	2 Hours
	Prepare for Combat	30 min
	Perform Passage of Line	1 Hour
	Conduct Tactical Movement	
	Break Contact	5 min
	Conduct Intermediate AAR	20 min
	Conduct Tactical Movement	
	Perform Overwatch/Support by Fire	1 Hours
	Assault a Building	1 Hours
	Consolidate and Reorganize	2 Hours
	Conduct Final AAR	1 Hours

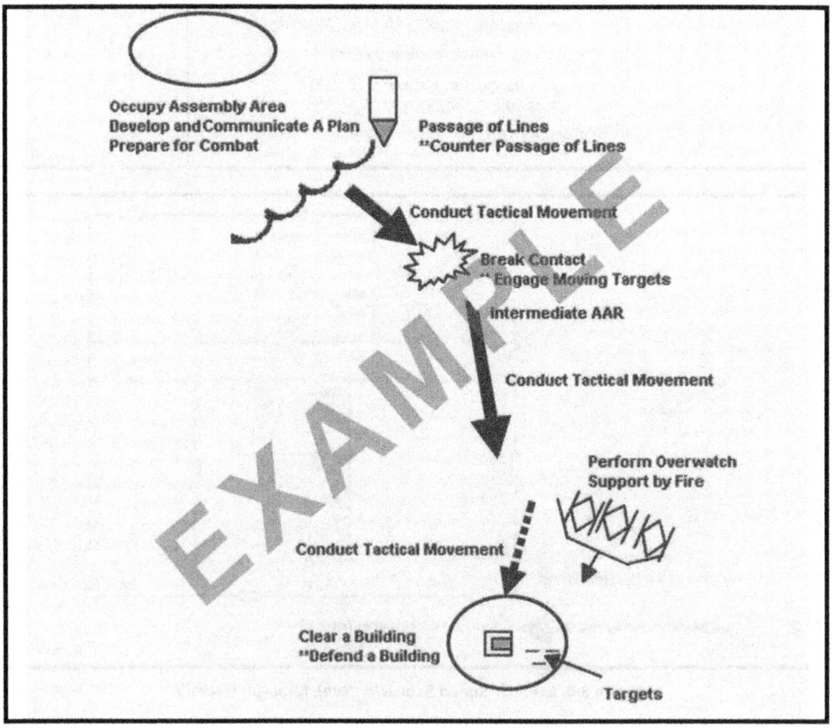

Figure 3-6. Example platoon tactical scenario (*Run*).

OPERATION ORDER

3-7. Figure 3-7 shows an example OPORD for the conduct of training.

```
                    ______________
                   (Classification)
                                             Copy ____ of ___ copies
                                         ____Company _____ Infantry
                                         (Unit Location)___________
                                          Date/time Group __________

(Message Reference Number)

OPORD: ____________
Reference Map, Series __________________
Time Zone used Throughout the Order:____________________

TASK ORGANIZATION:

______Platoon Platoon
______Platoon ____ 60-mm Section
______Antiarmor Section
```

1. SITUATION

a. **Enemy Forces**. Annex B (Intelligence Overlay [prepared by battalion S-2]).

(1) The company is opposed by elements of the Red Company of the Blue Brigade at about 60 percent strength. Within our zone, we have a platoon-sized element occupying a defensive position in a building in the vicinity of (______). The enemy has small arms weapons and may call on company mortars for support.

(2) Current indications are the enemy is regrouping to initiate small-unit offensive actions in the area. His most likely course of action is to move to the (direction) to link up with elements of the parent unit.

b. **Friendly Forces.**

(1) (___) Infantry attacks at (______) hours to seize Objective ALPHA (grid). The intent is to prevent the enemy element from moving (direction) to link up with other enemy units.

(2) Company (__) on our (left, right) makes the main attack at (____) hours to seize Objective RED (grid) in order to establish blocking positions to stop the enemy's movement to the (direction).

(Classification)

Figure 3-7. Example company OPORD.

(Classification)

2. **MISSION**

Company (__) attacks to seizes Objective BLUE (grid) at (date/time) hours in order to prevent the enemy from regrouping and engaging Company (__) (main effort) from Objective BLUE.

3. **EXECUTION**

a. **Concept of Operation**. Annex C (Operations Overlay). (This should include those steps listed in the task , Conduct Troop-Leading Procedures, paragraph 17. These should be written out when possible and not just placed on an overlay.)

(1) Maneuver. (___) Platoon establishes a defense in its zone, NLT (date/time) hours, to destroy small enemy elements moving or exfiltrating to the (direction), from the vicinity of Objective BLUE. This defense supports the attack on Objective BLUE by completing destruction of any enemy that may escape. (Evaluated) Platoon conducts the company main attack at (hours) to seize Objective BLUE to destroy the enemy or force them to withdraw into (___) Platoon's defense. (___) Platoon supports the attack of the (Evaluated) Platoon on Objective BLUE to suppress enemy fires and isolate the objective. On seizure of Objective BLUE, (Evaluated) Platoon and (___) Platoon will defend in sector to destroy the enemy elements withdrawing to avoid (___) Company's attack.

(2) **Fire support**.

(a) Priority of fires.

- Initially to (Evaluated) Platoon.
- Upon seizure of Objective BLUE, priority shifts to (Defensive) Platoon.

(b) Priority targets.

- 60-mm priority target initially AA4031.
- On initiation of assault on Objective BLUE, priority shifts to target AA4027 to isolate the objective.
- Upon seizure of Objective BLUE, priority shifts to target AA4025.

(Classification)

Figure 3-7. Example company OPORD (continued).

(Classification)

b. **Tasks to maneuver units.**

(1) (___) Platoon. Destroy enemy elements moving to the (direction), NLT (date/time).

(2) (Evaluated) Platoon. Consolidate and defend in sector to destroy enemy elements following the seizure of Objective BLUE.

(3) (___) Platoon. Be prepared to assault Objective BLUE, on order, consolidate and defend in sector to destroy enemy elements following the seizure of Objective BLUE.

c. **Tasks to Subordinate Units.**

(1) 60-mm Section. Initially in the vicinity of CP 7, priority of fires to the (Evaluated) Platoon. Move with the (Supporting) Platoon. On order, priority of fires to the (Defensive) Platoon.

(2) Antiarmor Section. Attached to the (Supporting) Platoon.

d. **Coordinating Instructions**:

(1) (Defensive) Platoon reports defense established and locations.
(2) Platoon reports the seizure of all objectives.
(3) Platoon reports the consolidation and establishment of the defense.
(4) Order of march from the assembly area to the objective: defense platoon, evaluated platoon, headquarters, support platoon, 60-mm mortar.

4. **SERVICE SUPPORT**

a. **General**. Company trains collocated with the battalion combat trains in the vicinity of (grid). Supply distribution at the company CP.

(Classification)

Figure 3-7. Example company OPORD (continued).

(Classification)

b. **Materiel and Services.**

(1) Class I: Ration cycle MRE-MRE-MRE.

(2) Class V: Pick up basic load at company CP, NLT (date/time).

(3) Medical: Company casualty collection point is __________.

(4) EPW: EPW collection point is ___________.

5. **COMMAND AND SIGNAL**

a. **Command.** Company command group initially follows the (Evaluated) Platoon.

b. **Signal.**

(1) Radio-listening silence is in effect until contact is made.

(2) SOI _____ is in effect.

JONES
Commanding

AnnexES:
B-Intelligence Overlay
C-Operations Overlay
D-Fire Support Overlay

(Classification)

Figure 3-7. Example company OPORD (continued).

TRAINING SUPPORT REQUIREMENTS

3-8. Minimum trainers and evaluators should be based on unit proficiency, training needs, and local safety requirements. Table 3-6 contains consolidated support requirements.

Table 3-6. Consolidated support requirements.

Ammunition	*Event 1*	*Event 2*	*Event 3*	*Event 4 (Optional)*
5.56-mm (per weapon) ball, blank, SRTA, or SESAMS	20 Blank	20 SESAMS	20 Ball/SRTA	20 Ball/SRTA
5.56-mm linked (per weapon)	30 Blank	30 SESAMS	30 Ball/SRTA	30 Ball/SRTA
Hand grenade body (practice)	0	--	4	4
Hand grenade fuzes (practice) per platoon	0	--	4	4
Hand grenade, smoke, per platoon	0	6	6	6
M84 hand grenade, stun, per platoon	0	4	4	0
TES (MILES Equipment)				
M16 (1 per weapon, 1 per OPFOR)	NO	YES	NO	NO
M249 (1 per weapon)	NO	YES	NO	NO
SAAF (1 per platoon)	NO	YES	NO	NO
Controller Gun (1 per OC)	NO	YES	NO	NO
Targetry	5	0	5	10
OPFOR	NO	6	NO	NO

TRAINING AND EVALUATION

3-9. The T&EOs are the foundation of the MTP and the collective training of the platoon. T&EOs are training objectives (task, conditions, and standards) for the collective tasks that support platoon critical wartime operations. T&EOs may be trained separately, in an STX, in an FTX, or in live-fire exercises. For collective live-fire standards, the trainer needs to refer to the applicable Infantry Gunnery Manual for the appropriate course of fire. Those standards and courses of fire need to be integrated into the training exercise. Table 3-8 provides a list of related collective and supporting individual tasks.

Table 3-7. Related collective and supporting individual tasks.

Task	FM 3-21.8	FM 3-06.11	STP 21-1-SMCT	STP 7-11B1-SM-TB	STP 21-24-SMCT
Conduct Consolidation and Reorganization Treat Casualties Evacuate Casualties Process Enemy Prisoners of War/Captured Materiel Defend MOUT Building Conduct Tactical Movement	X	X			
Supporting Individual Tasks					
Engage Targets with an M16-Series Rifle			071-311-2007		
Engage Targets with an M16-Series Rifle Using a Night Vision Sight AN/PVS-4			071-315-2007		
Engage Targets with an M249 Machine Gun				071-010-0006	
Employ Hand Grenades			071-325-4407		
Move Over, Through, or Around Obstacles (Except Minefields)			071-326-0503		
Neutralize Booby Traps					052-193-1013
Move as a Member of a Fire Team				071-326-0501	
Perform Movement Techniques during MOUT		X		071-326-0541	
Conduct the Maneuver of a Squad				071-326-5611	
Conduct an Explosive Breach				071-440-0030	
Execute a Mechanical Breach				071-440-0027	
Conduct a Breach				071-440-0029	
Operate Night Vision Goggles AN/PVS-7				071-710-0008	
Select Hasty Firing Positions during MOUT				071-326-0557	
Control Movement of a Fire Team				071-326-5605	

Chapter 4
Breach Facility

Urban operations require Soldiers to enter buildings through walls, doors, windows, and roofs. The Breach Facility is designed to train Soldiers using tasks and techniques on how to breach locked doors, windows, and create man-sized holes in walls. The facility can be used to train mechanical, ballistic, thermal, and explosive breaching.

PURPOSE

4-1. The Breach Facility (Figure 4-1) is used to train Soldiers semiannually on the technical aspects of breaching techniques. It is also used to train TTP and explosive techniques not trained on any other type range.

Figure 4-1. Breach facility.

DESCRIPTION

4-2. This facility has three stations:

STATION 1

4-3. This station is a facade that is 8 feet high and 88 feet wide with four sections of two doors for a total of eight doors. Soldiers can conduct mechanical, thermal, ballistic, and explosive breaching techniques on doors at this station. Half of the doors open from the exterior, and the other half of the doors open from the interior (Figure 4-2).

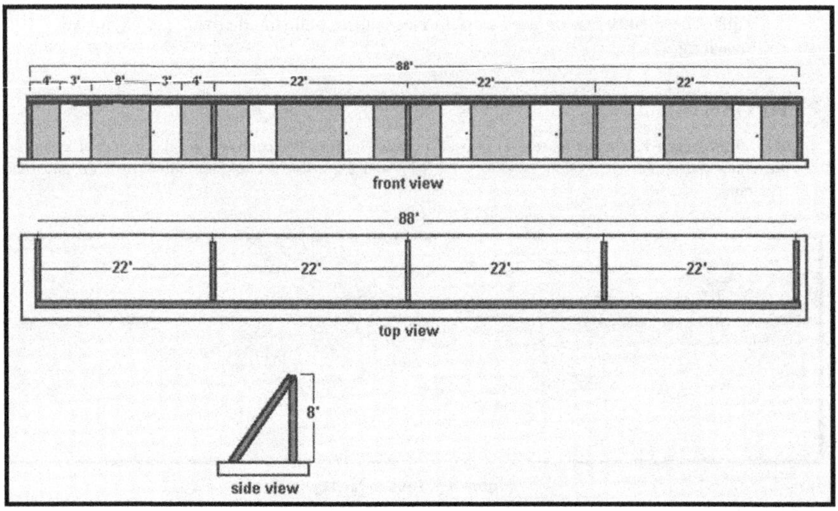

Figure 4-2. Station 1—Door breaching.

STATION 2

4-4. This station is a facade. It measures 10 feet high by 88 feet wide. It has four sections of two windows each, for a total of eight windows. Soldiers can conduct mechanical and explosive breaching techniques on windows at this station (Figure 4-3).

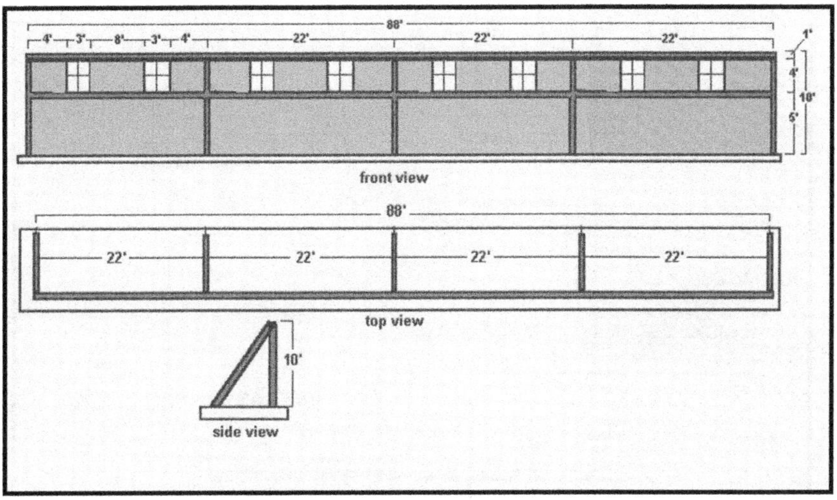

Figure 4-3. Station 2—Window breaching.

STATION 3

4-5. This station is a facade that looks like a wall of a building. It measures about 8 feet high by 28 feet wide. Precast panels will fit into each of the three sections. The panels measure 8 feet square and can be made of concrete, concrete blocks, stone, brick, wood or other material. The panels are designed for use with thermal and explosive breaching techniques (Figure 4-4).

TRAINING AUDIENCE

4-6. The facility provides Infantry, Engineer, MP, and Cavalry leaders with a facility to train and evaluate individual and leader breaching tasks. Unit training requirements determine the size of the unit and the tasks to be trained at this facility (Appendix B provides a unit training matrix). The individual tasks trained at the Breach Facility follow:

- 052-193-1003, *Prime Explosives Nonelectrically.*
- 071-440-0027, *Execute a Mechanical Breach.*
- 051-193-1055, *Construct a Nonelectric Initiating Assembly with Modernized Demolition Initiators (MDI).*
- 052-193-1312, *Construct Demolition Initiating Sets.*
- *Conduct a Thermal Breach* (Engineers only).

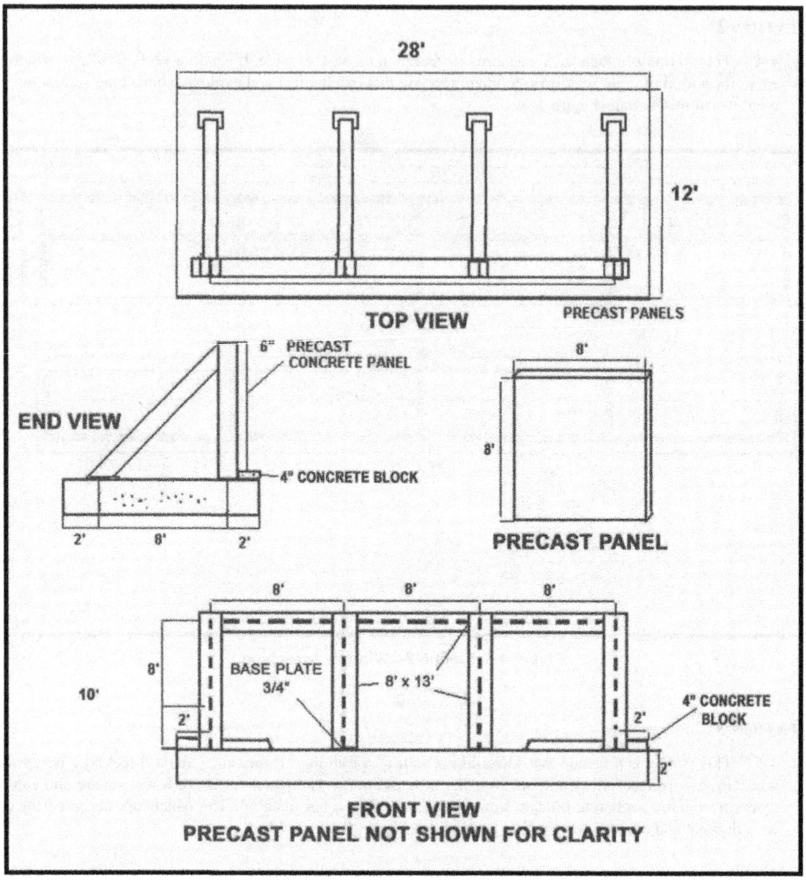

Figure 4-4. Station 3—Wall breaching.

TRAINING INFORMATION

4-7. Many types of organizations can train at this facility. Tactical operations in an urban environment may range from combat missions to stability and support actions. Commanders should note that this facility is an individual task and technique trainer. Training at the Breach Facility should be conducted prior to conducting collective training at the UAC, shoot house, and the CACTF. Trainers *must be certified* before conducting demolition training per local SOP or ACOM policy.

CONDUCT OF TRAINING

4-8. Trainers develop realistic time requirements for each task to be trained based on the *crawl-walk-run* concept and the size of the unit to be trained. The training event may be arranged in any order as long as it builds from simple to complex task and techniques. Table 4-1 contains an example sequence of events to assist units in developing their training plan. Leaders should do the following prior to conducting training. Before conducting ballistic breaching with automatic weapons, leaders should consult FM 3-06.11:

- Identify the unit's critical missions (see commander's training guidance) and select those tasks that support specific METL related to urban operations.
- Arrange the related tasks in the order of their importance to mission accomplishment.
- Assess the unit's current level of proficiency (strengths and weaknesses) and determine the degree of difficulty of each task. The training events should start from the less difficult tasks to the more difficult tasks. This instills confidence in the Soldiers conducting the training. Leaders should consult FM 3-06.11 for more information on explosive breach techniques.
- Identify the individual and leader tasks that are to be trained.
- Set up training conditions based on unit proficiency.

Table 4-1. Example platoon sequence of events.

Element	*Event Action*	*Estimated Time*
3d Sqd	Wall Breach: Explosive (crawl, walk, run)	3 hours
1st Sqd	Door Breach: Mechanical, ballistic, explosive (crawl, walk, run)	
2d Sqd	Window Breach: Mechanical, explosive (crawl, walk, run)	
2d Sqd	Wall Breach: Explosive (crawl, walk, run)	3 hours
3d Sqd	Door Breach: Mechanical, ballistic, explosive (crawl, walk, run)	
1st Sqd	Window Breach: Mechanical, explosive (crawl, walk, run)	
1st Sqd	Wall Breach: Explosive (crawl, walk, run)	3 hours
2d Sqd	Door Breach: Mechanical, ballistic, explosive (crawl, walk, run)	NOTE: Walls, doors, and windows require replacement after each run/iteration to train to standards.
3d Sqd	Window Breach: Mechanical, explosive (crawl, walk, run)	

TRAINING SUPPORT REQUIREMENTS

4-9. Table 4-2 shows minimum requirements based on three breach teams for each platoon-sized element. This facility is resource intensive, and Table 4-3 shows the total annual requirements per platoon-sized element.

Table 4-2. Consolidated minimum support requirements.

Ammunition/Equipment	*Crawl*	*Walk*	*Run*
Mechanical Breach: *Prying , Striking, or Cutting Tool*			
Door	(talk through)	1 per team (defeat one lock)	1 per team (defeat hinges)
Window	(talk through)	1 per team (defeat one lock)	1 per team (defeat hinges)
Ballistic Breach: Door			
Shotgun [1]	(talk through)	2 rds per weapon (defeat one lock)	5 rds per weapon (defeat hinges)
GREM (inert) [2]	(talk through)	1 per team – inert	1 per team - service
[1] If available to unit. [2] Grenade Rifle Entry Munition (GREM).			
Explosive Breach:			
Door: Det Cord (Flex Linear Charge) Nonelectric Firing System	(build)	(practice emplacement)	29 feet per team 1 per team
Wall: C4 (Satchel Charge) Nonelectric Firing System	(build)		4 blocks per team
		(practice emplacement)	23 feet per team 1 per team
Window: Det Cord (Flex Linear Charge) Nonelectric Firing System	(build)		29 feet per team 1 per team
		(practice emplacement)	
Thermal Breach Equipment Sappers (Only)	1	1	1

Table 4-3. Consolidated annual support requirements.

Ammunition/Equipment	*Total*
Doors	6
Windows	6
Wall Panels	6
Shotgun Rounds	42
GREM	6 inert/2 service
Det Cord (feet)	486
C4 (blocks)	24
Nonelectric Firing Systems	18

TRAINING AND EVALUATION

4-10. Individual training is the first step in the building-block approach and sets the basis for unit or collective training. Mastery of individual skills is an essential precondition for urban operations training and a good method for conducting individual training is through concurrent training. Concurrent training is designed to train groups of Soldiers on different tasks or techniques and these tasks or techniques may or may not be related. For example, a leader may subdivide the unit at the Breach Facility into breaching orders. One group of Soldiers may be conducting an explosive breach, while other groups are conducting mechanical and ballistic breaches. Ballistic breaching and explosive breaching does have a SDZ, an MSD, and ricochet considerations.

SAFETY CONSIDERATIONS

4-11. Safety requires effort on the part of everyone. It is not enough to teach safety rules; the rules must be practiced and enforced. Depending on the nature of the obstacle, the breacher may use mechanical, ballistic or explosive techniques. The following guidelines should be followed when training at the Breach Facility.

- Wear all appropriate personal protective equipment and follow appropriate guidelines:
  - -- Safety glasses.
  - -- Hearing protection.
  - -- Body armor.
  - -- Helmet.
  - -- Gloves.
  - -- Elbow and knee pads.
- Inspect equipment before use.
- Use appropriate tool.
- Use appropriate amount of explosive not to exceed four blocks of C4. Place the C4 on the wall only, not on the foundation, as this would cause irrepairable damage.
- Stay within the surface danger zone.
- Always use the minimum safe distance.
- Do not modify tool.
- Ensure proper use and maintenance of hydraulic and pneumatic tools; tools can burst under pressure when improperly used.

Chapter 5

Combined Arms Collective Training Facility

The home station CACTF is designed to conduct multiechelon, full-spectrum operations training up to battalion TF level. Example company and platoon STX lanes, and stability and support operations tasks, are provided to help units develop training plans. Leaders use training exercises to train, evaluate, and practice performance-oriented collective tasks. TEWT, STX, and FTX are the three types of exercises covered in this chapter. Each is designed to help the trainer develop urban operations training skills and increase unit proficiency in an urban environment. The urban operations collective training exercises were developed using the standard CACTF as described in this chapter and may be modified to fit available urban operations facilities.

PURPOSE

5-1. The leader or commander is provided with a CACTF to evaluate unit urban operations proficiency. The CACTF is intended to support blank fire, MILES/TES, SESAMS, STX, and FTX scenarios on a semiannual basis. The CACTF supports branch-specific lane training and combined arms training up to battalion level across the full spectrum of operations: offense, defense, stability, and support.

DESCRIPTION

5-2. The CACTF replicates an urban environment. The facility consists of 2.25 square kilometers of urban sprawl with 20 to 26 buildings, roads, alleys, parking areas, underground sewers, parks, athletic fields, and command and control building. The actual size and configuration of the CACTF depends on the local installation site requirements. The CACTF is designed to support heavy and light Infantry, Armor, artillery, and aviation positioning and maneuver. Table 5-1 shows the recommended types of buildings and features in the CACTF.

Table 5-1. CACTF characteristics.

20 to 26 Buildings (1.5 km by 1.5 km)	
Tunnel and sewer system Shantytown One 3-story building Three 2-story buildings Industrial area Electricity and potable water City dump	Props and furniture Targetry Audio and image capture with EDIT/REPLAY capability Control building with AAR facility Breachable walls

5-3. The buildings have one, two, or three stories (some with basements), sloped and flat roofs. CACTF building variations pose different tactical and technical training challenges. Figure 5-1 shows an example CACTF containing 24 buildings with the following characteristics:

- The hotel is a dominating structure typical of a central business district. The hotel has an elevator shaft, fire escapes, and a large first-floor lobby.
- The two-story office building contains a series of identical rooms.
- The school has a long, central corridor, large windows, and a single large room.
- The townhouse has multiple sections with a common attic.
- The service station accommodates combat vehicles in its service bay.

5-4. The buildings have other features that enhance the CACTF's training value.

- Constructed mouse-holes permit movement between rooms and floors.
- Loopholes in roofs and outside walls allow observation and fields of fire.
- Roof hatchways lead to the top floors of multiple story buildings.

5-5. Rooms vary in size, and in number of doorways and windows. These variations require Soldiers to identify rooms that provide suitable indoor fighting positions for TOW, Javelin, Dragon, and AT4 antitank weapon systems.

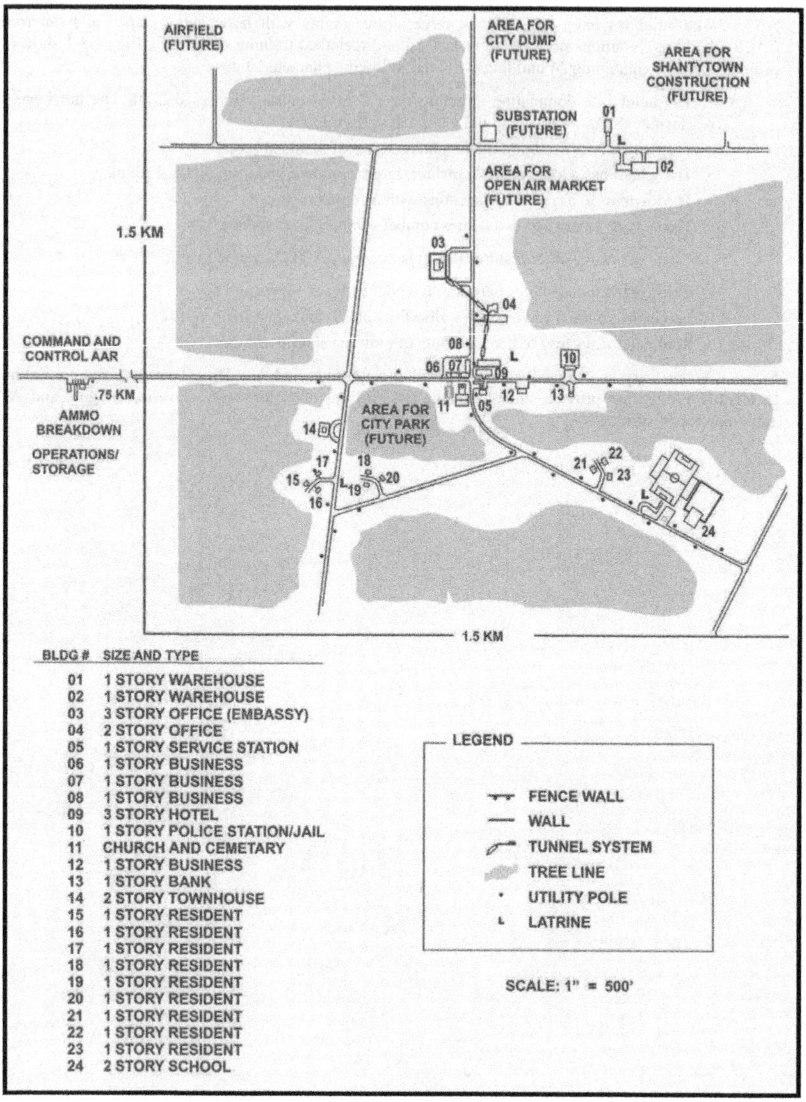

Figure 5-1. Example CACTF schematic.

5-6. Buildings are arranged in a realistic urban pattern. The CACTF represents an area consisting mainly of residential, commercial, public institutional, and light industrial buildings. Buildings and streets are arrayed with sufficient dispersion to allow up to nine platoon-sized STX sectors (Figure 5-2) or six company/team STX sectors (Figure 5-3).

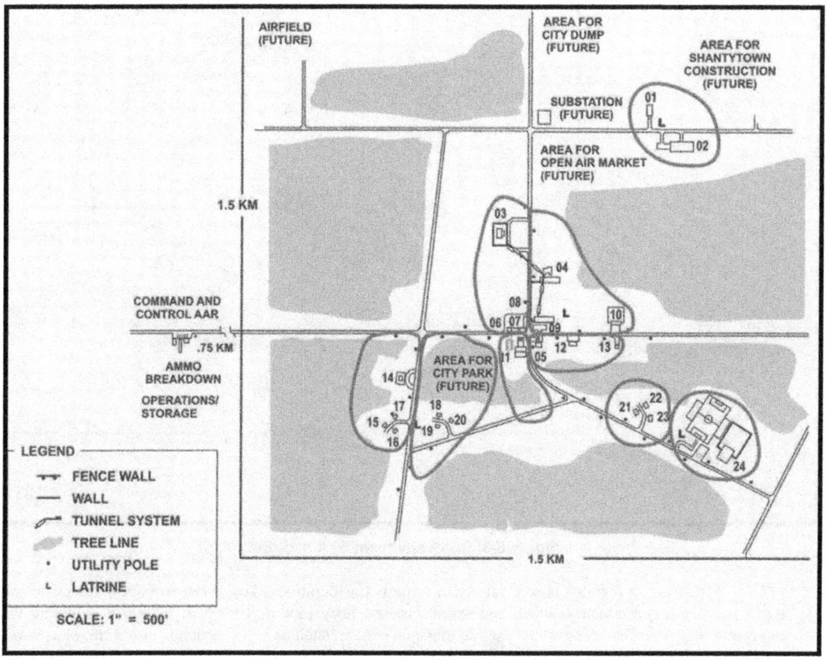

Figure 5-2. Platoon ODSS STX sectors.

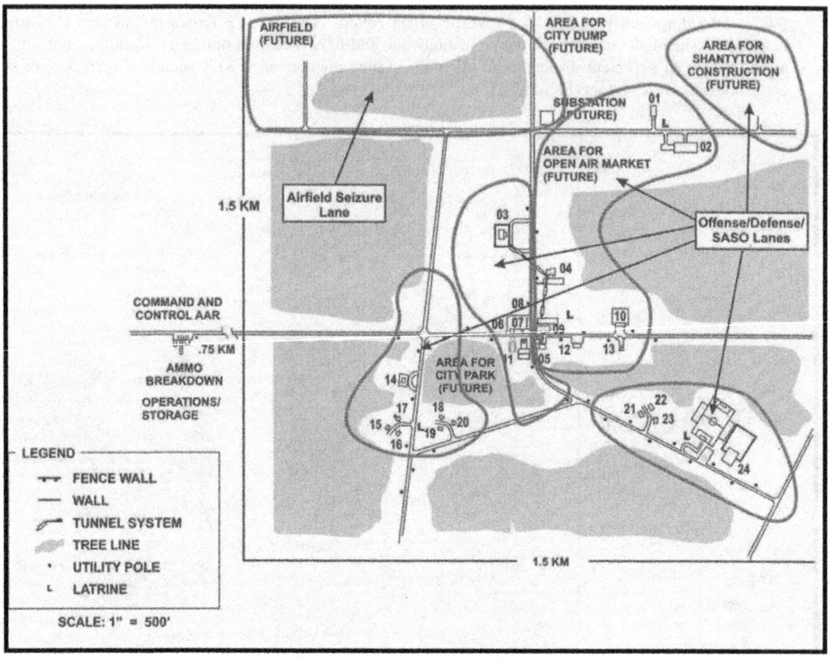

Figure 5-3. Company/team STX sectors.

5-7. Other urban features pose a variety of terrain considerations. The street network includes one and two-lane primary, secondary, local, and service roads. They join in T-shaped, L-shaped, and four-way intersections. Other features may be used to enhance realism such as signs, benches, furniture, newsstands, utility poles, street lights, vegetation, simulated rubble, live civilians on the battlefield, role players, and OPFOR. Open areas next to the CACTF can be used for maneuvering and staging areas. The specific details and layout of the CACTF may vary from one installation to another.

> **CAUTION**
>
> Trainers must ensure that fighting position reinforcement does not exceed the weight limits of the buildings in order to avoid structural damage and safety hazards to the Soldiers.

TACTICAL EXERCISE WITHOUT TROOPS

5-8. A terrain walk and a TEWT may be conducted in a local downtown area to prepare leaders and to take advantage of the available time in the CACTF. Terrain walks and TEWTs are conducted to teach leaders the tactical value of urban terrain. Appendix A provides training tips and additional information regarding the conduct of a TEWT.

COLLECTIVE TRAINING EXERCISES

5-9. The following exercises were developed IAW the guidance in FM 7-0, FM 7-1, FM 25-4, and TC 7-98-1. These publications include offensive and defensive scenarios for combat in urban areas, specifically, two platoon STXs and two company FTXs. Each can be conducted with blank ammunition; with MILES/TES; with 9-mm or 5.56-mm SESAMS for short-range precision target engagements (less than 50 meters); or force on force. Each was chosen to help the commander develop, sustain, and evaluate his unit's mission proficiency:

- Infantry Rifle Platoon Attack STX
- Mechanized Infantry Platoon Defend STX
- Infantry Rifle Company Attack FTX
- Mechanized Infantry Company Team Defend FTX

INFANTRY RIFLE PLATOON STX ATTACK

1. **Objective.**

This example STX trains collective, leader, and individual tasks for the platoon operation, Attack in a Built-Up Area.

2. **Interface.**

This STX is supported by two warrior drills, *React to Contact* and *Enter and Clear a Building*.

3. **Training.**

a. *Guidance*. FM 7-0 and FM 7-1 provide training guidance, tips, and enhancers for planning and executing STXs. The trainer should review the individual, leader, and collective tasks to be performed during the STX to determine the ones that require initial or refresher training.

b. *General Situation.*

(1) The exercise scenario follows: Contact with the enemy has been reestablished. Initial reports indicate he is at 65 to 70 percent strength, and that he has not received reinforcements. His defensive positions are in the CACTF and are not well established. He has indirect fire and CAS; he has already used chemicals, and will probably do so again. The platoon receives an order to prevent the enemy from establishing a heavily fortified defense in the CACTF. Such a defense could prevent friendly forces from continuing offensive operations.

(2) This exercise begins when the platoon leader receives the company FRAGO, and it ends after the platoon consolidates and reorganizes on the objective. An AAR should be held after completing the task, *React to Contact*, and at the completion of the exercise. If necessary, portions of the exercise should be repeated until the platoon performs them to standard. Figure 5-4 provides a graphic scenario of task performance for the STX *Attack*. Table 5-2 provides a recommended sequence for performance of T&EOs, and the time allotted for each portion of the STX.

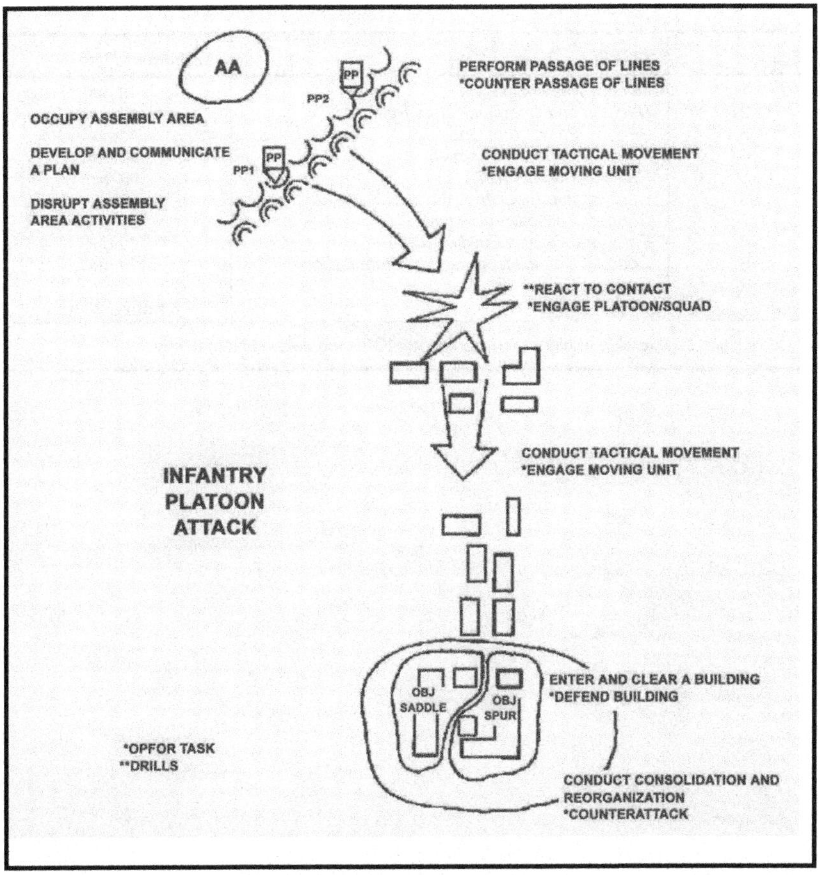

Figure 5-4. Attack STX scenario.

Table 5-2. Sequence of events and times for STX Attack.

Event	*Action*	*Estimated Time*
1	*Occupy Assembly Area*	1 hour
2	*Develop and Communicate a Plan*	1 hour
3	*Perform Passage of Lines*	1 hour
4	*Conduct Tactical Movement*	1 hour
5	*React to Contact (Drill)*	30 min
6	*Conduct Intermediate AAR*	30 min
7	*Conduct Tactical Movement*	30 min
8	*Enter and Clear a Building (Drill)*	1 hour
9	*Conduct Consolidation and Reorganization*	1 hour
10	*Conduct Final AAR*	1 hour
	* **Total Time**	**8 hours, 30 min**
* Conducting an exercise at night or in increased MOPP requires additional time.		

c. *Special Situation.* The platoon is part of a company in a secure assembly area when a FRAGO to attack is received (Figure 5-5).

1. **SITUATION.** The company is opposed by one platoon.

a. *Enemy Forces.* The enemy is at 65 to 70 percent strength. He is preparing a defense in CACTF (**MOUT site**) to prevent friendly forces from continuing to attack along the CACTF avenue of approach. He is expected to use artillery-delivered nonpersistent nerve agents in the defense of CACTF.

b. *Friendly Forces.* (Battalion designation) attacks (**date-time group**) to destroy enemy forces at OBJ COWBOY (**CACTF**) in order to pass follow-on forces, allowing friendly forces to continue the attack.

2. **MISSION.** (___) Company seizes OBJ HORSE (CACTF) vicinity (**grid**) NLT (**date-time group**) to prevent the enemy from establishing a heavily fortified defense.

3. **EXECUTION.**

a. *Concept of the Operation.* (See overlay.). The company moves in a wedge to Assault Position 1, assault CACTF to seize OBJ HORSE, consolidate, and reorganize.

(1) Maneuver. (**Evaluated**) Platoon will lead the company and be the main effort in the attack. They seize OBJ SADDLE vicinity (**grid**) to allow passage of friendly elements through CACTF. (___) Platoon moves on the company's left and seize OBJ SPUR vicinity (**grid**) to protect the left flank of the main effort. (___) Platoon moves on the company's right and establish a support-by-fire position located vicinity (**grid**).

(2) Fire support. fires are used to obscure movement to OBJ HORSE and to suppress the enemy during the assault. Priority of indirect fires is to (*evaluated*) Platoon, which is allocated one priority target (**120-mm mortar**).

Figure 5-5. Example FRAGO for STX Attack.

```
 b. Tasks to Maneuver Unit.
   (1)  (Evaluated) Platoon.
     (a)  Use Passage Point 1.
     (b)  Orient defense from TRP 2 to TRP 3.
     (c)  Be prepared to employ a wire-mined obstacle vicinity TRP 5
during consolidation.
   (2)  (___) Platoon.
     (a)  Use Passage Point 2.
     (b)  Orient defense from TRP 1 to TRP 2.
   (3)  (___) Platoon
     (a)  Use Passage Point 2.
     (b)  Orient defense from TRP 3 to TRP 4.
 c. Coordinating Instructions.
   (1)  Company RP is located (grid).
   (2)  Company linkup is (grid).
   (3)  MOPP1 is in effect (date and time).
   (4)  Selective rubbling is authorized. Destruction of an area
larger than one city block requires brigade commander approval.
```

Figure 5-5. Example FRAGO for STX Attack (continued).

4. **Support Requirements.**

a. *Minimum Trainers/Evaluators*: This exercise can be conducted with the company commander or platoon leader acting as the trainer or primary evaluator. At least one more controller or evaluator is required with the OPFOR. Other platoons being trained or evaluated will increase support personnel requirements.

b. *Vehicles/Communications*: All vehicles organic to the platoon should be included. The OPFOR should also have some combat vehicles.

c. *Opposing Force*:. The OPFOR should consist of at least a reinforced squad.

d. *Maneuver Area*: A training area at least 4 by 15 km is needed for cross-country movement and infiltration, and the CACTF is the ideal objective area. The terrain should offer multiple covered and concealed approaches to the objective.

5. **Evaluation.**

Table 5-3 lists the T&EOs from DTMS used to evaluate this STX.

Table 5-3. T&EOs used to evaluate STX Attack.

Task/Function	*Task Number*
Perform Passage of Lines	7-3-1125
Conduct Tactical Movement	7-2-1342
Occupy Assembly Area	7-3-1136
Develop and Communicate a Plan	7-3-1605
Conduct Consolidation and Reorganization (Company/Platoon)	7-2-5027

MECHANIZED INFANTRY PLATOON STX DEFEND

1. **Objective**. This example STX trains collective, leader, and individual tasks for the platoon operation, *Defend Urban Area/Building*.

2. **Interface.** This STX is supported by two warrior drills: *React to Contact* and *Enter and Clear a Building*.

3. **Training.**

a. *Guidance.* FM 7-0 and FM 7-1 provide training guidance, tips, and enhancers for planning and executing STXs. The trainer should review the individual, leader, and collective tasks to be performed during the STX to determine which require initial or refresher training.

b. *General Situation.*

(1) The exercise scenario is as follows: Latest INTSUMs indicate that enemy forces are expected to use a high-speed avenue of approach to counterattack. The approach is dominated by a small urban area containing strongly constructed buildings. The enemy has the capability for indirect fire, CAS, and chemicals. A defense in and around the town is necessary to deny the enemy the use of the avenue of approach. The platoon, as part of a larger force, is ordered to occupy and prepare defensive positions that overlook the avenue of approach. Indirect fire is available.

(2) Conduct of the exercise begins when the platoon receives an OPORD to move to establish a platoon defensive position. An AAR should be conducted after Soldiers enter a building and clear a room, and a final AAR should be conducted once all evaluation notes are compiled. If necessary, portions of the exercise should be repeated until the platoon's performance is satisfactory.

(3) When the platoon receives a warning order to prepare to move to establish a defense, the platoon is part of a company, which is part of a task force, and is located in an assembly area. Figure 5-6 provides a graphic scenario of task performance for the Defend STX. Table 5-4 provides a recommended sequence for performance of T&EOs and the time allotted for each portion of the STX.

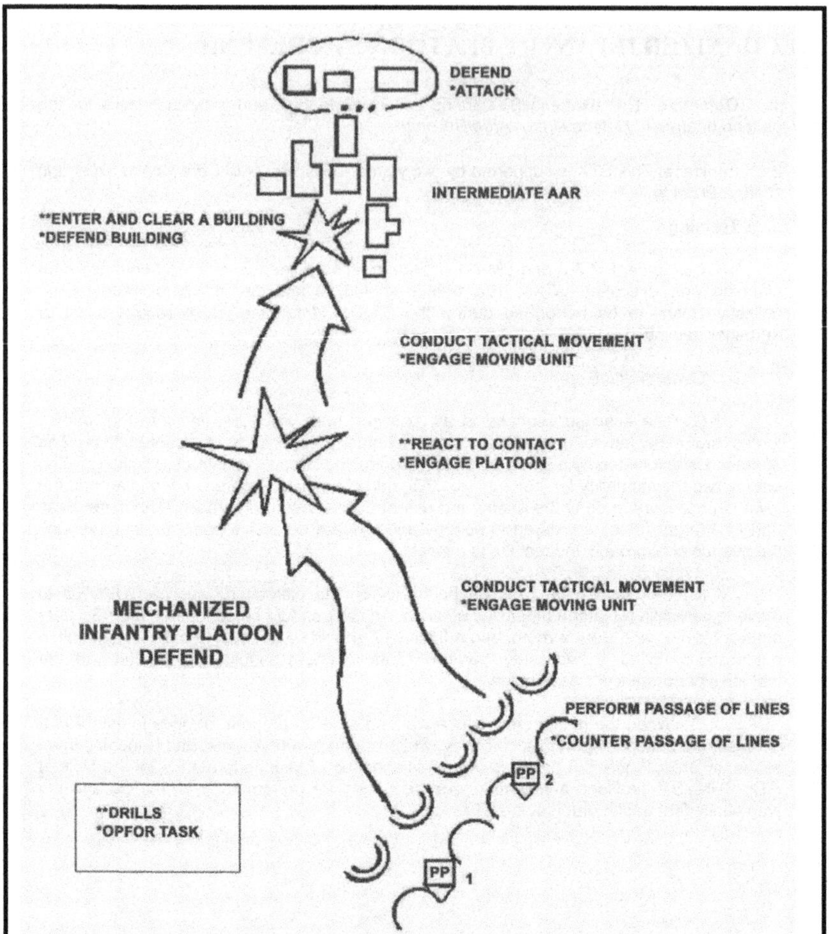

Figure 5-6. Defend STX scenario.

Table 5-4. Sequence of events and times for STX Defend.

Event	*Action*	*Estimated Time*
1	*Perform Passage of Lines*	1 hour
2	*Conduct Tactical Movement*	30 min
3	*React to Contact*	1 hours
4	*Conduct Tactical Movement*	30 min
5	*Enter and Clear a Building (Drill)*	1 hour
6	*Intermediate AAR*	30 min
7	*Defend MOUT/Building*	6 hours
8	*Conduct Final AAR*	1 hour
	***Total Time**	**11 hr 30 min**
*Conducting an exercise at night or in increased MOPP requires additional time.		

c. *Special Situation*. Thirty minutes after receiving the warning order, the platoon receives the company OPORD (Figure 5-7).

(Classification)

Copy no ___ of ___ copies
___ Battalion, ___ Infantry
Unit location or place of issue
Date-time group (with time zone)
Message reference number

OPERATION ORDER NO. _____
References: Map, series no. ___, sheet no. ___, edition ___
Time Zone Used Throughout the Order: ____
TASK ORGANIZATION:
___ Platoon, ___ Platoon(-), ___ Platoon, Company Control

Company HQ
1/1/C-52d Engr
1/2 ___, ___
(Reserve)
Company Trains

1. **SITUATION.**

a. **Enemy Forces.**

(1) *Composition, Distribution, and Strength*. We are opposed by elements of the Liberation Guards Militia, Red Brigade. Several platoon-sized outposts are located within the battalion sector and consist of about 30 to 35 Soldiers per platoon. They are equipped with an assortment of Soviet small arms, including AK-47s, RPKs, RPGs, and various wheeled and tracked vehicles including some BMPs. Enemy outposts are at 60 to 70 percent strength and have indirect fire assets available.

(2) *Capabilities*. The enemy units are well entrenched in their defensive positions. The strongpoints have dedicated fire support assets and are capable of using nonpersistent nerve agents to prevent coordinated attacks. No enemy air threat exists but some heavy armored vehicles have been sighted in the area.

(3) *Probable Course of Action*. Intelligence indicates the enemy strongpoints will continue to fortify their positions until relief arrives within 24 hours in the form of a main enemy counterattack.

b. **Friendly Forces**. (__) TF defends in sector NLT (______) along PL (____) to deny enemy forces use of avenue of approach (____). (__) TF must be prepared to continue movement forward of PL (____), on order.

(1) Missions of units on left and right are (as required).

(Classification)

Figure 5-7. Example OPORD for STX Defend.

(Classification)

(1) (__) TF defends in sector NLT (______) from (grid) to (grid) to (grid) to contain any penetrations of the brigade sector.

(1) (___) Battery, (___) Battalion provides DS to 1st Brigade.

c. **Attachments and Detachments.** 1/1/C-52d Engineers attached.

2. **MISSION.** (___) Company blocks enemy forces NLT (date-time group) to prevent them from using high-speed avenues of approach through (name of MOUT site), and to prevent the enemy from counterattacking into the battalion sector.

3. **EXECUTION.**

a. **Concept of the Operation.** (See overlay.) Company prevents the enemy from using the high-speed avenues of approach through (MOUT site) by moving to the site and establishing a blocking position, which overwatches the platoon BPs covering the main avenues. We will block in depth by placing two platoons forward and one back, and maintaining a reserve to block any penetration. We must ensure our sectors of fire are placed to mass fire at the decisive point.

(1) *Maneuver.* (Evaluated) Platoon acts as the main effort, blocking from BP (__), vicinity (grid), to prevent enemy penetration of PL (____). (___) Platoon occupies BP (__), vicinity (grid), to block enemy movement into the main effort's left flank. (___) Platoon occupies BP (__) to prevent the enemy from making a coordinated attack on the main effort.

(2) *Fire Support.* The purpose of fires is to disrupt the enemy when his counterattack reaches PL (____). Priority of indirect fires is to *(evaluated)* Platoon, which is allocated two 155-mm priority targets, one of which may be used as an FPF.

(3) *Intelligence.* Priority of collection is to identify and locate enemy combat reconnaissance patrols and the enemy main effort.

(4) *Engineering.* During movement to the battle position, obstacles are used to ensure mobility. During blocking actions, the purpose of obstacles is to prevent the enemy from advancing on larger streets and to canalize him into more restrictive terrain. Priority of effort is to countermobility. Priority of support is to *(evaluated)* Platoon.

b. **Tasks to Maneuver Units:**

(1) *(Evaluated) Platoon:*

(a) Conduct passage of lines using PP 1.

(b) Lead the company movement after crossing the LD. Be prepared to react to enemy contact when (MOUT site) is reached.

(c) Enter and clear building(s) (number determined by commander).

(Classification)

Figure 5-7. Example OPORD for STX Defend (continued).

(Classification)

(d) Occupy BP (__) and orient on TRPs 2 and 3.
(e) Establish an OP vicinity (grid) to observe main avenue of approach.
(2) *(___) Platoon*:
(a) Conduct passage of lines using PP 2.
(b) Move second in order of movement.
(c) Enter and clear building(s) (number determined by commander).
(d) Occupy BP (__) and orient on TRP 1.
(3) *(___) Platoon*:
(a) Conduct passage of lines using PP 3.
(b) Move last in order of movement.
(c) Enter and clear building(s) (number determined by commander).
(d) Occupy BP (__) and orient on TRPs 4 and 5.
(e) Establish an OP vicinity (grid).

c. **Tasks to Subordinate Units**: Engineer squad travels with the company headquarters behind (*evaluated*) Platoon and supports the main effort's emplacement of obstacles.

d. **Coordinating Instructions**.
(1) Timings.
(a) Brief back ________.
(b) Company rehearsal ________.
(c) SP time ________.
(d) In position NLT (time) ________.
(2) Report all enemy contact.
(3) Report crossing all phase lines.
(4) MOPP status ____; enemy chemical attack is expected.
(5) Air defense status is YELLOW, WEAPONS HOLD.
(6) Leaders briefback the movement and defense plan to the commander at the company CP at (date-time group).
(7) Destruction of subterranean approaches requires brigade commander's approval.
(8) Selective rubbling is authorized. Destruction of more than one building requires commander's approval.
(9) Marking of rooms and buildings is IAW CO SOP.

(Classification)

Figure 5-7. Example OPORD for STX Defend (continued).

(Classification)

(10) Rules of Engagement follow:
(a) Minimize collateral damage to structures.
(b) Civilians have been evacuated. Evacuate to company CP any other civilians encountered.

4. SERVICE SUPPORT. IAW the battalion SOP.

a. **Miscellaneous.**
(1) Battalion trains located vicinity (grid).
(2) Company trains located vicinity (grid).

b. **Materiel and Services.**
(1) *Supply.*
(a) Class I. Ration Cycle M-M-M. Platoons must maintain a three-day supply in each BP.
(b) Class II. Submit requirements for special equipment ASAP. Coordinate with 1SG.
(c) Class III. Refuel all vehicles before (date-time group).
(d) Class IV. Coordinate with S-4 for delivery.
(e) Class V.
- Platoons cache one additional basic load in BP.
- Platoons submit request for additional demolitions NLT (date-time group).

(2) *Services.* Battalion decontamination site is located vicinity (grid).
(3) *Maintenance.*
(a) Priority to (*evaluated*) Platoon M2 BFVs.
(b) UMCP located vicinity (grid).

c. **Medical Evacuation and Hospitalization.**
(1) Battalion aid station located vicinity (grid).
(2) Company CCP located with company CP.

d. **Personnel.**
(1) No replacements for next 72 to 96 hours.
(2) Brigade mortuary affairs point located in BSA.
(3) EPW collection sites located at (grid) for battalion and (grid) for company.

(Classification)

Figure 5-7. Example OPORD for STX Defend (continued).

```
                          _______________
                         (Classification)

5. COMMAND AND SIGNAL.

   a. Command. Initially, company CP will be traveling center of
company wedge. Location in BP to be determined.

   b. Signal.
      (1) Current SOI in effect.
      (2) Primary means of communication in BP is wire, FM, messenger,
in that order.
      (3) Radio listening silence is in effect until after crossing PL
(____) or until contact is made.
                          _______________
                         (Classification)
```

Figure 5-7. Example OPORD for STX Defend (continued).

4. **Support Requirements**.

a. *Minimum Trainers/Evaluators*: This exercise is conducted by one company grade officer who is the trainer and primary evaluator.

b. *Vehicles/Communications*: All vehicles organic to the platoon should be included.

c. *Opposing Force*:. The OPFOR should be platoon strength.

d. *Maneuver Area*: A training area at least 2 by 10 km, with hasty fighting positions, is desirable for cross-country movement. The CACTF serves as the objective area. The terrain, ideally, should offer multiple covered and concealed routes. Using terrain that limits the leader to a geographic or predictable route prevents evaluation of the unit's ability to conduct a terrain analysis and select covered and concealed routes.

e. *Barrier Materials*. Sufficient Class IV (sandbags, lumber, and so on) must be made available to fortify positions, IAW Chapter 3, FM 3-06.11.

5. **Evaluation**.

Table 5-5 lists the T&EOs from DTMS used to evaluate this STX, and Table 5-6 lists the supporting tasks.

Table 5-5. T&EOs used to evaluate STX Defend.

Task/Function	*Task Number*
Defend MOUT/Building	7-3-1118
Perform Passage of Lines	7-3-1125
Conduct Tactical Movement (Co/Plt)	7-2-1342

INFANTRY RIFLE COMPANY FTX ATTACK

Table 5-6. Supporting tasks for company FTX.

STX C-II-1	*STX C-II-2*	*STX C-II-3*
Conduct an Infiltration (Co/Plt) Perform Linkup Execute an Assault (MOUT) Conduct Consolidation and Reorganization (Co/Plt)	Perform Cordon and Search Take Action on Contact Conduct an Infiltration (Co/Plt) Perform Linkup Perform Passage of Lines Occupy Assembly Area	Disarm Belligerents Handle Captured Insurgents or Belligerents Apprehend/Detain Noncombatants

1. **Objective**. This FTX is designed to train the Infantry rifle company in conducting offensive operations. This exercise provides practice for the company commander, platoon leaders, squad leaders, and company headquarters personnel in planning, coordinating, and controlling combat operations.

2. **Interface**. This FTX is supported by STX C-II-1, C-II-2, and C-II-3 (Table 5-6). It is also supported by two warrior drills, *React to Contact* and *Break Contact*.

3. **Training**.

a. *Guidance*. FM 7-0 and FM 7-1 provide training guidance, tips, and enhancers for planning and executing STXs. The trainer should review the individual, leader, and collective tasks to be performed during the STX to determine which require initial or refresher training.

b. *General Situation*.

(1) The battalion is ordered to conduct offensive operations in support of the brigade mission. The company is ordered to infiltrate enemy lines and to destroy an insurgent operations base located in the CACTF. The company must be prepared to remain in the CACTF to conduct cordon and search missions.

(2) This exercise should be conducted under various conditions.

(3) When the company receives the battalion warning order the planning process for the attack mission begins. The exercise ends when the company completes a passage of lines, occupies an assembly area, and completes preparations to continue its mission. Figure 5-8 shows a graphic scenario of task performance for the FTX Attack. Table 5-7 provides a recommended sequence for performance of T&EOs and the time allotted for each portion of the STX, including AARs.

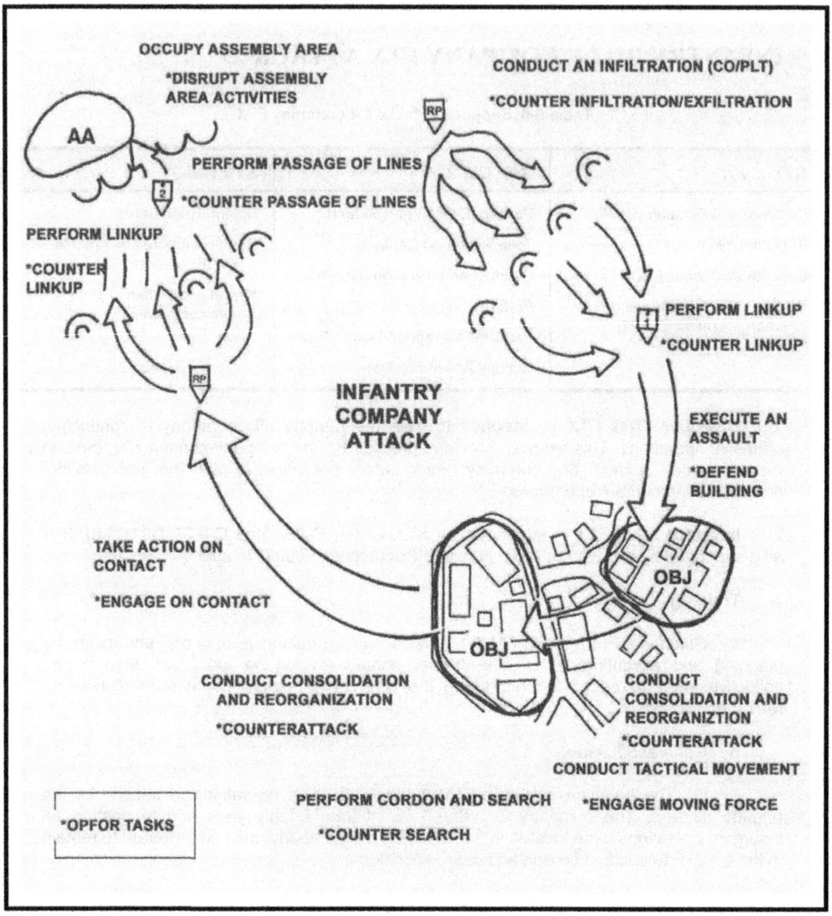

Figure 5-8. Attack FTX scenario.

Table 5-7. Sequence of events and times for FTX Attack.

Event	*Action*	*Estimated Time*
START STX C-II-1		
1	*Issue Company OPORD*	4 hours
2	*Conduct an Infiltration (Co/Plt)*	6 hours
3	*Perform Linkup*	3 hours
4	*Execute an Assault (MOUT) AAR*	3 hours
5	*Conduct Consolidation and Reorganization (Co/Plt)*	3 hours
ISSUE FRAGO, START STX C-II-2		
6	*Perform Cordon and Search*	6 hours
7	*Take Action on Contact / AAR*	3 hours
8	*Conduct an Infiltration (Co/Plt)*	6 hours
9	*Perform Linkup*	3 hours
10	*Perform Passage of Lines*	2 hours
11	*Occupy Assembly Area*	3 hours
START STX C-II-3		
12	*Disarm Belligerents*	3 hours
13	*Handle Captured Insurgents or Belligerents*	4 hours
14	*Apprehend/Detain Noncombatants*	4 hours
15	*Conduct Final AAR*	2 hours
	***Total Time**	**55 hours**

The time required to train an event varies based on the factors of METT-T and on the unit's training proficiency. Regardless, training is conducted to standards, not to time limitations. The following tasks are integrated and evaluated throughout the exercise:

- *Prepare for combat*
- *Maintain operations security*
- *Conduct tactical movement*
- *Perform consolidation*
- *Perform reorganization*
- *Perform stability and support operations*

* Time for intermediate AARs is excluded from this total; additional time is required if much of the exercise is conducted at night or during other limited visibility.

(4) The battalion commander issues an OPORD to the company commanders (Figure 5-9).

(Classification)

Copy no ___ of ___ copies
___ Battalion, ___ Infantry
Unit location or place of issue
Date-time group (with time zone)
Message reference number

OPERATION ORDER NO. _____
References: Map, series no. ___, sheet no. ___, edition ___
Time Zone Used Throughout the Order: ____

TASK ORGANIZATION:
___Company ___Company ___Company Battalion Control
Scout Platoon
81-mm Mortar Platoon
Antitank Platoon
1/C/58th ENGR (DS)
Battalion Trains

1. SITUATION.

a. **Enemy Forces**. Annex A (See intelligence overlay [prepared by brigade S-2].)

(1) *Disposition, Composition, and Strength*. The battalion is opposed by one company from the Dona Anna Militia Guards Battalion. This enemy company consists of 3 platoons with 35 to 40 Soldiers, equipped with M16s, AK-47s, and LMGs. The company headquarters is currently located in the town hall of CACTF. The company is at 90 percent strength and has indirect fire support assets available.

(2) *Capabilities*. The enemy possesses nonpersistent nerve agents and indirect fire assets capable of delivery. He also has some light armored vehicles in the area, but their location is undetermined. No enemy air threat exists. However, the enemy may be able to reinforce CACTF with an additional company, within two hours.

(3) *Most Probable Course of Action*. The enemy defends CACTF and continues using it as a base of operations and to consolidate weapons and material. His intent is to prevent the capture of CACTF and its support base, provided by its facilities and population.

(Classification)

Figure 5-9. Example battalion OPORD for FTX Attack.

(Classification)

b. **Friendly Forces.**

(1) (___) Brigade, (___) Division attacks at (date-time group) to seize the town of CACTF at (grid), to keep the enemy from using it as a base of operations. The intent is to secure the town and the road network in the area, to prevent the enemy from using it as a base of operations.

(2) (___) Battalion, (___) Infantry attacks to seize the road network at OBJ (____), vicinity (grid), at (date-time group) to prevent enemy forces from counterattacking along Route (____).

(3) (___) Battalion, (___) Infantry attacks at (date-time group), to seize the high ground at OBJ (____), vicinity (grid), and to establish defensive positions to block enemy routes in and out of CACTF.

2. **MISSION.** (___) Battalion, (___) Infantry attacks at (date-time group) to seize CACTF, vicinity (grid), NLT (date-time group), to prevent the enemy from using the town as an operations and logistics base.

3. **EXECUTION.**

a. **Concept of the Operation.** The battalion will accomplish its mission, which is to deny the enemy use of CACTF and its surrounding terrain and population, by seizing the town and controlling the key areas in the vicinity. We will conduct a feint in the west and then attack the decisive point within CACTF (OBJ [____]).

(1) *Maneuver.* (___) Company conducts a feint to deceive the enemy, then the main effort (*evaluated*) company seizes OBJ (____), to prevent the enemy from using CACTF as an operational and logistical base. (___) Company supports the main effort by seizing OBJ (____), in order to prevent the enemy from counterattacking the main effort. (___) Company seizes OBJ (____) to support the main effort's attack.

(2) *Fire Support.* Annex C. (See fire support.) Purpose of fires is to suppress the enemy and to obscure movement to CACTF. Priority of artillery fires is to (feinting) Company; on order, priority will shift to (supporting effort) Company. Priority of 81-mm mortar fires is to (*evaluated*) Company.

(3) *Engineering.* The purpose of the engineers is to breach obstacles during the attack, to help (*evaluated*) Company seize OBJ (____). Priority of support is to (*evaluated*) Company, (___) Company, and (___) Company in order. The battalion commander must authorize any booby traps, and point and hasty minefields.

(Classification)

Figure 5-9. Example battalion OPORD for FTX Attack (continued).

(Classification)

b. **Tasks to Maneuver Units.**

(1) *(Evaluated) Company.*

(a) Infiltrate using infiltration lane TANGO.

(b) Seize OBJ (____), vicinity (grid), NLT (date-time group).

(c) Be prepared to perform cordon and search of CACTF, after consolidation and reorganization.

(2) *(Supporting) Company.*

(a) Follow (main attack) Company and infiltrate on infiltration lane TANGO.

(b) Seize OBJ (____), vicinity (grid), NLT (date-time group).

(c) Coordinate the attack with (*evaluated*) Company, so that both units attack at the same time.

(d) Be prepared to perform cordon and search in the objective area after consolidation and reorganization.

(e) Be prepared to assume the mission of (*evaluated*) Company.

(3) *(Supporting) Company.*

(a) Infiltrate using infiltration lane ZEBRA.

(b) Conduct feint to OBJ (____), vicinity (grid), NLT (date-time group).

(c) Be prepared to establish defensive positions vicinity (grid) to prevent enemy movement to and from CACTF.

c. **Tasks to Subordinate Units.**

(1) *(Scout) Platoon.*

(a) During movement, determine location, strength, and disposition of enemy forces in CACTF.

(b) On order, screen the battalion flank from (grid) to (grid) to warn of the approach of enemy units from the southwest

(2) *Mortar Platoon.* Priority of mortar fire is to (*evaluated*) Company. Mortar platoon moves with (*evaluated*) Company, establishes a firing position vicinity (grid), and displaces on order.

(3) *Antitank Platoon.*

(a) Initially, over watch from (grid) to (grid) to secure the unit infiltration.

(b) On order, move vicinity (grid) to support (*evaluated*) Company's consolidation and reorganization.

(Classification)

Figure 5-9. Example battalion OPORD for FTX Attack (continued).

(Classification)

(4) *1/C/58 Engineers.*

(a) Move with (*evaluated*) Company to provide mobility support.

(b) Be prepared to reduce obstacles in CACTF during cordon and search mission.

d. **Coordinating Instructions.**

(1) Report when ready to execute an assault.

(2) Report any movement of enemy forces from the southwest.

(3) No rubbling is authorized, so minimize collateral damage to property.

(4) MOPP0 is in effect.

(5) Terrorist threat warning is red.

(6) Commanders brief back OPORD to battalion commander at battalion tactical CP at (date-time group).

(7) Rules of Engagement (ROE).

(a) Commanders take all steps necessary and appropriate for their units' protection.

(b) The minimum necessary force is used to control the situation.

(c) Leaders must take measures to minimize risk to civilians, without endangering the unit.

(d) Fire is returned directly to its source, not sprayed into a general area.

(e) Firing ceases when the threat is over.

(f) Anyone trying to surrender is allowed to do so.

(g) Civilians and property are treated with respect.

(h) WP can be used vicinity CACTF to aid in isolating the objectives. The requests for indirect fire, within the town, must be authorized by the battalion commander.

4. **SERVICE SUPPORT**. (See Annex D.)

5. **COMMAND AND SIGNAL**.

a. **Command**.

(1) The tactical CP moves with (*evaluated*) Company.

(2) The main CP is vicinity (grid) and displaces on order.

(3) The second in command remains in the battalion main CP.

(4) The rear CP is the alternate battalion main CP.

(Classification)

Figure 5-9. Example battalion OPORD for FTX Attack (continued).

(______________)
(Classification)

b. **Signal.**
(1) Current SOI is in effect.
(2) Radio-listening silence is in effect for all elements except scouts, until prepared to attack.

ACKNOWLEDGE:
JONES
LTC
OFFICIAL:
SMITH
S-3
Annexes: A–Intelligence
B–Operations Overlay
C–Fire Support
D–Service Support

(______________)
(Classification)

Figure 5-9. Example battalion OPORD for FTX Attack (continued).

(5) The battalion has destroyed the enemy operating in CACTF, and has seized the town. Two companies are defending against a limited counterattack. One company is in a blocking position. The (*evaluated*) company receives a FRAGO (Figure 5-10) to conduct a cordon and search operation. The purpose of the operation is to locate enemy personnel and equipment remaining in the CACTF.

Enemy personnel in squad-sized elements are reported by the civilian population to be hiding in the CACTF. Reports also indicate numerous weapons caches are being hidden by civilians collaborating with the enemy. Battalion has completed its mission and is preparing to conduct a cordon and search.

After completing this, the battalion exfiltrates to perform a passage of friendly lines. (Evaluated) Company moves to perform cordon and search vicinity OBJ (____), beginning (date-time group) to (date-time group), to locate and seize enemy personnel or equipment.

The (*evaluated*) company notifies the commander of the results of the search, and then notifies the commander when they (*evaluated* company) are ready to start movement. The company exfiltrates on order along Route Gold to perform linkup at (grid) and passage of lines at (grid). (See overlay for lanes.)

Figure 5-10. Example FRAGO for FTX Attack.

4. **Support Requirements**.

a. *Minimum Trainers/Evaluators*:. The company commander acts as the trainer-evaluator for the platoons. The battalion may direct the FTX internal evaluation along with a company evaluator and an FO evaluator. The platoon leaders and platoon sergeants are the primary trainers during the initial phases of the training. The company commander, the XO, the platoon leaders, and the platoon sergeants all act as trainers during the company phases of training. If the exercise is conducted by the battalion, either the battalion commander, XO, or S-3 acts as the trainer and evaluator, but additional battalion assets may be required to support the evaluation phase. The controllers or evaluators are positioned at critical locations throughout the exercise to observe the action. They must not interfere with the conduct of the operation, except to halt it when a safety violation occurs or to conduct an AAR.

b. *Vehicles/Communications*: Those organic to the company are used in the exercise. This includes one designated medical evacuation vehicle per platoon and at least one aidman on site.

c. *Opposing Force*:. The OPFOR ground force should be at least platoon-size. FTXs conducted by the company require internal resourcing; units should rotate to ensure total training. OPFOR tasks and standards are provided in each T&EO.

d. *Maneuver Area*: A training area 5 by 10 km which includes the CACTF is best for this exercise.

5. Evaluation.

Table 5-8 lists the T&EOs used to evaluate this STX.

Table 5-8. T&EOs used to evaluate STX Attack.

Task/Function	*Task Number*
Perform Cordon and Search	7-3-1105
Take Action on Contact	7-4-1107
Perform Passage of Lines	7-4-1125
Perform Linkup	7-4-1128
Conduct Tactical Movement	7-2-1342
Occupy Assembly Area	7-4-1136
Maintain Operations Security	7-4-1409
Prepare for Combat	7-4-1606
Conduct Consolidation and Reorganization (Co/Plt)	7-2-5027
Perform Cordon and Search	7-2-1105
Execute an Assault (MOUT)	7-2-1109
Perform Exfiltration	7-1-1146
Perform Infiltration	7-1-1147
Perform Reorganization	7-1-1608
Perform Consolidation	7-1-1609

MECHANIZED INFANTRY COMPANY OR TEAM FTX DEFEND

1. **Objective**. This FTX is designed to train the mechanized Infantry company or team in conducting defensive operations in urban terrain. It also provides platoons an opportunity to prepare for full-scale FTXs conducted and evaluated by the parent battalion. This exercise provides practice for the company commander, platoon leaders, squad leaders, and company headquarters personnel in planning, coordinating, and controlling combat operations.

2. **Interface**. This FTX is supported by the STX C-II-4, C-II-5, and C-II-6 tasks shown in Table 5-9. It is also supported by two warrior drills, *React to Contact* and *Break Contact*.

Table 5-9. Supporting tasks for company FTX.

STX C-II-4	*STX C-II-5*	*STX C-II-6*
Occupy Assembly Area	Perform Ambush	Operate a Checkpoint
Perform Tactical	Establish a Roadblock/Checkpoint	Interdict Smuggling Operations
Road March	Perform CBRN Operations	Defend Convoy
Perform Relief in Place	Conduct Tactical Movement	
Defend Built-Up Area/ Building	Perform Passage of Lines	
	Occupy Assembly Area	

3. **Training**.

a. *Guidance*. FM 7-0 and FM 7-1 provide training guidance, tips, and enhancers for planning and executing STXs. The trainer should review the individual, leader, and collective tasks to be performed during the STX to determine which require initial or refresher training.

b. *General Situation*.

(1) The battalion is ordered to conduct defensive operations in support of the brigade mission; the company is ordered to perform a relief operation and to prepare a defense in an urban area. The company remains in position for at least 36 hours before receiving a follow-on mission. Most civilians have been evacuated from the area, but a few remain. A terrorist threat exists as well as the possibility of enemy battalion-level offensive actions.

(2) This exercise should be conducted under various environmental conditions. The company also conducts operations under threat of CBRN attack.

c. *Special Situation.*

(1) The planning process for the defense mission begins when the company receives a warning order. The exercise ends when the company completes the passage of lines and occupies an assembly area. Figure 5-11 shows a graphic scenario of task performance for the FTX Defend.

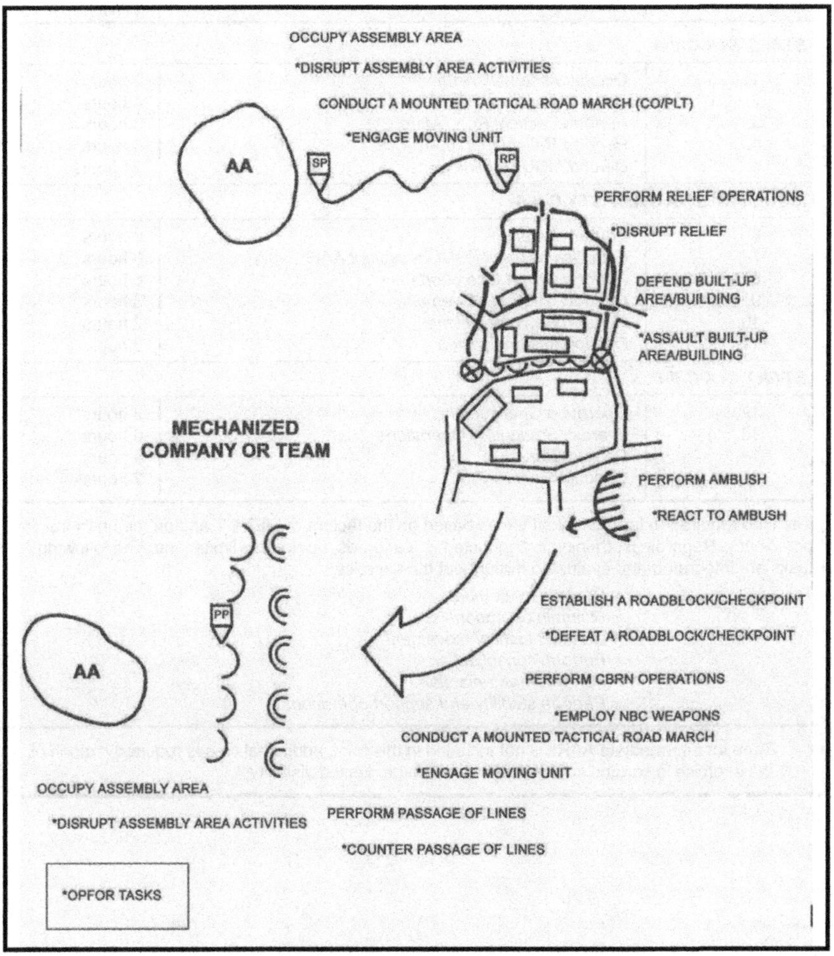

Figure 5-11. Graphic scenario of task performance for FTX Defend.

(2) Table 5-10 provides a recommended sequence for performance of T&EOs from DTMS and the time allotted for each portion of the STX, including AARs.

Table 5-10. Sequence of events and times for FTX Defend.

Event	*Action*	*Estimated Time*
START STX C-II-4		
1.	*Occupy Assembly Area*	2 hours
2.	*Issue Company OPORD*	4 hours
3.	*Perform Tactical Road March*	3 hours
4.	*Perform Relief Operations AAR*	6 hours
5.	*Defend MOUT/Building*	24 hours
ISSUE FRAGO, START STX C-II-5		
6.	*Perform Ambush*	6 hours
7.	*Establish a Roadblock/Checkpoint AAR*	6 hours
8.	*Perform CBRN Operations*	6 hours
9.	*Conduct Tactical Movement*	3 hours
10.	*Perform Passage of Lines*	2 hours
11.	*Occupy Assembly Area*	3 hours
START STX C-II-6		
12.	*Operate a Checkpoint*	2 hours
13.	*Interdict Smuggling Operations*	6 hours
14.	*Defend Convoy*	4 hours
15.	*Conduct Final AAR*	2 hours
The time required to train an event varies based on the factors of METT-T and on the unit's training proficiency. Regardless, training is conducted to standards, not to time limitations. The following tasks are integrated and evaluated throughout the exercise: ▪ *Prepare for combat* ▪ *Maintain operations security* ▪ *Conduct tactical movement* ▪ *Perform consolidation* ▪ *Perform reorganization* ▪ *Perform stability and support operations*		
* Time for intermediate AARs is not included in this total; additional time is required if much of the exercise is conducted at night or during other limited visibility.		

(3) The battalion commander issues the defense OPORD to the company (Figure 5-12).

```
                    ____________
                   (Classification)

                                         Copy no ___ of ___ copies
                                       ___ Battalion, ___ Infantry
                                      Unit location or place of issue
                                     Date-time group (with time zone)
                                          Message reference number
OPERATION ORDER NO. _____

Reference: Map, series no. ___, sheet no. ___, edition ___
Time Zone Used Throughout the Order: ____
TASK ORGANIZATION:
___Co ___Tm ___Tm (-)  Battalion Control
 1/__/__/AR 2/__/__/AR Scout Platoon
    120-mm Mortar Platoon
Team Tank E Co(-) 1/C/1-222 ADA (Stinger) (DS)
1/__/__/Inf  C/58 ENGR (DS)
2/__/__/Inf  1/1 (GSR) TM A/58 MI Bn
    ATK Force B (1/_/_/Inf)
    Battalion Trains
```

1. SITUATION.

a. **Enemy Forces**. Annex A (see intelligence overlay [prepared by brigade S-2]).

(1) *Composition, Disposition, Strength.* The company is opposed by the 1st Dragoons Regiment, 60th Guards Division. This regiment is at 85 percent strength and has received engineer and artillery reinforcements. The enemy is equipped with an assortment of former Soviet and Warsaw pact small arms, machine guns, and armored vehicles including T-62 tanks.

(2) *Capabilities.* The enemy has established defensive positions that follow a line from vicinity (grid) to vicinity (grid). Enemy armored reconnaissance units are operating in the more open terrain to the southwest of CACTF. The enemy has been observed moving forces in depth throughout the area in preparation for an offensive action.

(3) *Probable Course of Action.* The enemy's most probable course of action will be to begin offensive actions within the next 24 to 36 hours. Enemy encountered includes numerous reconnaissance elements trying to find the company's vulnerabilities and a main body trying to exploit any weakness.

____
(Classification)

Figure 5-12. Example battalion OPORD for mechanized company or team STX Defend.

(Classification)

b. **Friendly Forces.**

(1) (___) Brigade, (___) Division relieves (___) Brigade in place at (date-time group), to defend in sector from (grid) to (grid) to (grid) to (grid), in order to secure the corps counterattack axis of advance, (grid). The intent is to secure the road network around CACTF to prevent the enemy from impeding the movement of the (___) Brigade, (___) Division.

(2) TF (__) is the main effort in the defense and, NLT (date-time group), holds key terrain vicinity (grid) on our left, to prevent an enemy bypass of the brigade left flank.

(3) TF (__), on our right, defends along Route 360, vicinity (grid) to (grid) to (grid) to (grid), NLT (date-time group), to prevent an envelopment of the right flank. They conduct a supporting attack at (date-time group) to seize the high ground at OBJ OAK, vicinity (grid), to establish defensive positions that prevent the enemy armored reconnaissance units from interfering with the seizure of the pass at (grid) or the attack of the (___) Battalion, (___) Infantry, Mechanized.

2. **MISSION.** TF (____) defends NLT (date-time group), to retain the town of CACTF and its surrounding road network and allow brigade to prevent enemy interference with corps counterattack.

3. **EXECUTION.**

a. **Concept of the Operation.** Annex B (see operations overlay).

(1) Maneuver. Company (main effort) retains the main intersection (traffic loop) in CACTF, to prevent enemy movement through the sector. Team (*evaluated*) blocks movement west of 3d Street, to prevent the enemy from enveloping the main effort. The counterattack force is committed as needed. Team (__) blocks movement east of Oak Street, to canalize the enemy into EA (____). Team (__) defends from BP1, to prevent a coordinated attack against the main effort. E Company occupies BPs, to destroy enemy vehicles and protect the TF flanks.

(2) Fire Support. (See Annex C.) Fires are used to suppress the enemy and obscure our movement during the relief operation. Once positions are occupied, fires are used to suppress enemy overwatch positions, disrupt attempts to breach obstacles, and separate enemy Infantry from their supporting carriers. Priority of CAS and indirect fire is initially to Team (*evaluated*), during the relief operation, and then to Team (main effort). Teams (main effort) and (*evaluated*) each have two priority 155-mm targets. Each team may make one priority target an FPF.

(Classification)

Figure 5-12. Example battalion OPORD for mechanized company or team STX Defend (continued).

____
(Classification)

(3) Counter-Air Operations. Priority of protection is to the TF trains and main CP. Air defense weapons status is free; air defense warning status is yellow.

(4) Intelligence. Priority of collection is to identify and locate the C2 elements of the MRR and MRB, enemy tank battalions, and the enemy assault detachments' actual axes of advance.

(5) Engineering. The purpose of obstacles is to turn the enemy into restrictive terrain and engagement areas. Priority of support is to Team (main effort), Team (*evaluated*), Team (__), and Team (__), in that order.

b. **Tasks to Maneuver Units**.

(1) Team (Main Effort).

(a) Sector of defense includes the east side of Front Street, the south side of 1st Street, the west side of 3d Avenue area, and the north side of 3d Street.

(b) Team (main effort) relieves Team (__), TF (__).

(2) Team (Evaluated).

(a) Sector includes the east side of 3d Avenue, the south side of 1st Street, the west side of 6th Avenue, and the north side of 3d Street.

(b) Team (*evaluated*) relieves Team (__), TF (__).

(3) Team (__).

(a) Sector includes the east side of Oak Street, the south side of 1st Street, the east side of Front Street, and the north side of 3d Street.

(b) Team (__) relieves Team (__), TF (__).

(c) Team (__) coordinates directly with right flank team of TF (__), to ensure mutually supporting fires.

(4) Team (__)

(a) Occupy BP 1 vicinity (grid)

(b) Be prepared to occupy BP 2 (grid) to engage enemy in EA (____).

(5) Company E.

(a) Occupy BP 2-1 (grid) and move on order to BP 2-2 (grid).

(b) Provide antiarmor fires to destroy enemy tanks in EA (____).

(6) Counterattack B (Reserve).

(a) Occupy AA Black, vicinity (grid).

(b) Reconnoiter routes into TF (main effort) sector.

____
(Classification)

Figure 5-12. Example battalion OPORD for mechanized company or team STX Defend (continued).

(Classification)

(3) Counter-Air Operations. Priority of protection is to the TF trains and main CP. Air defense weapons status is free; air defense warning status is yellow.

(4) Intelligence. Priority of collection is to identify and locate the C2 elements of the MRR and MRB, enemy tank battalions, and the enemy assault detachments' actual axes of advance.

(5) Engineering. The purpose of obstacles is to turn the enemy into restrictive terrain and engagement areas. Priority of support is to Team (main effort), Team (*evaluated*), Team (__), and Team (__), in that order.

b. **Tasks to Maneuver Units.**

(1) Team (Main Effort).

(a) Sector of defense includes the east side of Front Street, the south side of 1st Street, the west side of 3d Avenue area, and the north side of 3d Street.

(b) Team (main effort) relieves Team (__), TF (__).

(2) Team (Evaluated).

(a) Sector includes the east side of 3d Avenue, the south side of 1st Street, the west side of 6th Avenue, and the north side of 3d Street.

(b) Team (*evaluated*) relieves Team (__), TF (__).

(3) Team (__).

(a) Sector includes the east side of Oak Street, the south side of 1st Street, the east side of Front Street, and the north side of 3d Street.

(b) Team (__) relieves Team (__), TF (__).

(c) Team (__) coordinates directly with right flank team of TF (__), to ensure mutually supporting fires.

(4) Team (__)

(a) Occupy BP 1 vicinity (grid)

(b) Be prepared to occupy BP 2 (grid) to engage enemy in EA (____).

(5) Company E.

(a) Occupy BP 2-1 (grid) and move on order to BP 2-2 (grid).

(b) Provide antiarmor fires to destroy enemy tanks in EA (____).

(6) Counterattack B (Reserve).

(a) Occupy AA Black, vicinity (grid).

(b) Reconnoiter routes into TF (main effort) sector.

(Classification)

Figure 5-12. Example battalion OPORD for mechanized company or team STX Defend (continued).

(Classification)

(c) Priority for planning:
- Battle position to reinforce main effort.
- Battle position to counterattack into Team (main effort) sector.
- Battle position to counterattack into Team (*evaluated*) sector.

(1) Scout Platoon.
(a) Determine location, strength, and disposition of enemy forces.
(b) Perform a screen of the TF's right flank from (grid) to (grid) beginning (date time group) to prevent enemy infiltration of CACTF.

c. **Tasks to Subordinate Units.**
(1) Mortar Platoon.
(a) Priority of fire to TF (main effort). Move with TF (main effort) and establish firing position vicinity (grid).
(b) Displace on order to (grid).
(2) 1/C/1 222 ADA. Coordinate locations with S-3 NLT (date time group).
(3) C/58 Engineers.
(a) Coordinate the engineer effort with the team commanders NLT (date time group).
(b) Provide an obstacle clearing team to Counterattack Force B, NLT (date time group).
(4) 1/1 (GSR) TM A/58 MI. Coordinate location of assets with the task force S-3 NLT (date time group).

d. **Coordinating Instructions.**
(1) Selective rubbling is authorized, but destruction of more than one building requires TF commander approval.
(2) Destruction of subterranean approaches requires TF commander approval.
(3) Commanders brief back OPORD to TF commander at the TF tactical CP, vicinity (grid), at (date time group).
(4) MOPP1 is in effect as of (date time group).
(5) Decontamination site established vicinity Miller Pond (grid).

(Classification)

Figure 5-12. Example battalion OPORD for mechanized company or team STX Defend (continued).

(Classification)

4. SERVICE SUPPORT. (See Annex D.)

a. **General.**
(1) Combat trains located vicinity (grid).
(2) Field trains located in the BSA, vicinity (grid).

b. **Materiel and Services.**
(1) Supply.
(a) Class I. Ration cycle C-A-C. Team establishes MRE cache sufficient for 48 hours in sector.
(b) Class III. Resupply available on request. Class III located in combat trains.
(c) Class IV. Materials are allocated as shown in Annex D.
(d) Class V. Each team establishes ammunition caches in sector, but should also have two basic loads of ammunition on hand.
(2) Transportation. The main streets are capable of two way heavy vehicle traffic (class 60). Secondary streets may be one way and of lighter construction (class 50).
(3) Maintenance.
(a) Priority by unit: Team (main effort), Team (*evaluated*), Team (__), Company E, Team (__).
(b) Priority by vehicle: M1A1, M2/M3, M113.

c. **Medical Evacuation and Hospitalization.**
(1) TF aid station located vicinity (grid).
(2) Civilian casualties should be transported by nonmilitary means to clinic, located vicinity (grid).

d. Personnel.
(1) No replacements for the next 96 hours.
(2) EPW collection point located vicinity (grid).
(3) Mortuary affairs point located in BSA, vicinity (grid).

e. **Civil-Military Cooperation.**
(1) Curfew in effect 1800 to 0800 daily.
(2) Majority of civilians have been evacuated. Those remaining have been notified of the risk involved.

(Classification)

Figure 5-12. Example battalion OPORD for mechanized company or team STX Defend (continued).

(Classification)

5. COMMAND AND SIGNAL.

a. **Command.**
(1) Brigade main CP located vicinity (grid).
(2) TF tactical CP located vicinity (grid).
(3) TF main CP located vicinity (grid).
(4) TF commander located at tactical CP.
(5) TF XO located at main CP.

b. **Signal.**
(1) SOI index 1 9 in effect.
(2) Primary means of communication is wire, FM, messenger in that order.
(3) Building and obstacle marking IAW TF SOP.

ACKNOWLEDGE:

WILLIAMS
LTC

OFFICIAL:

SMITH
S-3

Annexes: A-Intelligence
B-Operations Overlay
C-Fire Support
D-Service Support

(Classification)

Figure 5-12. Example battalion OPORD for mechanized company or team STX Defend (continued).

(3) The TF has retained the north half of the CACTF after the enemy attack. The enemy was defeated but has established squad-sized to platoon-sized strongpoints in the southern half of the CACTF. The enemy is expected to receive reinforcements within the next 12 to 24 hours. Theprevent enemy reinforcements from moving to the strongpoints to strengthen and resupply them, TF headquarters orders the team to conduct platoon-sized ambushes. They also order the team to prepare to establish a roadblock, to control refugee flow through the area of operations, before the team moves out of the area. Headquarters issues a FRAGO (Figure 5-13).

```
Numerous enemy strongpoints are located to your front, vicinity (grid), (grid), and (grid). We expect these strongpoints to receive reinforcements within the next 12 to 24 hours.

Your mission is to establish platoon ambushes on the likely avenues of approach to the strongpoints, NLT (date-time group), in order to prevent reinforcements or resupply to these strongpoints.

Notify the TF commander when you are prepared to move.

Coordinate with TF scouts for guides. On order, after return from ambush missions, establish roadblocks at (grid) and (grid). Be prepared to move within the next 36 to 48 hours to conduct further follow-on missions.
```

Figure 5-13. Example FRAGO for FTX Defend.

4. **Support Requirements.**

a. *Minimum Trainers/Evaluators*: This exercise should be conducted for practice with the company commander as the trainer-evaluator for his platoons. The battalion may direct the FTX (internal evaluation) using a company evaluator and an FO evaluator. The platoon leaders and platoon sergeants are the primary trainers during the initial phases of the training. The company commander, the XO, the platoon leaders, and the platoon sergeants act as trainers during the company phases of training. If the exercise is conducted by the battalion, either the battalion commander, the XO, or the S-3 acts as the trainer and evaluator, but may require additional assets from the battalion to support the evaluation phase. The controllers or evaluators must be positioned at critical locations throughout the exercise, to observe the actions. They must not interfere with the conduct of the operation, except to halt it when a safety violation occurs or to conduct an AAR.

b. *Vehicles/Communications*: Those organic to the company are used.

c. *Opposing Force*:. The OPFOR ground force should be at least platoon-size. FTXs conducted by the company require internal resourcing, and units should be rotated to ensure total training. T&EOs provide training tasks and standards for the OPFOR.

d. *Maneuver Area*: A training area at least 5 by 15 km is best for this exercise.

e. *Consolidated Support Requirements*: This exercise requires the items shown in Table 5-11.

f. *Barrier materials.* Sufficient Class IV must be made available to fortify positions, IAW FM 3-06.11.

Table 5-11. Ammunition and pyrotechnics support requirements for a six-day FTX.

Ammunition	*DODAC*	*Per Plt/Co/Bn FTX*
5.56-mm blank	1305A080	120 rounds per rifle
5.56-mm SESAMS		120 rounds per rifle
5.56-mm blank	1305A075	200 rounds per M249 LMG
5.56-mm SESAMS		200 rounds per M249 LMG
7.62mm	1305A111	200 rounds per BFV/M1
		400 rounds per M60/240B
ATWESS cartridge	1370L367	6 (Viper) (for LAW) 18 per Dragon 14 per TOW system 9 per OH-58D/AH-64
Pyrotechnic device	1370L602	6 per M1 2 per target
Hand grenade fuze (M228)	1330G878	150 per battalion
Stun grenade M84	1330GG09	150 per battalion
Smoke grenade, HC		150 per battalion
Smoke pot		6 per battalion

5. **Evaluation.**

Table 5-12 lists the DTMS T&EOs used to evaluate this STX.

Table 5-12. T&EOs used to evaluate STX Attack.

Task/Function	*Task Number*
Defend MOUT/Building	7-3/4-1118
Conduct a Dismounted Tactical Road March (Co/Plt)	7-2-1189
Conduct a Mounted Tactical Road March (Co/Plt)	7-2-1198
Perform Relief Operations	7-3/4-1124
Perform Passage of Lines	7-3/4-1125
Conduct Tactical Movement	7-2-1342
Occupy Assembly Area	7-3/4-1136
Establish a Roadblock/Checkpoint	7-3/4-1401
Perform CBRN Operations	7-3/4-1406
Maintain Operations Security	7-3/4-1409
Prepare for Combat	7-3/4-1606
Conduct Consolidation and Reorganization	7-2-5027
Perform Ambush	7-2-1106

TARGETRY

5-10. Table 5-13 lists the combination of precision and nonprecision targets to support the diversity of scenarios required to support combined arms and branch specific lane training. Precision targets, less than 50 meters, are used in and around buildings within the CACTF and reinforce precision marksmanship techniques described in FM 3-22.9 (Table 5-13).

Table 5-13. CACTF target requirements.

Event	*Number Of Lanes*	*Human Targets*		*Vehicle Targets*	*Totals*	
					HT	*VT*
Company/Platoon STX Offense	3/9	Precision	Nonprecision	9	45	9
		15	30			

CONSIDERATIONS FOR TRAINING WITH ARMOR

5-11. Built-up areas consist mainly of man-made features such as buildings, streets, and subterranean systems. These features of urban terrain create a variety of tactical problems and possibilities. In order to ensure that the tank platoon can operate effectively in the urban environment, the platoon observation and direct-fire plans must address the ground-level or surface fight (in streets and on the ground floor of buildings), the aboveground fight (intrasurface and supersurface), and the subterranean fight (Figure 5-14). Infantry can assist the tank platoon by—

- Locating targets for tanks to engage.
- Destroying antitank weapons.
- Assaulting enemy positions and clearing buildings with tank support.
- Protecting tanks from antitank fires.

Figure 5-14. Urban battlespace.

TRANSPORTING INFANTRY

5-12. At times, the tank platoon must transport Infantrymen (Figure 5-15). They only do so when not expecting contact. If the platoon is moving as part of a larger force, and if it is tasked to provide security for the move, then the lead section or element does not carry Infantry.

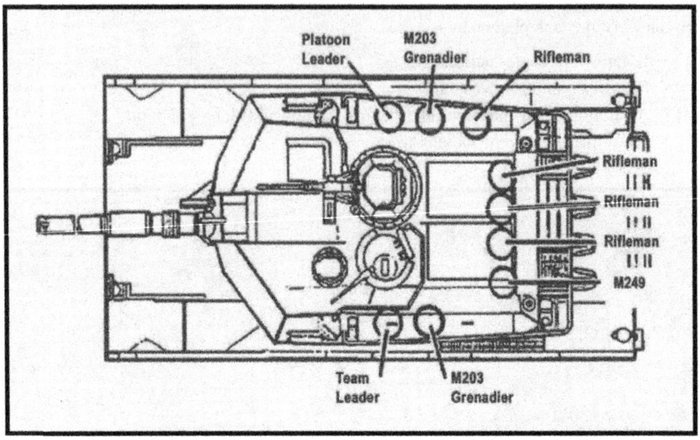

Figure 5-15. Example positions for Infantry riding on a tank.

5-13. Infantry and Armor leaders must observe the procedures, precautions, and considerations in Figure 5-16 when Infantrymen ride on tanks:

- Infantry should thoroughly practice mounting and dismounting procedures and actions on contact.
- Infantry must always alert the TC before mounting or dismounting. They must follow the commands of the TC.
- Infantry platoons should be broken down into squad-sized groups, similar to air assault chalks, with the Infantry platoon leader on the Armor platoon leader's vehicle and the Infantry PSG on the Armor PSG's vehicle.
- Platoon leaders, PSGs, and team leaders should position themselves near the TC's hatch, using the external phone (if available) to talk to the TC and relay signals to the unit.
- If possible, the lead vehicle should not carry Infantrymen. Riders restrict turret movement and are more likely to be injured or killed on initial contact.
- Whenever possible, Infantrymen should mount and dismount over the left front slope of the vehicle. This procedure ensures that the driver can see the Infantrymen and that the Infantrymen DO NOT pass in front of the coax machine gun. Infantrymen must ensure that they remain behind the vehicle's smoke grenade launchers. This will automatically keep them clear of all weapon systems.
- Infantrymen must always have three points of contact with the vehicle; they must watch for low-hanging objects such as tree branches.
- All Infantrymen should wear hearing protection.
- Infantrymen should not ride with anything more than their battle gear. Rucksacks and B-bags should be transported by other means.
- Infantrymen should scan in all directions. They may be able to spot a target the vehicle crew does not see.

Figure 5-16. Precautions for Infantrymen riding on a tank.

5-14. Passengers should be prepared to take the actions on contact shown in Figures 5-17 and Figure 5-18.

- Wait for the vehicle to stop.
- At the TC's command, dismount immediately (one fire team on each side). ***Do not*** move forward of the turret.
- Move at least 5 meters to the sides of the vehicle.
  Do not move behind or forward of the vehicle.
- ***Do not*** move in front of vehicles unless ordered to do so. Main gun discharge overpressure can inflict severe injury or death to forward dismounts.
- ***Do not*** dismount a vehicle unless ordered or given permission to do so.
- ***Do not*** dangle arms or legs, equipment, or anything else off the side of a vehicle; they could get caught in the tracks, causing death, injury, or damage to the equipment or vehicle.
- ***Do not*** place too many riders on the vehicle.
- ***Do not*** fall asleep when riding. The warm engine may induce drowsiness; a fall could be fatal.
- ***Do not*** smoke when mounted on a vehicle.
- ***Do not*** stand near a moving or turning vehicle at any time. Tanks have a deceptively short turning radius.

Figure 5-17. Actions on contact for Infantrymen riding on a tank.

DANGER

1. The overpressure from the tank 120-mm cannon can kill a dismounted Infantryman within a 90-degree arc extending from the muzzle of the gun tube out to 200 meters.
2. From 200 to 1,000 meters along the line of fire, on a frontage of about 400 meters dismounted Infantry must be aware of the danger from discarding sabot petals, which can kill or seriously injure Soldiers.

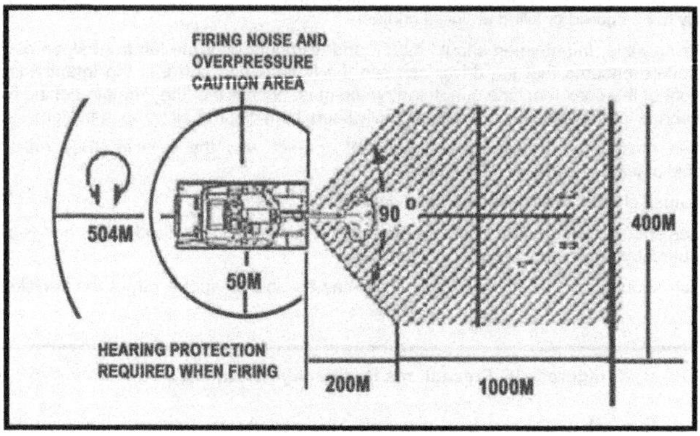

Figure 5-18. Danger areas around a tank when firing the 120-mm main gun.

ADDITIONAL CONSIDERATIONS

5-15. Additional considerations and preparations for transporting Infantrymen include the following:

- Armor--
  - Uses main gun fire to reduce obstacles or entrenched positions for the Infantry.
  - Takes directions from the Infantry ground commander (platoon leader/PSG/squad leader) to support their fire and maneuver.
  - Provides reconnaissance by fire for the Infantry.
  - -- Should know and understand how the Infantry clears buildings, how they mark cleared buildings, casualty evacuation plan, signal methods, engagement criteria for tank main gun, front line trace reporting, ground communication from the tank with the dismounted Infantry.
  - -- Uses its night vision capability to augment and supplement the Infantry's night vision capability.
- Infantry--.
  - -- Provides real time information for the tank crewmen to help them overcome tank noise and the lack of ground situational understanding.
  - -- Provides reconnaissance and fire direction of enemy positions for main gun attack.

- Considerations for dismounted tank security include--
  - -- Tank crewman should rehearse the mounting and dismounting of Infantry from their vehicle, briefing the Infantrymen on safety procedures for the vehicle and weapon systems.
  - -- Tank commanders need to rehearse communicating with dismounted Soldiers by way of TA-1 and DR-8 in the bustle rack.
- Vehicle preparation for combat in urban terrain should cover--
  - -- Keep at least one ballistic shield to the "Dog House" closed. Most engagements will be under boresight range, and the battlesight technique will suffice.
  - -- Place sandbags around antenna connections and electrical wiring on the turret top.
  - -- Place extra coax ammunition inside the turret.
  - -- Remove any highly flammable products from the outside of the vehicle and from the sponson boxes.

VEHICLES, WEAPONS, AND MUNITIONS

5-16. Numerous factors related to vehicles and their organic weapons and munitions affect the tank platoon's urban operation planning and execution, including the following:

- The preferred main gun rounds in the urban environment are HEAT, MPAT (ground mode), and MPAT-OR. These all perform much better than sabot rounds against bunkers and buildings.
- HEAT ammunition will open a larger hole in reinforced concrete or masonry structures than MPAT or MPAT-OR. Both MPAT and MPAT-OR, however, offer greater incapacitation capability inside the structure.
- HEAT ammunition arms approximately 60 feet from the gun muzzle. It loses most of its effectiveness against urban targets at ranges of less than 60 feet.
- MPAT and MPAT-OR rounds arm approximately 100 feet from the muzzle of the gun. Because of the shape and metal components of the projectiles, however, this ammunition remains effective at ranges of less than 100 feet.
- Sabot petals, including those on MPAT and MPAT-OR, endanger accompanying Infantry elements. They create a hazard area extending 70 meters on either side of the gun-target line out to a range of 1 kilometer.
- The tank's main gun can depress only to -10 degrees and can elevate only to +20 degrees. This creates considerable dead space for the crew at the close ranges that are typical in the urban environment.
- The external M2 HB machine gun can deliver a heavy volume of suppressive fire and penetrate light construction, buildings and most barricades. The M2 HB MG can elevate to +36 degrees; however, the TC must be unbuttoned to fire the M2 on the M1A2 or M1A2 SEP.
- The M240 coax machine gun can effectively deliver suppressive fires against enemy personnel and against enemy positions that are behind light cover.
- The loader's M240 machine gun can effectively deliver suppressive fire against enemy personnel and against enemy positions that are behind light cover; however, the loader must be unbuttoned to operate it. This weapon may be dismounted and used in a ground role if units are equipped with the M240 dismount kit.

- When buttoned up, the tank crew has limited visibility to the sides and rear and no visibility to the top. Figures 5-19 and 5-20 show the dead spaces associated with tank operations in an urban environment.
- FM 3-20.12 explains special uses for tank-mounted machine guns in the urban environment.

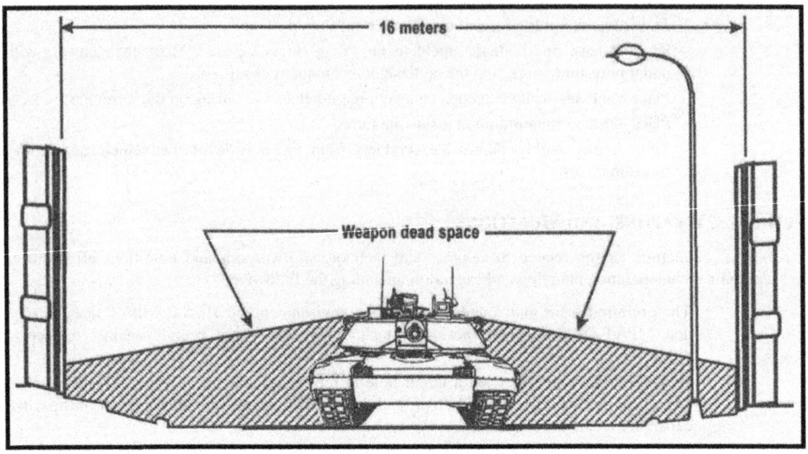

Figure 5-19. Tank weapon dead space at street level.

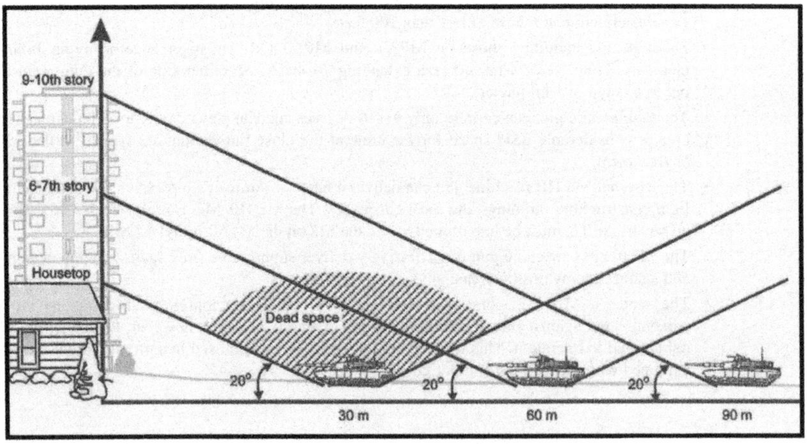

Figure 5-20. Tank main gun and coax dead space above street level.

Tank Platoon Command and Control in Urban Operations

5-17. Communications problems, fire control, proximity and visibility, personnel factors, ROE, ROI, civilians, and slow-pace urban operations all affect the tank platoon's urban operations planning and execution:

Communications Problems

5-18. The low-level task organization that may take place during urban operations requires elements to establish additional communications links, which can be disrupted by buildings and other urban terrain features.

Fire Control

5-19. Extensive direct fire planning and restrictive fire control measures are an absolute requirement in urban operations.

Proximity and Visibility

5-20. Friendly elements often must operate in confined and restrictive areas during urban operations, and they may not be able to see other nearby friendly forces. These factors significantly increase the danger of fratricide.

Personnel Factors

5-21. Urban operations impose significant, and often extreme, physical and psychological demands on Soldiers and leaders.

ROE, ROI, and Civilian Considerations

5-22. The ROE or ROI may restrict the use of certain weapon systems and TTP. As an integral part of urban operations, noncombatants create special operational problems. Thedeal with these concerns, units operating in urban terrain must know how to effectively employ linguists and counterintelligence and civil affairs teams.

Slow Pace Urban Operations

5-23. This operation usually prevents the tank platoon from taking full advantage of the speed and mobility of its tanks.

Appendix A

Training Tips

The effectiveness of urban operations training depends on the imagination and thoroughness applied to a training plan. All trainers, from training managers to junior leaders, should prepare their units to fulfill their urban operations mission. They can employ procedures that are based on training doctrine and proven techniques. This appendix discusses several techniques for conducting urban operations training. Trainers make the choice based on the training needs and available resources.

CONCURRENT TRAINING

A-1. Concurrent training allows groups of Soldiers to train on different tasks at the same time. This training requires a plan for rotating Soldiers between the stations that teach different tasks. This procedure is referred to as round-robin training. Concurrent training is an efficient training technique since it makes the best use of training time and facilities. Concurrent training in the urban operations training facilities works best with performance-oriented training. It also requires detailed planning, control, coordination, and assistant trainers. Each station should conduct training that is relevant to, and supports the training the unit is conducting.

URBAN ASSAULT COURSE AND BREACH FACILITY

A-2. These facilities are well suited for conducting concurrent training. After mastering the tasks at one station, a group of Soldiers rotates to the next station in the sequence. Theavoid bottlenecks at any of the stations, trainers plan instructional blocks that last about the same length of time. Concurrent training is often conducted with dedicated instructors at each station. The *advantage* of this technique is that it ensures uniformity of performance standards and saves on trainer preparation time. The *disadvantage* is that the Soldier is not always trained by his leader. Therefore, when feasible, small-unit leaders should rotate with their Soldiers from station-to-station so they can take part in the training. This technique demands prior planning, preparation, and instructors that can teach.

URBAN ASSAULT COURSE AND COMBINED ARMS COLLECTIVE TRAINING FACILITY

A-3. These facilities are well suited for conducting concurrent exercises. Concurrent exercises are variants of concurrent training. In a concurrent exercise, elements of the same unit receive different missions. These missions correspond to each element's training objective for that exercise. This technique can be used during FTXs, STXs, or CFXs. For example, one unit attacks; a second unit supports by fire; a third unit acts as a reserve. Also, elements of the same unit can challenge each other. For example, if a trainer has been training his unit in urban offense and defense operations, he can create a scenario in which one force is required to attack the other. This technique allows the unit to train with an OPFOR without an extra investment of resources. Leaders must remember to carefully plan the force ratios between the OPFOR and counterattacking forces.

DEMONSTRATIONS

A-4. Trainers must demonstrate the correct way to perform the tasks that the Soldiers will be required to perform. A demonstration is more effective when the trainer uses hand-on techniques. A live demonstration is an effective teaching method since it holds Soldiers' interest and they are able to learn with their eyes. The UAC and the Breach Facility are ideal facilities for teaching individual and collective tasks using demonstrations. The trainer must ensure that each Soldier. Performs the task being taught to guarantee that the soldier has learned. Key points should be explained to the satisfaction of each Soldier before progressing to the next step. The trainers can use video cameras to film the action occurring inside the buildings to show the Soldiers the correct way to perform a task or use the same action to explain a better technique to accomplish the same task. This technique is very effective because the Soldiers can watch the demonstrations on a remote monitor, and trainers can replay the videotape as often as required to reinforce his training objective.

EXERCISES

A-5. Training exercises are used to develop tactical SOPs, teamwork among the participants, and Soldier confidence in his leaders and equipment. Table A-1 matches training exercise types to appropriate training facilities.

Table A-1. Training exercise matrix.

EXERCISE	*CACTF*	*SHOOT HOUSE*	*SIMULATIONS*	*LOCAL TOWN*
MAPEX	X		X	X
TEWT	X			X
CPX	X		X	X
CFX	X			X
BSX			X	
LFX		X		
STX	X	X		X
FTX	X			X

MAP EXERCISE

A-6. A MAPEX allows participants to plan and execute urban operations using maps, overlays, sand tables, photographs, or models. MAPEXs develop teamwork and skills in terrain analysis, planning, coordination, order writing, and employment of urban operation tactics and are more suited for battle staff and leaders.

- Using a model of the CACTF is an excellent training aid for instructing Soldiers and leaders. Gather leaders around the map or model and have them devise a tactical plan. Use figures or markers to represent friendly units and the OPFOR. Create tactical situations that challenge participants to react as they would in real combat.
- Incorporate direct fire, casualties, CBRN, logistics, and civilians into the exercise. Ensure the ROE for each exercise is clear to participants.
- Have plenty of detailed maps available for use in the MAPEX and in the planning and execution of subsequent CFX operations.

Tactical Exercise Without Troops

A-7. A TEWT focuses on leader and staff tasks. The senior trainer selects the terrain for the urban TEWT. Either the CACTF or a real village, town, or city is appropriate. The procedure for conducting a TEWT is as follows:

- Create a scenario and enemy situation, and assign missions.
- Allow the leaders to devise their reconnaissance plans and implement them on the terrain.
- Require the leader to brief subordinates while walking the terrain.
- Ensure their plans include the following tasks:
  - -- Analyze the terrain.
  - -- Employ units according to the terrain analysis.
  - -- Emplace weapons.
  - -- Devise a unit plan.
- Conduct an AAR.

Command Post Exercise and Battle Simulation Exercise

A-8. CPXs and BSXs allow leaders and battle staffs to perform their tasks in a unit CP, where they receive information as reports requiring certain reactions. This type of training can be achieved through the constructive simulations training portion of the urban operations training strategy using JANUS or JCATS. A company CP is usually associated with a battalion level or higher CPX, but a company can conduct its own CPX. The headquarters section can set up a CP in a CACTF building, displace to another building, and repeat the exercise until they meet the required standards. The section can practice hiding wire on pavement, stringing it between buildings, and running it through sewers. The CP can practice receiving reports, posting situation maps, and sending reports forward to higher headquarters. The CPX/BSXs are most effective in achieving the following general training objectives:

-- Preparing plans.
-- Issuing orders.
-- Building teams.
-- Reconnoitering, selecting, and tactically occupying CP locations.
-- Establishing and using communications.
-- Preparing and sending reports.
-- Coordinating between subordinate elements.

Command Field Exercise

A-9. The CFX is used to train commanders, staffs, and junior leaders. The CACTF or a local town can accommodate this type of exercise. However, the unit commander must go through the correct procedures when conducting training off federal property. The CFX employs all of a unit's command and control assets, with a number of subordinate elements reduced according to a set scale. For example, one Soldier might represent an entire squad, one vehicle a platoon, or one gun an entire battery.

Preliminary Training

A-10. Preliminary training, such as CPXs, TEWTs, MAPEXs, and BSXs, should precede the CFX. Units conduct operations, and controllers resolve battle outcomes. The exercise ends with an AAR. Fewer controllers are needed in a CFX than in an FTX. However, if the number of unit functions increases, the number of controllers must increase also.

Planning Procedures

A-11. Combat in urban areas entails centralized planning with decentralized execution.

Centralized

A-12. Centralized planning can be practiced in a CFX; however, it can be practiced more efficiently in a CPX, TEWT, MAPEX, or BSX.

Decentralized

A-13. Decentralized execution means that platoon, squad, and team leaders lead the battle in an urban area. However, because element participation is scaled down in a CFX, trainers usually find STXs, FTXs, and TEWTs more useful.

LIVE-FIRE EXERCISE

A-14. Chapter 3 and local range regulations provide specific guidance for conducting LFXs at the shoot house. Live-fire training allows Soldiers and leaders to maneuver while firing service ammunition or SRTA. An LFX is most productive when preceded by non-live-fire practice (dry fire or dry run). However, dry fire should *always* precede live fire, with either TES or SESAM. An AAR should follow each repetition, and units should only proceed to the next level of difficulty after mastering the current step. The training in an LFX occurs in the following sequence, which can be shortened if time and ammunition dictate.

1. Dry fire to practice unit drills and individual tasks.
2. Conduct an AAR.
3. Practice drills, tasks a second time with limited use of ammunition to show the complex coordination of fire and maneuver required.
4. Conduct an AAR.
5. Practice exercise a third time with full use of ammunition to enhance realism and build confidence.
6. Conduct an AAR.

SITUATIONAL TRAINING EXERCISE

A-15. STXs are mission-oriented, limited exercises used to train a unit to perform a single collective task or a series of related tasks and drills. With their distinct start and stop points, STXs represent a segment of battle. An STX is more flexible and complex than a single T&EO or drill. In fact, an STX typically includes drills, leader tasks, and separate individual tasks, and involves a unit's "slice" of external combat assets. An STX often precedes an FTX or CFX, and can be conducted in the CACTF by creating a brief situation and mission. For example, the MP company commanders might want to teach his Soldiers about straggler control. They can string the tasks together into small scenarios such as STXs, describe a tactical situation that produces stragglers from nearby units, use other Soldiers to represent the stragglers, and train Soldiers to process stragglers at straggler collection and transport points.

FIELD TRAINING EXERCISE

A-16. FTXs allow units to execute collective tasks under conditions that closely resemble actual urban terrain. FTXs are suitable for the CACTF.

- An FTX requires significant time, resources, and planning. The time is well spent if units master their collective tasks. An FTX is an excellent way to conduct the *run* stage of collective training. Additionally, OPFOR and MILES greatly enhance the training value of the FTX.
- A unit's urban operations FTX training plan addresses ROE, maneuver restrictions, and the controller plan. Soldiers must know the exercise's training objectives before they begin, so they can focus on the correct tasks. Every participant, regardless of position or rank, must fully understand all safety requirements.

- To make the most of their training time in the field, leaders must plan in great detail. The concept of multiechelon training demands that they consider what training can occur at each level in each event.

EVALUATORS

A-17. Soldiers from the training unit or from another unit can evaluate urban operations training, especially during the *run* stage of training in an STX, FTX, or CFX. A dedicated evaluator provides objective and thorough comments on the unit's performance throughout all stages of the exercise. Evaluators monitor the critical point of the action but must avoid interfering. They must not disrupt unit performance or undermine realism. They must be trained and given standardized evaluation guides if their feedback is to be meaningful; MTPs and T&EOs are a good start. Evaluators facilitate AARs alone or along with the unit leader.

TACTICAL ENGAGEMENT SYSTEM

A-18. The TES realistically replicates the effects of actual fire in training. It uses laser adapters on the weapons and laser-sensitive equipment on the Soldiers and vehicles. A TES weapon emits a laser beam when fired. Accurate fire strikes the laser-sensitive equipment and scores a hit. The TES requires Soldiers to aim accurately and to use cover and concealment effectively. When used with an OPFOR, TES offers realistic training. It is an excellent training aid for an FTX or STX. It is also effective at the shoot house as a substitute for live fire. The two major shortcomings of TES are that it allows Soldiers to seek cover behind structures that normally would not protect them, and that it does not replicate suppression.

NAVIGATION

A-19. Navigational techniques for urban areas differ from those needed on other terrain.

- A typical 1:50,000-scale military map provides insufficient detail for accurate navigation. Therefore, Soldiers must learn to navigate using commercial maps, sketches, and aerial photographs. Leaders must learn how to plan operations using these items.
- Underground navigation is a special challenge. Soldiers must know subway plans and be able to identify tunnels, elevated tracks, stairways, and other critical features. They must learn how to measure distance, remain oriented, and mark their routes.
- Destruction often erases urban features that could aid in navigation. Soldiers must learn to orient using distinctive, difficult-to-erase features such as bridges, waterways, towers, and intersections.
- Leaders conduct navigational exercises in an actual urban area. Soldiers must move from point to point while controllers at each point monitor and verify Soldiers' progress.
- The GPS is a navigational aid that triangulates with satellites to calculate position. However, this system may be affected by nearby tall buildings, and its effectiveness decreases inside buildings or underground.

LIMITED VISIBILITY TRAINING

A-20. Urban operations involve day and night operations. FTXs, STXs, CFXs, and CPXs are the most valuable exercises for night training.

- Field-expedient lighting, NVDs, illumination, and electrical lighting must be exploited when the tactical situation allows. Soldiers must learn discipline in their use of light, and also learn the lighting techniques that allow them to carry out their urban training program.

- Situational awareness is the Soldier's knowledge of his location, the locations of friendly forces, and the locations of the enemy on the battlefield. Situational awareness is the key to avoiding fratricide, especially during limited visibility operations. Units must plan to use all the equipment and devices available to mark friendly units and to avoid fratricide, without exposing themselves to enemy observation.
- Reverse-cycle training should be part of an urban training program. Limited visibility fighting skills are perishable, but they can be sustained and improved by effective training. The goal is for every Soldier to operate as effectively at night as during the day.

SAFETY AND AMMUNITION CONSIDERATIONS

A-21. When using ammunition and pyrotechnics, all associated safety requirements must be followed. This paragraph describes some of the safety equipment and training procedures to consider:

SPECIAL EFFECT SMALL-ARMS MARKING SYSTEM

A-22. SESAMS is 5.56-mm and 9-mm color-marking training ammunition. It is designed for close combat training. It is used with the close combat mission capability kit (CCMCK) adapter kit, which is installed on the M4/16 rifle, M9/M11 pistol, and M249 SAW. Different colors can be assigned to OPFORs for engagement identification and fratricide tracking. SESAMS has a 30-meter effective range and activates TES (MILES), so the two are used together. Anytime during a SESAMS exercise that a Soldier loses any of his protective gear, he must immediately cover his face and eyes with his hands. He must stay put until the engagement ends, or until he is told that it is safe to replace his safety equipment. Everyone, to include observer-controllers, must wear the protection shown in Table A-2.

Table A-2. SESAMS protection.

SKIN	Wear gloves, and cover all exposed skin. At close ranges, the SESAMS projectile can break skin.
HEAD	Take every precaution to protect your head.
EYES	Keep your eyes completely enclosed within eye protection. You can use standard Army sun, wind, and dust goggles.
EARS	Wear hearing protection to prevent hearing loss.
MOUTH	Keep your mouth covered.

DANGER

Failure to wear the required protection during SESAMS exercises can result in injury or death.

M84 STUN GRENADE

A-23. The hand-deployed M84 stun grenade serves as a reliable, effective, nonlethal means of neutralizing, diverting, and disorienting enemy personnel. It is used tactically as well as in training.

A-24. The M84 produces an intense (1 to 2.5-million candlepower) flash and, at 1.5 meters, a 170 dBs bang. Even though this is too low to produce permanent hearing damage, one-time use Soldiers are required to wear hearing protection anytime noise levels exceed 85 dBs.

A-25. The M84 stun grenade has an M201A1 fuze and a pyrotechnic output charge in a synthetic casing with a one- to two-second delay fuze, which cannot be cooked off. The grenade body has a nonfragmenting metal hexagonal casing, which, once deployed, resists rolling.

SMOKE GRENADES

A-26. Smoke grenades are used in urban combat to cover movement, deceive enemy forces as to the exact location of the attacking force, and for signaling. Smoke grenades should not be used in buildings and subterranean passages as they displace oxygen and protective masks do not filter the smoke or provide oxygen. Buildings in urban areas channel air currents, which influence emplacement of smoke grenades. One grenade should be used to test the airflow before employing additional grenades.

WARNING

Avoid using smoke grenades in buildings and in subterranean passages.

OPPOSING FORCES

A-27. OPFOR should be used as often as possible to enhance the realism of training, especially for FTXs, CFXs, and STXs. Thebe realistic, the OPFOR must have defined countertasks. These countertasks must support "Blue Force" training, and should prevent accidental contact between the Blue Force and the OPFOR. Leaders should develop a thorough control plan for OPFOR scenarios. This plan should describe uniform and vehicle markings, ammunition, safety instructions, controller guidance, and guidelines for handling enemy prisoners of war (EPW) and noncombatants. The ROE must clearly specify the limits of close combat between Soldiers.

- OPFOR heighten Soldiers' interest by giving them a real opponent. This fosters unit competition and realism, especially when the OPFOR is used in addition to TES and SESAM.
- OPFOR challenges leaders' ability to improvise. However, this free-play scenario should be monitored to ensure the unit meets the training objectives.
- OPFOR can use threat urban operations and doctrine to determine their course of action. If available, OPFOR uniforms, mock documents, overlays, and equipment add realism and intelligence to the scenario. Leaders conduct preliminary training for the OPFOR, which helps the OPFOR recreate the threat's urban tactics.
- Noncombatants should be integrated into training scenarios.

PRECISION MARKSMANSHIP TRAINING

A-28. Precision marksmanship training should be conducted before training at the UAC, CACTF, and the Shoot House, which accommodate training in target discrimination (FM 3-22.9). This is the best use of the unit's time and resources.

PHYSICAL READINESS TRAINING

A-29. Combat in urban areas places extraordinary physical demands on Soldiers. Therefore, physical readiness training, with emphasis on upper body development, is paramount to any urban operations training program.

VARIED SCENARIOS

A-30. Leaders must vary training scenarios to prevent redundant and predictable tactical behavior. Using predictable avenues of approach into the facility produces stereotypical tactical solutions.

BUILDING FACADE WALLS, WINDOWS, AND DOORS

A-31. The use of façade walls, windows and doors, in conjunction with SESAMS in the interior of some stations at the UAC, provides for more realistic training. This method of training reinforces the fact that walls, windows, and doors in most cases *do not* stop the penetration of service ammunition. Plastic wrap sheeting, tar paper, or some other type of material can be used to allow SESAMS penetration (Figure A-1).

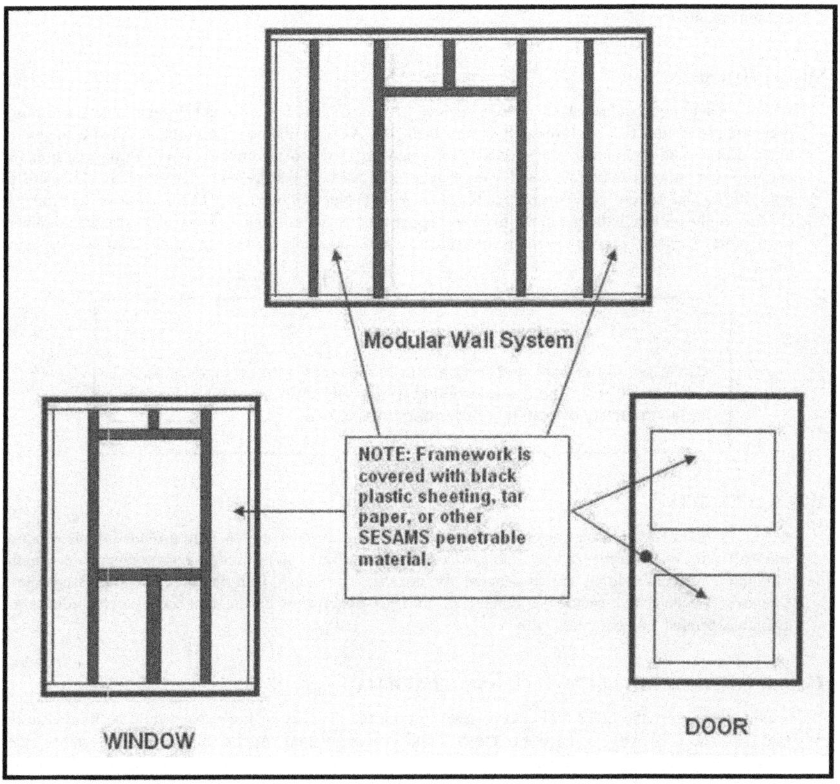

Figure A-1. Shoot-through walls and doors.

USE OF OLDER OR NONSTANDARD FACILITIES

A-32. Until installations receive funding for the UAC, SH, BF, and CACTF, they can use the older MOUT Assault Course (MAC) and Collective Training Facility (CTF) as described in the 2002 edition of TC 90-1, which this edition supersedes. Installations may also use nonstandard shoot houses and other nonstandard urban facilities. These facilities may be adapted as follows:

URBAN ASSAULT COURSE

A-33. Generally, the MAC can be used to train most of the UAC tasks. Of course, the MAC lacks targetry and instrumentation. Instead, units can record training with portable video cameras. During the AARs, they can play back the videos on VCRs hooked to TVs or on TV/VCR combos. OPFOR combatants will have SESAMS or MILES. People in civilian clothes can play the parts of the noncombatants.

SHOOT HOUSE

A-34. Many nonstandard shoot houses, such as tire houses, exist in different installations. Local range regulations and unit risk analyses will govern their use. Again, units can use portable video cameras for video capture and VCRs and television sets for video play back to facilitate an AAR. Both precision and nonprecision targets can be fabricated using E-type silhouettes on a stake, old BDUs/civilian clothes filled with straw, and so on. Simulate noncombatants with E-type silhouettes painted white or dummies in civilian clothes. Place balloons in the head and upper torso areas to simulate lethal shot placement. Where local policy permits, convert portions of collective training facilities into live-fire shoot houses using SRTA.

CAUTION

Units should ensure that they are in compliance with all local policies affecting live fire before attempting to use SRTA in an older collective training facility or nonstandard urban training site.

BREACH FACILITY

A-35. Existing demolition ranges may be used to conduct explosive breaching provided door, window, and wall units, or facsimiles, are available. Mechanical breaching can be trained in cantonment areas on the installation where building are designated for demolition, if such facilities are available. Regardless, whenever mechanical breacking is conducted, unit personnel must ensure that local safety policies are available, briefed, and complied with.

COMBINED ARMS COLLECTIVE TRAINING FACILITY

A-36. Generally the older CTF can be used to train the majority of tasks that would be trained in the CACTF. The CTF will not have any targetry and instrumentation. Again, units can use portable video cameras for video capture and VCRs and television sets for video play back to facilitate an AAR. Targets can be OPFOR with MILES. Role players in civilian clothes can be introduced to play the part of noncombatants. If armored vehicles are used, units should ensure that local range policies allow for their use. Older CTFs are normally either a 16 or 32 building design. Appropriate unit-level training, platoon, company, or battalion, should be planned based on the size of the facility.

Appendix B

Unit Training Matrix

Unit training plans are prepared IAW FM 7-1, which states that "Planning links the unit METL and the execution of battle-focused training.". The commander begins his planning with guidance from a higher echelon and an in-depth assessment of METL tasks (in this case, mission-essential urban tasks). Based on his own observations, feedback from his subordinates, and other sources the commander determines which tasks should be trained, which should be trained first, and how much time should be allotted for each. Tables B-1 through B-5 list training tasks by the WFF to be trained at each facility.

Table B-1. Army unit training matrix, brigade tasks.

Brigade Tasks	*SHOOT HOUSE* Semiannual	*BREACH HOUSE* Semiannual	*UAC* Quarterly	*CA/CTF* Semiannual	*CTC* *As Scheduled	*SIMULATION* Annual
Develop Intelligence						
Perform Intelligence Operations					X	X
Perform S-2 Operations					X	X
Deploy/Conduct Maneuver						
Attack					X	X
Perform Air Assault					X	X
Defend					X	X
Lodgment					X	X
Exercise Command and Control						
Command and Control the Brigade					X	X
Maintain Communications					X	X
Coordinate Air Defense Operation					X	X
Conduct Noncombatant Evacuation					X	X
Develop a Media Plan					X	
Conduct Mediation and Negotiation					X	
Employ Firepower						
Execute Fires					X	X
Employ Air Support					X	X
Employ Attack Helicopters					X	X
Synchronize Fire Support					X	X
Synchronize CAS					X	X
Conduct Targeting Process					X	X
Protect the Force						
Perform Mobility/Survivability Operations					X	X
Provide Engineer Support					X	X
Coordinate CBRN Operations					X	X
Conduct Operations Security					X	X
Perform Sustainment						
Evacuate Casualties					X	X
Treat Casualties					X	X
Sustain the Force					X	X
Conduct a Civil-Military Operation					X	X
Provide Humanitarian Support					X	X

Table B-2. Army unit training matrix, battalion tasks.

Battalion Tasks	SHOOT HOUSE Semiannual	BREACH HOUSE Semiannual	UAC Quarterly	CA/CTF Semiannual	CTC As Scheduled	SIMULATION Annual
Develop Intelligence						
Conduct S-2 Operations				X	X	X
Deploy/Conduct Maneuver						
Attack a Built-Up Area				X	X	X
Plan for Urban Operation				X	X	X
Defend an Urban Area				X	X	X
Conduct an Infiltration/Exfiltration				X	X	X
Perform Cordon and Search Operations in an Urban Area				X	X	X
Plan for Urban Operations				X	X	X
Conduct Tactical Movement				X	X	X
Conduct Presence Operation in a Stability Environment				X	X	X
Defend an Urban Area				X	X	X
Traverse a Built-up Area				X	X	X
Conduct a Mobile Defense				X	X	X
Employ Firepower						
Integrate Fires and Effects Support				X	X	X
Establish the Fire Support Element				X	X	X
Protect the Force						
Conduct Mobility or Survivability Operations				X	X	X
Secure Civilians During Operation				X	X	X
Conduct Passive Air Defense Measures				X	X	X
Conduct Active Air Defense Measures				X	X	X
Employ Operations Security Measures				X	X	X
Integrate Air Defense Support				X	X	X
Perform Sustainment						
Conduct S-1 Operations				X	X	X
Conduct S-4 Operations				X	X	X
Conduct a Civil-Military Operation				X	X	X
Exercise Command and Control						
Conduct S-6 Operations				X	X	X
Conduct Noncombatant Evacuation				X	X	X
Conduct S-3 Operations				X	X	X
Conduct Mediation and Negotiations				X	X	X
Plan Roadblocks or Checkpoints				X	X	X
Establish Liaison				X	X	X

Table B-3. Army unit training matrix, company tasks.

Company Tasks	*SHOOT HOUSE Semiannual*	*BREACH HOUSE Semiannual*	*UAC Quarterly*	*CA/CTF Semiannual*
Develop Intelligence				
Conduct an Area or Zone Reconnaissance				X
Reconnoiter a Built-Up Area				X
Deploy/Conduct Maneuver				
Conduct Cordon and Search				
in a Built-Up Area				X
Attack a Built-Up Area				X
Conduct Presence Operations				X
Conduct Tactical Movement				
in a Built-Up Area				X
Defend in Urban Environment				X
Conduct an Infiltration (Co/Plt)				X
Conduct Operations with Armored or				
Mechanized Infantry Vehicles				
in an Urban Environment				X
Conduct Civil Disturbance Operations				X
Establish Checkpoints				X
Employ Firepower				
Integrate Direct Fires				X
Integrate Indirect Fires				X
Protect the Force				
Maintain Operations Security				X
Conduct Passive Air Defense Measures				X
Conduct Passive Air Defense Measures				X
Perform Sustainment				
Treat Casualties				X
Evacuate Casualties				X
Secure Civilians during Operations				X
Conduct Resupply Operations				
Process Captured Documents and				X
Equipment				X
Handle Enemy Prisoners of War				
Exercise Command and Control				
Prepare for Combat				X
Establish Communications				X
Conduct Consolidation and Reorganization				X
Conduct Negotiation				X
Conduct Troop-Leading Procedures				X

Table B-4. Army unit training matrix, platoon/squad tasks.

Platoon/Squad Tasks	*SHOOT HOUSE Semiannual*	*BREACH HOUSE Semiannual*	*UAC Quarterly*	*CA/CTF Semiannual*
Develop Intelligence				
Report Tactical Information	X		X	X
Conduct an Area or Zone Reconnaissance	X		X	X
Reconnoiter a Built-Up Area	X		X	X
Deploy/Conduct Maneuver				
Conduct Tactical Movement in a Built-Up Area	X		X	X
Assault a Building	X		X	X
Conduct a Strongpoint Defense of a Building				X
Conduct an Infiltration or Exfiltration				X
Conduct Operations with Armored or Mechanized Vehicles in an Urban Environment	X			X
Conduct a Presence Patrol	X			X
React to a Civil Disturbance				X
Establish a Checkpoint				X
Search a Building	X		X	X
Employ Firepower				
Employ Fire Support				X
Protect the Force				
Maintain Operations Security				X
Conduct Passive Air Defense Measures				X
Conduct Active Air Defense Measures				X
Perform Sustainment				
Secure Civilians During Operations	X		X	X
Perform Resupply Operations				X
Treat Casualties	X		X	X
Evacuate Casualties	X		X	X
Handle Enemy Prisoners of War				
Exercise Command and Control				
Conduct Troop-Leading Procedures	X		X	X
Establish Radio Communication	X		X	X
Conduct Consolidation and Reorganization	X		X	X
Prepare for Combat	X		X	X

Table B-5. Army unit training matrix, individual tasks/techniques.

Individual Tasks/Techniques	*SHOOT HOUSE Semiannual*	*BREACH HOUSE Semiannual*	*UAC Quarterly*	*CA/CTF Semiannual*
Skill Level 1				
Perform Movement Techniques in an Urban Environment, 071-326-0541	X	X	X	X
Prepare Positions for Individual and Crew-Served Weapons in an Urban Environment, 071-326-0550	X	X	X	X
Select Hasty Fighting Positions in an Urban Environment, 071-326-0557	X	X	X	X
Employ Hand Grenades, 071-325-4407	X	X	X	X
Engage Targets with an M16-Series or M16A2 Rifle, 071-311-2007	X	X	X	X
Execute a Mechanical Breach, 071-440-0027			X	X
Skill Level 2				
Conduct an Explosive Breach	X	X	X	X
Conduct a Breach, 071-440-0029	X	X	X	X
Prime Explosive Nonelectrically, 051-193-1003	X	X	X	X
Construct a Nonelectric Initiating Assembly with Modernized Demolition	X	X	X	X

Appendix C

Stability and Support Operations Unit Training Matrix

When assigned a stability or support mission, a well-trained unit must be able to rapidly shift its focus from war fighting to stability and support and also from stability and support to war fighting. During a stability or support operation, the unit performs numerous activities. Essentially, the unit accomplishes these activities through execution of tactical missions and tasks. While stability and support operations can occur anywhere, it is most probable that they will occur in an urban environment. Tables C-1 through C-4 list training tasks by the WFF to be trained at each facility.

Table C-1. Stability and support operations unit training matrix, brigade tasks.

Brigade Tasks	*SHOOT HOUSE*	*BREACH HOUSE*	*UAC*	*CA/CTF*	*CTC*	*SIMULATION*
Develop Intelligence						
Collect and Disseminate Information					X	X
Deploy/Conduct Maneuver						
Demonstrate a Show of Force					X	
Employ Psychological Operation					X	X
Employ Firepower						
Provide Fire Support					X	X
Protect the Force						
Protect the Force					X	
Perform Sustainment						
Restore Law and Order					X	
Exercise Command and Control						
Establish an Evacuation Control Center/Conduct Evacuation Control Center Operations					X	
Negotiate					X	
Perform Noncombatant Evacuation Operations					X	
Plan a Media Visit					X	
Develop Intelligence						
Collect and Disseminate Information				X	X	X

Table C-2. Stability and support operations unit training matrix, battalion tasks.

Battalion Tasks	*SHOOT HOUSE*	*BREACH HOUSE*	*UAC*	*CA/CTF*	*CTC*	*SIMULATION*
Deploy/Conduct Maneuver						
Interdict Smuggling Operations				X	X	
Disarm Belligerents				X	X	
Handle Captured Insurgents or Belligerents				X	X	
Employ Psychological Operation				X	X	
React to Civil Disturbance				X	X	
Employ Firepower						
Provide Fire Support				X	X	X
Protect the Force						
Protect the Force				X	X	
Restore Law and Order						
Identify and Process Detainees				X	X	
Deliver Supplies or Humanitarian Aid				X	X	
Control Civilian Movement				X	X	
Prepare Traffic Control Plan				X	X	
Exercise Command and Control						
Negotiate				X	X	
Perform Noncombatant Evacuation Operations				X	X	
Conduct Marshalling Force Operations				X	X	
Negotiate a Belligerent Force Checkpoint				X	X	
Plan a Media Visit				X	X	
Perform a Cordon and Search				X	X	
Coordinate and Monitor Convoy Security				X	X	

Table C-3. Stability and support operations unit training matrix, company tasks.

Company Tasks	*SHOOT HOUSE*	*BREACH HOUSE*	*UAC*	*CA/CTF*	*CTC*	*SIMULATION*
Develop Intelligence						
Collect and Disseminate Information				X	X	
Deploy/Conduct Maneuver						
Interdict Smuggling Operations				X	X	
Disarm Belligerents				X	X	
Handle Captured Insurgents or Belligerents				X	X	
Defend Convoy				X	X	
Apprehend/Detain Noncombatants				X	X	
Perform Cordon and Search				X	X	
Employ Firepower						
Provide Fire Support				X	X	
Protect the Force						
Establish a Checkpoint				X	X	
Operate a Checkpoint				X	X	
Protect the Force				X	X	
Perform Sustainment						
Process Captured Documents and Equipment				X	X	
Exercise Command and Control						
Negotiate a Belligerent Force Checkpoint				X	X	
Conduct Marshaling Force Operations				X	X	
Negotiate a Belligerent Force Checkpoint				X	X	
Plan a Media Visit				X	X	
Coordinate and Monitor Convoy Security				X	X	
Link Up with a Convoy				X	X	

Table C-4. Stability and support operations unit training matrix, platoon/squad tasks.

Platoon/Squad Tasks	*SHOOT HOUSE*	*BREACH HOUSE*	*UAC*	*CA/CTF*	*CTC*	*SIMULATION*
Deploy/Conduct Maneuver						
Disarm Belligerents				X	X	
Handle Captured Insurgents or Belligerents				X	X	
Apprehend/Detain Noncombatants				X	X	
Search a Building	X			X	X	
React to a Sniper			X	X	X	
Conduct Platoon Riot Control Formation				X	X	
Protect the Force						
Establish a Checkpoint				X	X	
Operate a Checkpoint				X	X	
Protect the Force				X	X	
Exercise Command and Control						
Negotiate a Belligerent Force Checkpoint				X	X	
Exercise Command and Control						
React to a Media Interview				X	X	

Appendix D
Leader Development Training

To be successful in combat, the Army must continually train to develop and maintain combat-ready Soldiers, leaders, and units that can perform assigned tasks to specific standards.

SIMULATIONS TRAINING

D-1. Commanders train their staffs to plan, coordinate, and synchronize their units' operations from initial mobilization and deployment. These operations include the full spectrum of Army operations; offense, defense, stability, and support. Current simulations lack adequate replication of urban scenarios, terrain, and weapons effects. The introduction of WARSIM 2000 and OneSAF should help to solve these simulations modeling deficiencies. For lower echelon units, battalion and below, constructive simulations complement collective battle staff training that occurs during virtual simulation similar to the close combat tactical trainer and live events. For echelons above battalion, it becomes difficult to use simulators or live maneuver to train. Studies to date indicate the crossover point for simulation versus simulator is at the battalion level. Some training events from battalion and below can be trained effectively using various types of simulators. Above battalion level, effective training almost has to be conducted with a simulation (Figure D-1).

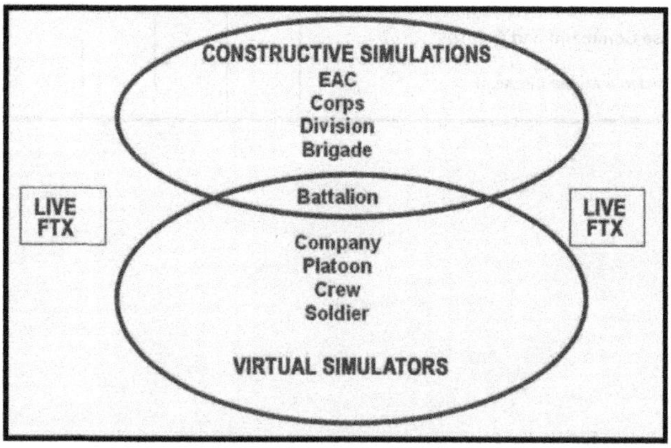

Figure D-1. Echelons and uses.

D-2. Simulations that support leader and battle staff training focus on collective tasks rather than individual leader tasks. Simulations, used in this context, rely on the human dynamics in the command setting and use real-time actions in a tactical environment with a unit's go-to-war systems. This process forces commander and staff interaction that provides feedback and helps refine the integration process. Battle staff training encourages team building, task coordination, and standardization of staff actions.

- Constructive simulations can provide commanders large areas for conducting training, access to sensitive areas through computer and mapping technology, and futuristically provide a much better level of fidelity in terms of urban operations terrain and scenario replication.
- Simulations can portray large, capable, and doctrinally correct OPFORs. All of the Soldiers portrayed in the units in a simulation do not have to be present in order for the training to be realistic. An attractive feature of simulations has always been the reduced staffing necessary to conduct this type of training exercise.
- Simulations can stress CPs and provide the realistic conditions under which commanders must make decisions. Brigade and below simulation (BBS) and corp battle simulation (CBS) are well known for the stressful environments created during an exercise. The stress is realistic and can assist in conditioning the training audience before CTC rotations.
- Simulations allow different units to train under the same conditions and to the same standards. This technique is a unique feature of simulations as the exact weather, opposing force organization, location of units, and equipment can be used to train many different units on the same terrain with the same conditions.
- The size and number of personnel in a battle staff vary depending on the echelon and other demands unique to that commander. However, the WFF are represented by the following staff officers: intelligence, operations, fire support, communications, logistics, air defense, and engineer.
- Characteristics and benefits of simulations used to support battle staff training follow:
  -- Build and sustain staff teamwork.
  -- Build and sustain effective and efficient organizations.
  -- Provide training and validation of command and staff procedures (SOPs, TACSOPs, FSOPs).
  -- Provide proficiency to the individual staff officer, while building proficiency in a fully integrated staff.
  -- Allow practice of battle drills and tactical missions (movement to contact, hasty defense).
  -- Allow commanders to determine if their staff and subordinate commanders understand the commander's intent and concept of operations.

D-3. The major development efforts for FAMSIM is WARSIM 2000. WARSIM 2000, in conjunction with OneSAF, exploits new technologies to enable commanders and staffs at battalion through EAC to train in a realistic simulation environment (Table D-1). The increased realism of WARSIM 2000 over existing models allows units to use Army Battle Command Systems (ABCS) to synchronize across the complete range of operations. WARSIM 2000's design allows warfighting CPs to use organic ABCS equipment to interact with the simulation from training sites in the field, as well as from simulation centers. WARSIM 2000 will also be capable of showing a joint and combined environment across the operational continuum.

Table D-1. Urban operations simulations crosswalk.

CAMTF Training Tasks	*WARSIM*	*OneSAF*
Corps	X	
Division	X	
Brigade		X
Battalion		X
Company		X
Platoon		X

D-4. Future simulation capabilities simultaneously integrate live, virtual, and constructive environments. Constructive simulation is a viable solution for battalion and above training and the federation of WARSIM and OneSAF models should alleviate the following training simulation problems:

- Eliminate the need for swivel chair interface.
- Allows focus on subordinate units fighting the fight.
- Digital and analog feeds to C4I systems including FBCB2.
- May provide for distance learning opportunities.
- Facilitates simultaneous multi-echelon urban operations training

D-5. At company and below, the best training solution is live and virtual. This is accomplished by linking the Close Combat Tactical Trainer (CCTT) to OneSAF and VLET to Engagement Skills Trainer (EST) technology. Entity-based models like OneSAF can allow the leader to fight from a TOC or vehicle, if linked to virtual or live environments. It also has the potential for distance learning.

PHASES OF TRAINING

D-6. The Army Urban Operations Training Strategy consists of initial, intermediate, and culminating stages. These phases of training, shown in Figure D-2, describe the sequence in which urban operations training tasks should be conducted.

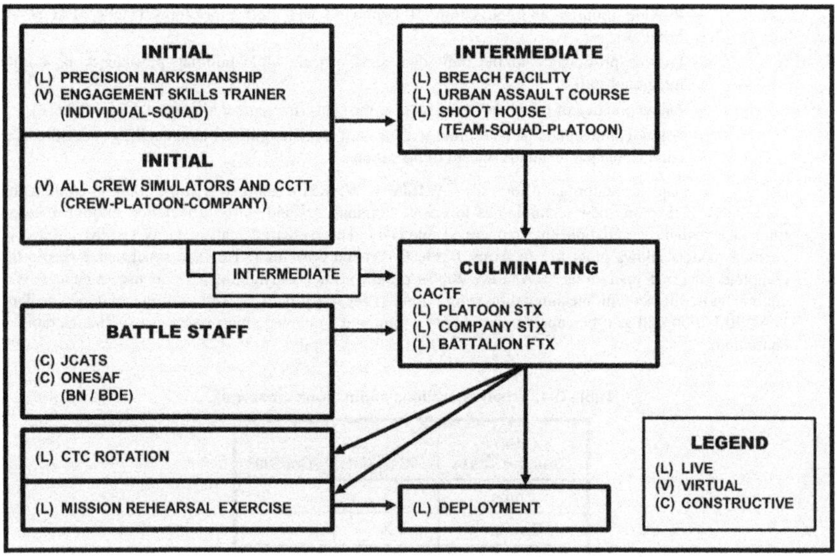

Figure D-2. Phases in the Live, Virtual, and Constructive Training Strategy.

Initial Phase

D-7. The initial phase of training covers those individual and collective tasks that Soldiers and leaders should be proficient at before arriving at any of the urban operations training sites. They consist primarily of basic marksmanship skills and Soldier tasks needed to perform any tactical operation.

Intermediate Phase

D-8. The intermediate phase consists of training conducted at the breach facility, shoot house, and the urban assault course. Urban operations training tasks should be conducted here before advancing to the Combined Arms Collective Training Facility with the exception of Aviation, Armor, and other branches that have no practical use for the Shoot House, Breach Facility, or Urban Assault Course.

Culminating Phase

D-9. The combined arms collective training facility is the culminating phase of urban operations training and focuses on multiechelon collective skills for all branches of the combined arms team. Urban operations leader training currently exists in our leader professional development courses. Units have developed internal urban warfare training programs and TRADOC schools have assembled technical mobile training teams (MTTs) for marksmanship, combatives, and breaching techniques.

Education Phase

- Urban warfare reading/video list.
- Urban operations equipment.
- Review of doctrinal publications (FM 3-21.10, FM 3-06.11, for Infantry) and corresponding publications for other branches.
- Small-unit urban movement techniques and considerations.

Planning and Execution Phase

- Prepare company OPORD or WFF-related products.
- Brief platoon/company OPORD/WFF-related products to commander or staff representative.
- Participate as briefer in urban environment TEWT.

Sustainment Phase

- Participate in company/platoon TEWT or other urban operations event.
- Execute/facilitate a company or platoon TEWT.

EXAMPLE LEADER'S COURSE POI

D-10. An example urban leaders' course POI, which is intended to provide unit trainers with a tailorable urban operations training program, is shown in Table D-2. It is a *train-the-trainer* course for team leaders through platoon leaders to train tactics, techniques, and procedures of advanced urban operations skills. The length and focus of the POI can be tailored to support the commander's training objectives and unit METL. The target audience is newly arrived, or soon to be, small unit leaders. The recommended course length is about two weeks or 120 to 140 hours of instruction.

REALISTIC TRAINING OFF FEDERAL FACILITIES

D-11. Realistic training for urban operations is critical to force readiness. Training opportunities that are both realistic and challenging may be unavailable at federal installations, or at facilities specifically designed for urban operations training. Training in off-post civilian urban settings may become necessary in order to support readiness and training requirements.

D-12. The Department of Defense has developed the following SOP for planning and approving such events. The SOP assists in balancing the needs of the military training and of the civilian community/ies to be used for realistic urban training events.

Table D-2. POI for urban leader's train-the-trainer course.

Course Overview	• Introduction	
Precision Marksmanship	• Firing techniques. • Target discrimination.	
Breaching Techniques	• Explosive. • Ballistic. • Mechanical.	

Table D-2. POI for urban leader's train-the-trainer course (continued).

Offensive Operations	• Planning considerations	-- METT-TC. -- Snipers. -- CASEVAC. -- Communications. -- ROE.
	• Platoon attack a built-up area (task organization)..	-- Assault (isolate, enter, clear). -- Rooms, hallway, stairs, superstructure, subsurface. -- Markings. -- Movement techniques. -- Streets. -- Alley ways. -- Obstacles. -- Other danger areas. -- Open areas. -- Roof tops.
	• Combined arms	-- Armor/mechanized platforms. -- Aviation assets. -- Artillery. -- Engineer.
	• Limited visibility	
Defensive Operations	• Planning considerations	-- METT-TC. -- Snipers. -- CASEVAC. -- Communications. -- ROE.
	• Characteristics of urban areas	
	• Hasty defense	-- Occupation and preparation of positions. -- Hasty firing positions. -- Improving the defense.
	• Deliberate defense, including priorities of work and other considerations	
	• Defensive plan at platoon level.	-- Defense of a strongpoint. -- Defense against armor.
	• Combined arms	-- Armor/mechanized. -- Aviation assets. -- Artillery. -- Engineer.
	• Limited visibility • Fundamental combat skills	

Table D-2. POI for urban leader's train-the-trainer course (continued).

Stability and Support Operations	• Principles of stability and support.	
	• Stability and support activities	-- NEO. -- Humanitarian assistance. -- Disaster relief. -- Show of force. -- Attack and raids.
	• Rules of engagement application • Convoy operations • Checkpoints • Fire support • Sniper employment/counter sniper measures • Civil disturbance	
Urban Training Facilities	• Breach facility • Urban assault course • Shoot house • CACTF	
New Technologies	• Rifle launch entry munition	
	• Unmanned vehicles	-- UAV. -- UGV. -- Ladders. -- Night vision devices.

STANDING OPERATING PROCEDURES FOR
PLANNING AND APPROVING REALISTIC URBAN TRAINING

1. **REFERENCES**.

   10 USC § 167, *Unified Combatant Command for Special Operations Forces.*

   10 USC §§ 3013, 5013, 8013.

   Department of Defense Directive 5111.10.

   Unified Command Plan 99.

2. **PURPOSE**. This SOP establishes uniform planning and approval procedures for realistic urban training (RUT) events.

3. **APPLICABILITY**. This SOP applies to all RUT events in civilian urban settings within the United States, its territories, and its possessions; and to Active or Reserve Forces, including National Guard Forces in Federal service. This policy does not apply to National Guard Forces under in-state status (Title 32), or to aviation operations that do not require an exemption from standard Federal Aviation Regulations. The training event should not be located on federal state, local, or private installations or facilities.

4. **DEFINITIONS**. For the purposes of this SOP, the following definitions apply:

a. **Urban Operations. This term** includes (see FM 3-06) all military actions planned and conducted on a topographical complex, and on adjacent natural terrain, where man-made construction is the dominant feature. It includes combat in towns and cities, including house-to-house and street-by-street fighting.

b. **Realistic Urban Training.** RUT is high-intensity, close-quarter battle training using live or nonlethal fires; demolitions or explosives such as breaching charges; and air and naval supporting platforms at the objective, in ***civilian urban settings***.

c. **Civilian Urban Setting.** A civilian urban setting is any area(s) located in or near a civilian community. Civilian urban settings will not be developed or established for training events, especially where training activities would affect civilian residences or commercial areas due to noise, traffic, and so on.

5. **PROCEDURES**. The following procedures apply to the planning and approval of RUT in civilian urban settings:

a. The combatant commander or the service ensures compliance with this SOP. Approval authority for RUT events may be delegated to a general or flag officer within the chain of command, or to the combatant common service conducting the training.

b. The approval authority reviews proposed training events to determine whether they are necessary and appropriate for unit readiness. Training must support unit METs and JMETs.

c. On approval of the proposed training, and at least 14 days before training starts, the combatant common service notifies the **SECR/PF** through the CJCS (by way of message) of the approved training. At a minimum, this notice includes the following:

- Name and location of the unit conducting training.
- Scope of training and activities planned such as live fire, breaches, or use of aircraft.
- Start and end dates.
- Training location.
- A statement that all necessary coordination has been accomplished IAW this SOP. The statement should include a list of the contacted official's names and their positions.

d. **Coordination with Civilian Officials**.

- While planning the training event, the commander coordinates with and obtains written approval to conduct the event from the appropriate federal, state, and local civilian authorities. Commanders will consult their legal and public affairs officers to identify the civilian officials, for example the mayor, borough chief, county commissioners or supervisors, with whom they must coordinate the training event. To do this, they consult local and Federal law enforcement agencies. At a minimum, they must consult a senior official with responsibility for each affected civilian urban setting.
- During coordination, the commander covers the details of the proposed training activities, such as whether the event involves live fire, rotary- or fixed-wing aircraft, or close quarters battle training; and whether the exercises will be conducted during the day, night, or both: and so on. It will also list actions to ensure public and participant safety, proposed public or resident notification and outreach procedures, and any other issues that might cause public interest or concern.
- As much as possible, the commander briefs civilian officials in person on the exercise concept.

- The US congressional representative or staff for the training district, and the US senators for the state, or their staffs, are advised of the proposed training event and given whatever information they request. Neither written approval nor concurrence from the representative or senators is required. However, if these officials raise objections or issues, and if those issues cannot be resolved, they are immediately elevated to the Joint Staff, J3, and the service or combatant commander point of contact.
- For each training site, license agreements with civilian or government property owners or managers are executed. Agreements address liability and conditions of acceptable usage of property. Property owners are informed of the nature of the training including written notification as to whether the activities will include use of breaching charges and, if so, how much; or live ammunition and, if so, which weapons and what ammunition. This notice will also state *if* rotary or fixed-wing aircraft will be used *and* whether the exercises will be conducted during the day, night, or both. Before the exercise can begin, all license agreements must be reviewed for legality. Copies of the license agreements will be provided to the senior civilian officials responsible for training area.
- A record of coordination activities is maintained for each training event. The record includes the names and positions of the officials *who* coordinate and approve the event. It also provides officials with a summary of the events such as date, time, and location of the training; types of activities; and planned public notification and outreach. A follow-on letter is distributed to the senior civilian officials responsible for the affected civilian urban settings. The letter confirms the training event *and* lists the names of the civilian officials who plan the training event.
- The unit may schedule recurring training events at a particular area or location by submitting a single MOA to the authorizing civilian officials. The terms of the MOA must comply with the requirements of this SOP. The service or the combatant commander reviews the MOA, which authorizes the recurring training, to ensure it complies with the procedures. Once the commander approves the MOA, the unit may conduct individual training events under it without further Joint Staff or OSD review. However, services and combatant commanders notify SO/LIC *and* the Joint Staff 14 days *before* the training in this MOA begins. Thereafter, services and combatant commanders must review these MOAs periodically to ensure *that* they continue to fulfil the intent.
- *If* nondisclosure agreements are required, command legal representatives must review them. Nondisclosure statements should clearly explain *that* the agreement does not preclude persons from informing others of general, nonsensitive information about the training exercise.
- Depending on OPSEC and safety considerations, commanders must adhere to the recommendations of civilian officials regarding community notification, outreach, and press procedures. Community notification and outreach procedures may include door-to-door notification by the local law enforcement agency. Commanders will advise civilian officials when they cannot follow their recommendations.

e. Other instructions include--

- Legal and PA representatives attend final planning meetings, and they remain on-site throughout all training activities.
- The on-site PAO handles all press queries after coordinating with the on-scene commander.
- All requested training must follow applicable statutes, law, DOD directives, other guidance, and implementing service directives.

Glossary

Section I. ACRONYMS AND ABBREVIATIONS

1SG first sergeant

A

AA assembly area
AAR after-action review/report
ABCS Army Battle Command System
ACU Army combat uniform
ADA air defense artillery
ANCOC advanced noncommissioned officer course
AO area of operation
AR Army Reserve
AT4 a shoulder-fired munition, formerly called a "light antiarmor weapon"
atk attack
ATWESS antitank weapon effect signature simulator
AVCATT armored vehicle combined arms tactical trainer

B

BBS brigade and below simulation
BCT brigade combat team
bde brigade
BDU (obsolete; *see* ACU)
BF Breach Facility
bn battalion
BNCOC basic noncommissioned officer course
BOLC Basic Officer Leadership Course
BOS (obsolete; *see* WFF)
BP battle position
BSA brigade support area
BSX battle simulation exercise

C

C2 command and control
CACTF Combined Arms Collective Training Facility
CASEVAC casualty evacuation
CBRN chemical, biological, radiological, nuclear
CBS corps battle simulation
CCMCK close combat mission capability kit
CCOC Captain's Career Officer Course
CCTT close combat tactical trainer
CCTV closed-circuit television
CFX command field exercise
CMTC Citizens Military Training Corps
co company
commo communications
CP command post
CPX command post exercise
CTC Combat Training Center
CTF collective training facility

D

DES demolition effects simulator
det cord detonation cord
div division
DLIC detachment left in contact
DS Digital Training Management System
DTMS direct support

E

EA engagement area
EAC echelons above corps
ENCATT engineer combined arms tactical trainer
engr engineer
EPW enemy prisoners of war
EST Engagement Skills Trainer

F

FAMSIM family of simulations
FO forward observer
FRAGO fragmentary order
FSCATT fire support combined arms tactical trainer
FSOP field standing operating procedures
FTX field training exercise

G

GREM	grenade rifle entry munition
GSR	ground surveillance radar

H

HC	herachloroethane zinc
HEAT	high-explosive antitank

I

IAW	in accordance with
Inf	Infantry

J

JANUS	Joint Army-Navy Uniform Simulation
JCATS	Joint Conflict and Tactical Simulation
JMET	joint mission-essential task
JRTC	Joint Readiness Training Center
JTF	joint task force
JTX	joint training exercise

L

LAW	light antiarmor weapon
LFX	live-fire exercise
LMG	light machine gun

M

MAC	MOUT assault course
MACOM	major command
MAPEX	map exercise
M/C/S	mobility/countermobility/ survivability
MET	mission-essential task
METL	mission-essential task list
METT-TC	mission, enemy, terrain, troops, time available, and civilians
MI	military intelligence
MILES	Multiple Integrated Laser-Engagement System
MOPP	mission-oriented protective posture
MPAT	multipurpose antitank (ammunition)
MPAT-OR	multipurpose antitank-obstacle reducing
MRE	meals ready to eat
MTT	mobile training team

N

NBC	(obsolete; *see* CBRN)
NEO	noncombatant evacuation operations
NLT	not later than
NTC	National Training Center

O

OneSAF	one semiautomated force (software used to create modeling and simulation programs with military applications)
op	operation
OPFOR	opposing force
OPORD	operation order
OPSEC	operations security

P

plt	platoon
POI	program of instruction
PPP	power projection platform
PSP	power support platform

R

ROE	rules of engagement
ROI	rules of interaction
RUT	realistic urban training

S

S-1	personnel staff officer (adjutant)
S-2	intelligence staff officer
S-3	operations staff officer
S-4	logistics staff officer
S-5	civil affairs officer
S-6	command, control, communications and computer operations (C4 operations) officer
SAAF	small-arms alignment fixture
SBCT	Stryker brigade combat team
SDZ	surface danger zone
SEP	system enhancement program: a version of the M1A2 machine gun with enhanced digital command and control and second-generation forward-looking infrared

SESAMS	Special Effects Small Arms Marking System
SH	Shoot House
SOI	signal operation instructions
SOP	standing operating procedure
sqd	squad
SRTA	short-range training ammunition
STAFFEX	staff exercise
STX	situational training exercise

T

T&EO	training and evaluation outline
TACSOP	tactical standing operating procedure
TES	Tactical Engagement System
TEWT	tactical exercise without troops
TF	task force
TOW	tube-launched, optically tracked, wire-guided (missile)
TTP	tactics, techniques, and procedures

U

UAC	urban assault course
UO	urban operations

V

VCR	videocassette recorder
VLET	virtual leader effects trainer

W

WARSIM	warfighting simulation
WFF	warfighting function
WFX	warfighting exercise

X

XO	executive officer

Section II. TERMS

bustle rack	storage rack mounted on turrets of some armored vehicles
entity-based model	a representation of an actual or conceptual system that includes mathematical, logistical, or computer simulation
warfighting functions	group of tasks and systems (people, organization, information, and processes) united by a common purpose that commanders use to accomplish missions and training objectives. The warfighting functions are intelligence, movement and maneuver, fire support, protection, sustainment, and command and control. These warfighting functions replace the battlefield operating systems (BOSs).

References

SOURCES USED/DOCUMENTS NEEDED

These are the sources quoted or paraphrased in this publication. All of them are needed by and must be available to the intended user.

DOD DIRECTIVE

DOD Directive 2311.01E *DOD Law of War Program.* 9 May 2006.

FIELD MANUALS

FM 3-06	*Urban Operations.* 26 October 2006.
FM 3-06.11	*Combined Arms Operations in Urban Terrain.* 28 February 2002.
FM 3-20.12	*Tank Gunnery (Abrams).* 15 August 2005.
FM 3-21.8	*The Infantry Platoon and Squad.* 28 March 2007.
FM 3-21.9	*SBCT Infantry Rifle Platoon and Squad.* 2 December 2002.
FM 3-21.10	*The Infantry Rifle Company.* 27 July 2006.
FM 3-21.11	*The SBCT Infantry Rifle Company.* 23 January 2003.
FM 3-21.20	*The Infantry Battalion.* 13 December 2006.
FM 3-21.21	*The Stryker Brigade Combat Team Infantry Battalion.* 8 April 2003.
FM 3-21.71	*Mechanized Infantry Platoon and Squad (Bradley).* 20 August 2002.
FM 3-22.9	*Rifle Marksmanship M16A1, M16A2/3, M16A4, and M4 Carbine.* 24 April 2003.
FM 7-0	*Training the Force.* 22 October 2002.
FM 7-1	*Battle Focused Training.* 15 September 2003.
FM 25-4	*How to Conduct Training Exercises.* 10 September 1984.

SOLDIER TRAINING PUBLICATIONS

STP 21-1-SMCT	*Soldier's Manual of Common Tasks Skill Level 1.* 14 December 2007.
STP 21-24-SMCT	*Soldier's Manual of Common Tasks (SMCT) Warrior Leader Skill Levels 2-4.* 2 October 2006.

TRAINING CIRCULAR

TC 7-98-1	*Stability and Support Operations Training Support Package.* 5 June 1997.

Index

Made in the USA
Monee, IL
05 November 2026

67807972R00216